An Introduction to the

Earth–Life System

This undergraduate textbook brings together Earth and biological sciences to explore the co-evolution of the Earth and life over geological time. It examines the interactions and feedback processes between the geosphere, atmosphere, hydrosphere and biosphere. It also explains how the Earth's surface environment involves a complex interplay between these systems.

The book opens with an investigation of the physical controls that combine to make Earth a habitable planet. It explores the emergence and persistence of life and the role life plays in global water and carbon cycles. Subsequent chapters demonstrate the interactions and feedback processes between climate, life and geological processes, such as plate tectonics, mountain building and erosion. The evolution of complex life during the Phanerozoic is examined within the context of environmental conditions that lead to patterns of radiation and extinction. The book concludes with an assessment of how and why the Earth's climate has varied over geological time from icehouse to greenhouse conditions, and considers whether life itself is passive or a force for change.

An Introduction to the Earth–Life System provides a concise overview of the complete Earth system at and above the surface of the Earth throughout its geological history. It is designed for use on intermediate undergraduate courses and incorporates a wealth of features to support student learning.

CHARLES COCKELL is Professor of Microbiology at The Open University, Milton Keynes, UK. His academic interests lie in geomicrobiology, astrobiology and space exploration and he has undertaken expeditions to the Arctic and Antarctic, among other places, to study life in extreme environments. Professor Cockell has written and edited six other books including *Impossible Extinction* (Cambridge University Press, 2003).

Cover image: Stromatolites. These are mineralised microbial communities, formed from blue–green algae (also called cyanobacteria). Over the last 4000 years, algae growing in this area have trapped detritus and sediment, forming large, living rafts known as microbial mats. The secretion of calcium carbonate by the algae has caused the mats to mineralise, forming the rock-like structures seen here. Stromatolites are known as living fossils because the process of mat formation and mineralisation continues today. Photographed in Hamelin Pool Marine Nature Reserve, Shark Bay, Western Australia.

An Introduction to the Earth–Life System

Edited by Charles Cockell

Authors:

Charles Cockell

Richard Corfield

Neil Edwards

Nigel Harris

CAMBRIDGE
UNIVERSITY PRESS

University Printing House, Cambridge CB2 8BS, United Kingdom

Cambridge University Press is part of the University of Cambridge.

It furthers the University's mission by disseminating knowledge in the pursuit of education, learning and research at the highest international levels of excellence.

www.cambridge.org
Information on this title: www.cambridge.org/9781107123456

First published 2007
Reprinted 2016

This co-published edition first published 2008

Edited and designed by The Open University.

Printed in the United Kingdom by Bell and Bain Ltd, Glasgow

This book forms part of an Open University course S279 *Our dynamic planet: Earth and life*. Details of this and other Open University courses can be obtained from the Student Registration and Enquiry Service, The Open University, PO Box 197, Milton Keynes MK7 6BJ, United Kingdom: tel. +44 (0)845 300 60 90, email general-enquiries@open.ac.uk

http://www.open.ac.uk

British Library Cataloguing in Publication Data available on request

Library of Congress Cataloguing in Publication Data available on request

ISBN 978 0 521 493918 hardback; ISBN 978 0 521 729536 paperback

1.1

Contents

CHAPTER 1 A HABITABLE PLANET 1

Neil Edwards

1.1 How does the Earth differ from other planets? 1
1.2 Energy from the Sun 5
1.3 The Earth's surface temperature pattern 20
1.4 The Earth's air-conditioning and heating systems 24
1.5 Earth–ocean–atmosphere: the support system for life 47
Summary of Chapter 1 57
Learning outcomes for Chapter 1 59

CHAPTER 2 THE CARBON CYCLE 61

Richard Corfield

2.1 Carbon and life 61
2.2 Carbon and climate 63
2.3 The natural carbon cycle: a question of timescale 64
2.4 A system in balance? 93
Summary of Chapter 2 100
Learning outcomes for Chapter 2 102

CHAPTER 3 PLATE TECTONICS, CLIMATE AND LIFE 103

Nigel Harris

3.1 Volcanism and the Earth system 103
3.2 Volcanic aerosols and climatic change 108
3.3 Flood basalts and their effects on climate and life 112
3.4 Continental drift and climate 117
3.5 Sea-level changes: causes and consequences 130
Summary of Chapter 3 136
Learning outcomes for Chapter 3 138

CHAPTER 4 MOUNTAINS AND CLIMATE CHANGE 139

Nigel Harris

4.1 Mountain building and the carbon cycle 139
4.2 The uplift of Tibet and the monsoon 142

4.3 Global climate change during the Tertiary 153
Summary of Chapter 4 164
Learning outcomes for Chapter 4 165

CHAPTER 5 THE EMERGENCE AND PERSISTENCE OF LIFE 167
Charles Cockell
5.1 Former worlds 167
5.2 Mat world 170
5.3 Empire of the eukaryotes 172
5.4 The carnival of the animals 185
5.5 The rules of the new evolutionary game 187
Summary of Chapter 5 192
Learning outcomes for Chapter 5 193

CHAPTER 6 LIFE IN THE PHANEROZOIC 195
Richard Corfield
6.1 The Proterozoic–Phanerozoic transition 198
6.2 Radiations and extinctions 206
6.3 The greening of the land 217
Summary of Chapter 6 227
Learning outcomes for Chapter 6 228

CHAPTER 7 THE EARTH AT EXTREMES 229
Charles Cockell
7.1 The icehouse world 229
7.2 Permo-Carboniferous glaciation and subsequent warming 234
7.3 The impact of land vegetation 239
7.4 A synthesis for the icehouse 244
7.5 The Cretaceous greenhouse world 245
7.6 Polar climate 246
7.7 Low-latitude vegetation and climate 253
7.8 Climate reconstructions 256
7.9 Geographical framework 259
7.10 A surfeit of carbon: the key to the Cretaceous greenhouse 265
7.11 Conclusions from the case studies 267
Summary of Chapter 7 267
Learning outcomes for Chapter 7 269

CHAPTER 8 END-OF-BOOK SUMMARY 271

Charles Cockell

8.1 Possible worlds 271

8.2 Review of the options 271

8.3 Conclusions 274

ANSWERS TO QUESTIONS 275

APPENDICES 294

A The elements 294

B SI fundamental and derived units 295

C The Greek alphabet 296

D Additional figures 297

GLOSSARY 299

FURTHER READING 308

ACKNOWLEDGEMENTS 310

SOURCES OF FIGURES, DATA AND TABLES 313

INDEX 315

A habitable planet

No one knows whether life flourishes on planets orbiting stars more distant than our own Sun. Despite the excitement in 1996 about possible fossils in Martian meteorites, it is still not known whether life exists, or has ever existed, on Mars. It is known, however, that within the Solar System, the profusion of life in the forms that are recognised and understood on Earth could not be supported on any of the other planets. Why is this? In this introductory chapter you will explore some of the conditions that make the Earth habitable to life.

1.1 How does the Earth differ from other planets?

If you could observe the Solar System from far off, you would see the Sun and its orbiting planets, all virtually in the same plane. This is a legacy of their common origin from the same spinning nebular disc. If you were to look more closely at the four planets nearest to the Sun, you would see a small planet, Mercury, then a very bright one, Venus, then our own blue Earth with its swirls of white cloud, and finally the reddish globe of Mars. These four planets have very different surface environments, yet formed in more or less the same part of the solar nebula, and so are likely to be made up of the same elements in very similar proportions. Looking closer still at the blue planet, it has a thin envelope of hazy atmosphere. Through this, below the clouds, not only the blue of oceans and seas and the bright white of the ice caps can be seen, but also the greens, greys and browns of land. Even the night hemisphere is illuminated by billions of tiny light sources grouped into cities, by flashes of bluish lightning, and by the occasional red glow of fire or erupting magma.

The habitability of this mysterious and beautiful environment is the result of the complex interplay of many processes – physical, chemical, biological and geological – acting over a vast range of temporal and spatial scales. Looked at simply, however, the Earth is hospitable principally as a result of its position in relation to the Sun. Earth is half as far again from the Sun as Venus, where the average surface temperature is 460 °C. The Earth is about two-thirds of the distance of Mars, where the average surface temperature is –50 °C. By contrast, the average temperature at the surface of the Earth is a moderate 15 °C. Averages can, of course, conceal enormous ranges. On the Earth, at the present time, surface temperatures rarely rise above 50 °C, and rarely fall below –50 °C, although there are geographical variations, such as with latitude. However, some microorganisms on Earth can grow at temperature extremes much greater than even these (see Box 1.1 overleaf). Most life on Earth, particularly multicellular life, is confined to a much narrower range. The relatively small temperature range of most of the Earth's surface is a result of the form and content of the Earth's atmosphere and ocean. As you will see, the full story is very complicated (and by no means completely understood) but, in terms of regulating the Earth's surface temperature, the two most important atmospheric constituents are carbon dioxide (CO_2) and water.

Box 1.1 The diversity of microbial life in Earth's environments

The habitability of the Earth can be defined by the boundaries of environmental conditions in which life can grow. Many environments, once thought to be too extreme for life, are now known to support microbial life.

1 Currently, the highest temperature limit for life is set by the microorganism known as Strain 121, which was isolated from hot fluid emanating from a hydrothermal vent on the floor of the northeast Pacific Ocean. The previous record holder was *Pyrolobus fumarii*, which can grow at 114 °C. Strain 121 can grow at a temperature of 121 °C (the temperature used for sterilising objects). It could survive at 130 °C, but it needs to be at the lower temperature of 121 °C to grow. Although organisms with higher growth temperatures might eventually be found, it is likely that, at much higher temperatures, the energy imparted to molecules makes the energetic cost of repairing or synthesising them prohibitive to growth.

   Many of the microorganisms that can grow at high temperatures (known as **hyperthermophiles**) belong to the domain **Archaea**. Some of these organisms have been recovered from the deep subsurface, for example at 3 km depth in African mines, showing that life is not limited to the surface of the Earth and the oceans. It has been postulated that the characteristics of many of these groups of microorganisms reflect the conditions on the early Earth, when both volcanism and asteroid and comet impacts, and thus hot environments, were more common.

2 At the other end of the scale, frozen permafrost can provide an environment for metabolising bacteria. At temperatures below 0 °C, saline liquid water can exist because salt depresses the freezing point of water and at temperatures of −10 °C, bacteria that can metabolise and grow, albeit slowly, have been recovered from Siberian permafrost.

3 Microorganisms have also been found growing at extremes of pH. *Ferroplasma acidarmanus* can grow at a pH of 0, living in acidic waters at Iron Mountain in California, USA. Similarly, organisms have been found growing at high pH in alkaline soda lakes.

4 Some microorganisms, such as *Deinococcus radiodurans*, also have high radiation resistance, using highly efficient DNA repair processes to reverse radiation damage caused to their DNA. This organism can tolerate a radiation dose at least three orders of magnitude higher than that which is lethal to a human.

These **extremophiles** help define the envelope of life on Earth, and thus the extremes beyond which the planet becomes uninhabitable.

On the Earth, surface temperatures are such that most of the planet's surface water is in liquid form, with the remainder in the ice caps and in the atmosphere. The atmosphere contains a small amount of CO_2, the oceans a good deal more, and large amounts are effectively 'locked up' in crustal rocks. By contrast, on Venus the atmosphere is largely CO_2, there is a minute amount of atmospheric water vapour (i.e. H_2O gas) and, at the prevailing temperatures, none of it can condense on the planet's surface. The Martian atmosphere, like the Venusian one, largely consists of CO_2, but some CO_2 and most of the planet's water is in the form of ice. Were it not for the presence of liquid water on the surface of the Earth, life – at least in its familiar forms – could not exist here.

Look at Figure 1.1. The cloud cover, which obscures much of the Earth's surface, has been removed. Immediately obvious are bright areas of ice cover, the enormous area of ocean in the Southern Hemisphere, and the green of forests and crops.

■ To what extent does this map tell you about life on Earth?

□ It indicates where life is concentrated because much of the biologically productive life on Earth, and on land in particular, is based on plants and other **phototrophs** – organisms that can build their own organic material by harnessing the energy of sunlight by **photosynthesis**.

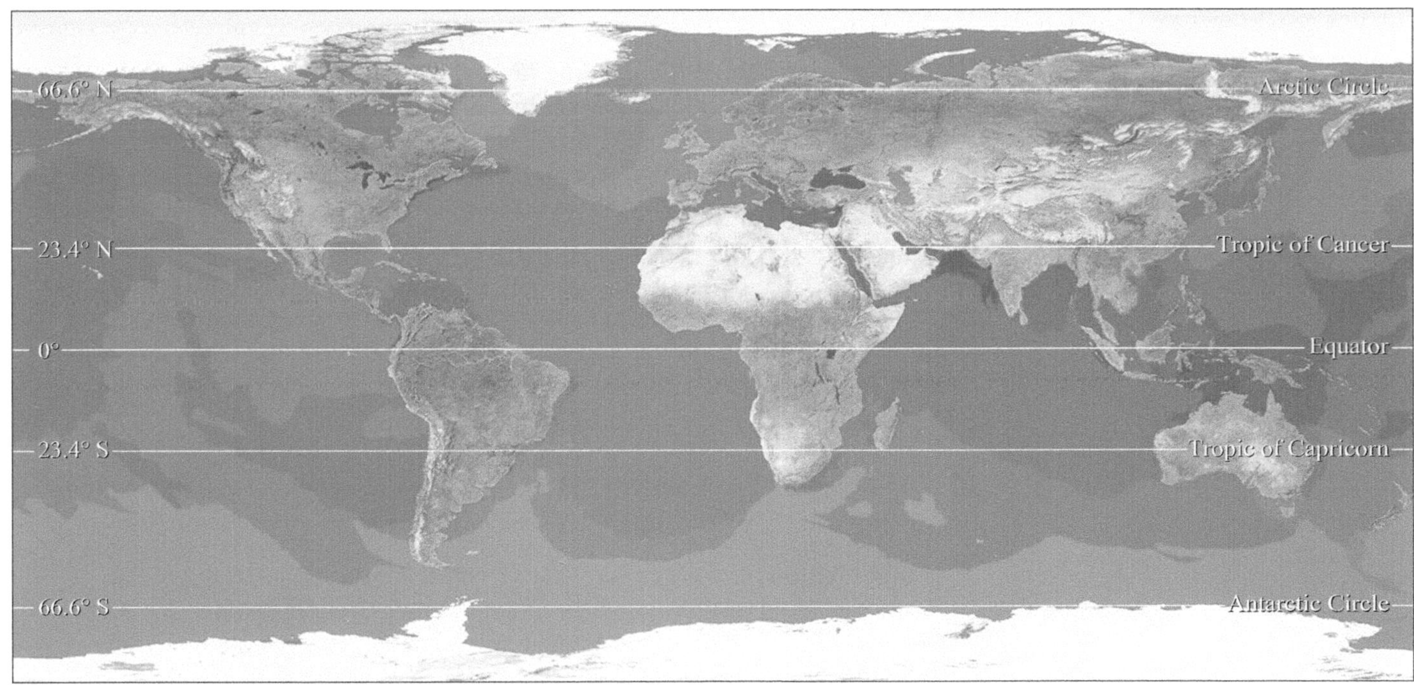

Figure 1.1 Composite satellite image of the Earth's surface. The green areas (e.g. Central Brazil and Western Europe) are forests and crops; the light-coloured areas in the tropics are mostly savannah, semi-arid scrub and desert; and the brown areas at high northern latitudes are tundra. At high latitudes, and at lower latitudes where there are mountains (e.g. the Rockies, Andes and Himalaya), white corresponds to ice and snow. *Note*: the area of ice at high latitudes is grossly exaggerated by this type of projection. (NASA)

Expressed simply, the chemical equation for photosynthesis can be written as follows:

$$\underset{\text{carbon dioxide from atmosphere}}{nCO_2(g)} + nH_2O \xrightarrow{\text{light energy}} \underset{\text{organic matter}}{(CH_2O)_n} + \underset{\text{oxygen}}{nO_2(g)} \quad (1.1)$$

where n can have various values, $(CH_2O)_n$ represents a range of carbohydrate materials of which glucose ($C_6H_{12}O_6$) is the simplest, and (g) is gas. You could rewrite Equation 1.1, for instance, with glucose, $C_6H_{12}O_6$, representing organic matter in the form:

$$\underset{\text{carbon dioxide from atmosphere}}{6CO_2(g)} + 6H_2O \longrightarrow \underset{\text{organic matter}}{C_6H_{12}O_6} + \underset{\text{oxygen}}{6O_2(g)} \quad (1.2)$$

This process of taking free carbon from the atmosphere and combining it into living organic material is referred to as **'fixing' carbon**, and the process of building living material by fixing carbon is known as **primary production**. You will look at this in much more detail in Chapter 2. Animals cannot fix carbon and so can live only by consuming **primary producers**, either directly or indirectly.

But Figure 1.1 tells only a small part of the story; for one thing, the oceans like the land also support abundant phototrophs. Figure 1.2 shows not only the geographical variation of the potential for primary production on land (i.e. the potential for carbon to be incorporated into terrestrial living material), but also the average concentration of chlorophyll in algae living in surface waters. Chlorophylls are the light-collecting pigments found in photosynthesising organisms which give them their green colouration. Although both Figures 1.1 and 1.2 are composites of many satellite images, they are akin to snapshots in time. They provide information about the standing stock of plant material, but by themselves they do not reveal anything about the *rates* at which plant material is being made – the primary *productivity* – in different environments. Nor do they indicate anything about the rates at which the plant material is being eaten by animals, decomposing or being recycled. As you will see, because organic material is essentially carbon, almost all of which can eventually find its way back into the atmosphere as gaseous CO_2, all these processes are important influences on the Earth's climate and thus its habitability.

■ Returning to Figure 1.1 for a moment, suggest why the vegetation patterns shown are not wholly reliable as indicators of local climatic conditions.

□ (i) Humans have direct effects on vegetation through for example the removal of forests (particularly in temperate latitudes) and irrigation of land in arid regions; their domesticated animals graze the vegetation. (ii) Seasonal changes in local climatic conditions are not reflected in the figure.

Nevertheless, the patterns of primary production seen in Figures 1.1 and 1.2 are to a large extent determined by the movement of air and water over the surface of the Earth – including the swirling clouds. You will see *how* shortly, but first one of the primary influences on the continual motion of the Earth's atmosphere and oceans, i.e. energy from the Sun, is discussed.

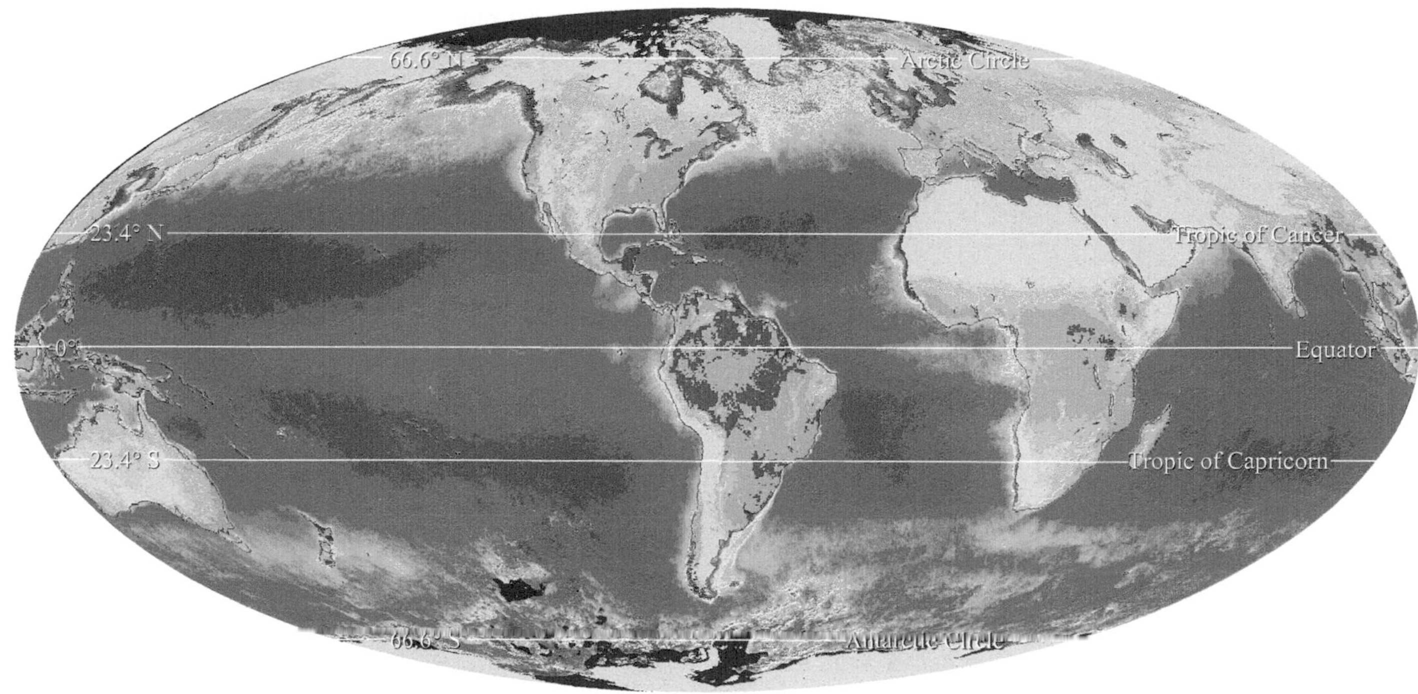

Figure 1.2 Global distribution of the potential for primary production on land and in surface waters, as indicated by chlorophyll concentration (determined using satellite-borne sensors). On land, the darkest green areas (e.g. Brazil) correspond to the greatest potential for production of new plant material; decreasing production potential is indicated by increasingly paler greens. Least productive of all are the paler deserts, high mountains and arctic regions shown in yellow. In surface waters, regions of highest productivity (bright red, e.g. Canadian coast) are mostly around the coasts, followed by yellow, green and blue. Least productive oceanic regions (e.g. large areas in the tropics) are shown in purplish-red. (NSF/NASA)

1.2 Energy from the Sun

The Earth is, on average, about 150×10^6 km from the Sun. From this, the average amount of solar energy reaching the top of the atmosphere can be calculated. The amount of solar energy that would fall on a surface at right angles to the Sun's rays, known as the **solar flux** (or solar irradiance, or solar constant), is ~1370 W m^{-2}. This means that the amount of solar radiation that the Earth intercepts is ~1370 W m^{-2} $\times \pi r^2$ W, where πr^2 is the area of a disc with the same radius as the Earth. However, the Earth is roughly *spherical*, so the area presented to the incoming solar radiation by the rotating Earth (over any period longer than a day) is $4\pi r^2$, i.e. four times as great. The average flux of solar energy is therefore effectively only a quarter of the solar flux:

1 W = 1 J s^{-1}

Surface area of a sphere $A = 4\pi r^2$, where r is the radius.

$$\frac{1370 \text{ W m}^{-2}}{4} \approx 343 \text{ W m}^{-2}.$$

So the average amount of solar energy reaching the top of the atmosphere, i.e. the *effective solar flux*, is ~343 W m^{-2} (Figure 1.3 overleaf).

Not all of this incoming solar energy is available to heat the Earth–atmosphere system: about 30% of it is reflected back into space, mainly from the tops of clouds. In other words, the **albedo** of the Earth as a whole (i.e. the fraction of

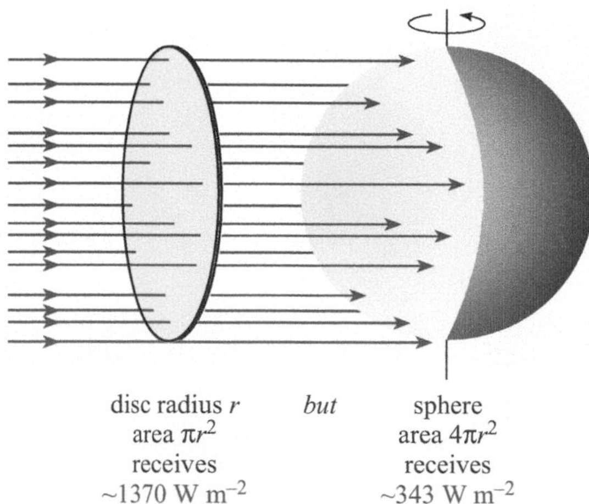

Figure 1.3 Diagram of the average flux of solar energy reaching a disc and the Earth (with the same radii).

incoming solar radiation that is reflected from it) is about 30%. This means that the Earth–atmosphere system receives ~70% of the solar flux, i.e. 240 watts of solar energy per square metre (240 W m^{-2}). But this number of 240 W m^{-2} is an average, and it is the uneven distribution of this heating that drives the Earth's climate engine.

■ Look at Figure 1.4 below. Bearing in mind that the atmosphere absorbs a proportion of incoming solar energy, suggest two reasons why the intensity of solar radiation at the Earth's surface, and hence the surface temperature, is generally lower at high latitudes than at low latitudes.

□ The intensity of solar radiation at the Earth's surface (on the diagram, the number of rays per unit area) depends on the angle of the rays with respect to the surface: the more oblique the angle, the larger the area over which the solar energy will be spread. Furthermore, the more oblique the rays, the greater the thickness of atmosphere through which the rays will have to travel.

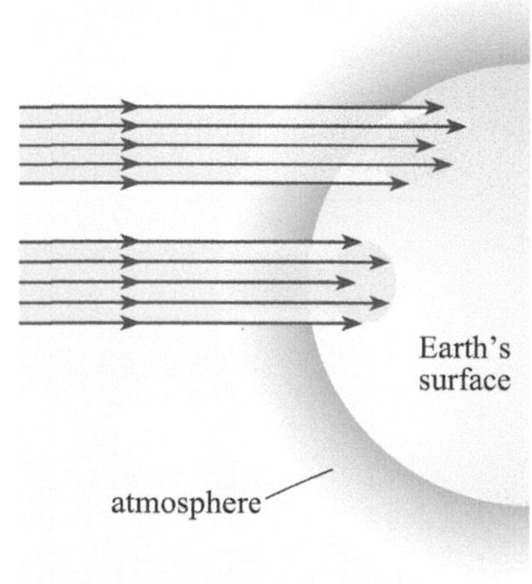

Figure 1.4 Schematic diagram to show why there is a difference in the intensity of solar radiation reaching the Earth's surface at different latitudes (the atmosphere is not to scale).

The relationship between the angles the Sun's rays make with the ground and their ability to warm it is, in fact, the origin of the word 'climate' (from the Greek *klima*, meaning slope). Slopes that face equatorwards (i.e. southwards in the Northern Hemisphere) – where the Sun's rays meet the ground at a steeper angle – are warmer.

If the Earth's axis of rotation were at right angles to the plane of its orbit, for any given latitude, the angle at which the rays of the noonday Sun fell upon the surface would remain constant, with higher latitudes in both hemispheres always receiving less solar radiation than lower latitudes. In other words, the Earth's surface at, for example, 10° N and 10° S would always receive the same amount of solar radiation, with this always being more than that received at, for example, 40° N and 40° S. The Sun would be overhead at noon only at the Equator, the poles would have perpetual twilight, and night and day would always be the same length (i.e. each 12 hours long) everywhere around the globe.

But the Earth's axis of rotation is *tilted* with respect to the plane of its orbit, currently at an angle of 23.4°. As a result, the latitude at which the noonday Sun is overhead migrates between 23.4° N (the Tropic of Cancer) and 23.4° S (the Tropic of Capricorn), passing the Equator only twice a year, at the equinoxes (when the lengths of night and day are equal). This change in the position of the noonday Sun throughout the year is the cause of the seasons (Box 1.2).

Box 1.2 The cause of seasons

Figure 1.5a shows the passage of the seasons for the Northern Hemisphere.

- Along the Tropic of Cancer (23.4° N), the noonday Sun is overhead, and maximum solar radiation is received during the summer solstice (the longest day), which is 21 June.

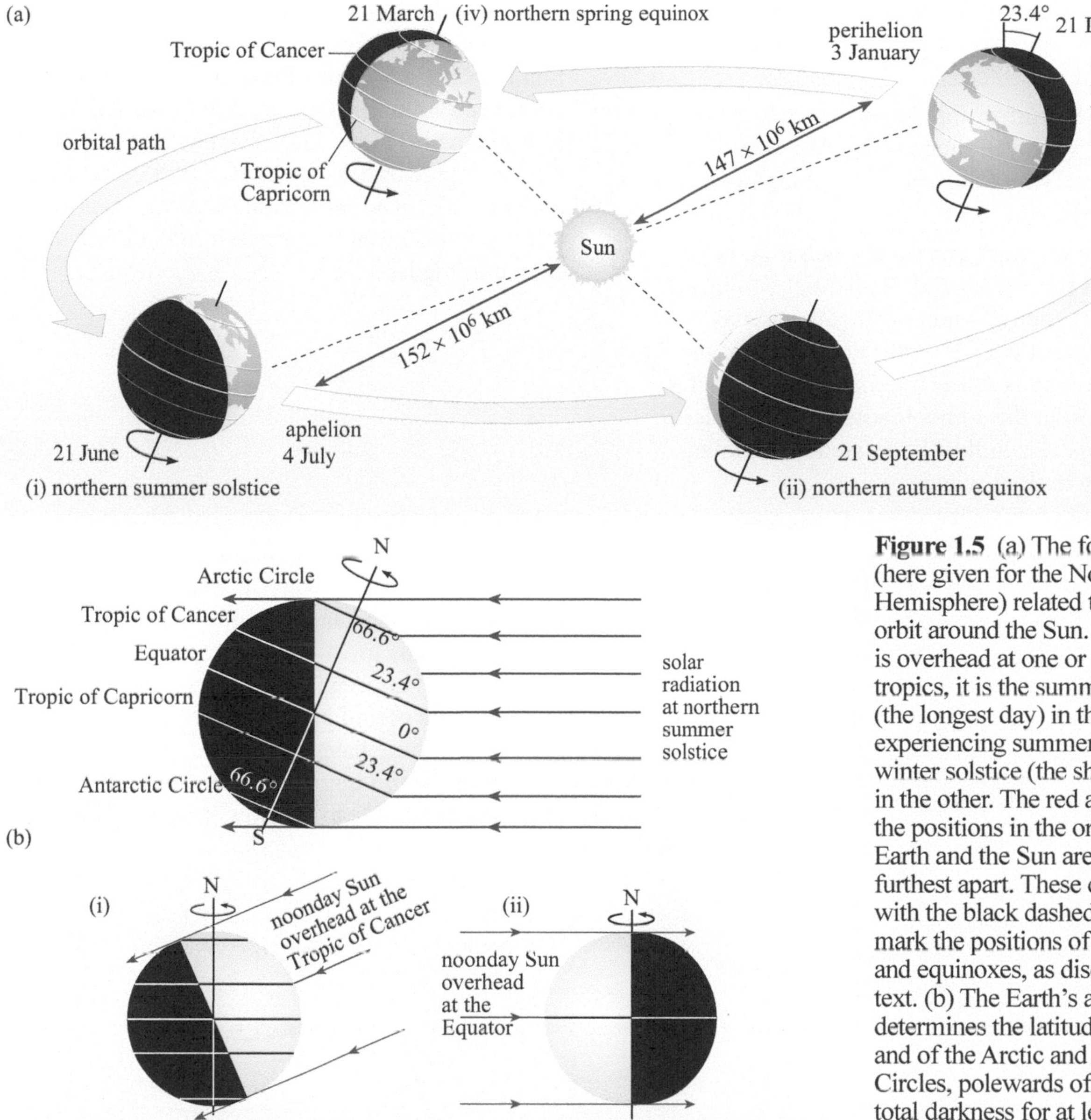

Figure 1.5 (a) The four seasons (here given for the Northern Hemisphere) related to the Earth's orbit around the Sun. When the Sun is overhead at one or other of the tropics, it is the summer solstice (the longest day) in the hemisphere experiencing summer, and the winter solstice (the shortest day) in the other. The red arrows mark the positions in the orbit where the Earth and the Sun are closest and furthest apart. These do not coincide with the black dashed lines, which mark the positions of solstices and equinoxes, as discussed in the text. (b) The Earth's angle of tilt determines the latitude of the tropics and of the Arctic and Antarctic Circles, polewards of which there is total darkness for at least part of the year. (c) The passage of the seasons shown in terms of the position of the noonday Sun with respect to the Earth: (i) the northern summer solstice and (iii) the southern summer solstice; (ii) and (iv) at the equinoxes, the Sun is overhead at the Equator, and the Northern and Southern Hemispheres are illuminated equally; days and nights are the same duration at all latitudes, except at the poles, which are grazed by the Sun's rays for 24 hours.

- After 21 June, the days begin to shorten, until at the autumn equinox, on 21 September, day and night are of equal length.
- After 21 September, day lengths continue to shorten, until the shortest day (the winter solstice, on 21 December), after which the days begin to lengthen again.
- At the Equator, maximum solar radiation is received at the March and September equinoxes, when the noonday Sun is overhead, and day and night are of equal length. Polewards of the tropics, the Sun is *never* overhead, although it is at its highest at the summer solstice. The poles themselves are wholly illuminated in summer and wholly dark in winter.

The Earth's angle of tilt (presently 23.4°) determines the latitude of the tropics (where the Sun is overhead at one of the solstices) and of the Arctic and Antarctic Circles at 66.6° N and 66.6° S respectively (90° – 23.4°), polewards of which there is total darkness for at least part of the year (Figure 1.5b). The seasons in terms of the position of the noonday Sun with respect to the Earth are shown in Figure 1.5c.

Figure 1.6 shows the seasonal variation in the amount of solar radiation received daily at the Earth's surface (i.e. taking into account the amount absorbed in the atmosphere). The zero contour corresponds to 24-hour darkness. At the North Pole (90° N), it encompasses the period between 21 September and 21 March, and at the South Pole it encompasses the period between 21 March and 21 September. In each case, the first of these dates is the autumn equinox and the second is the spring equinox (compare with Figure 1.5). Don't worry if this type of diagram seems strange – you should be able to see how it works when you have tried Question 1.1.

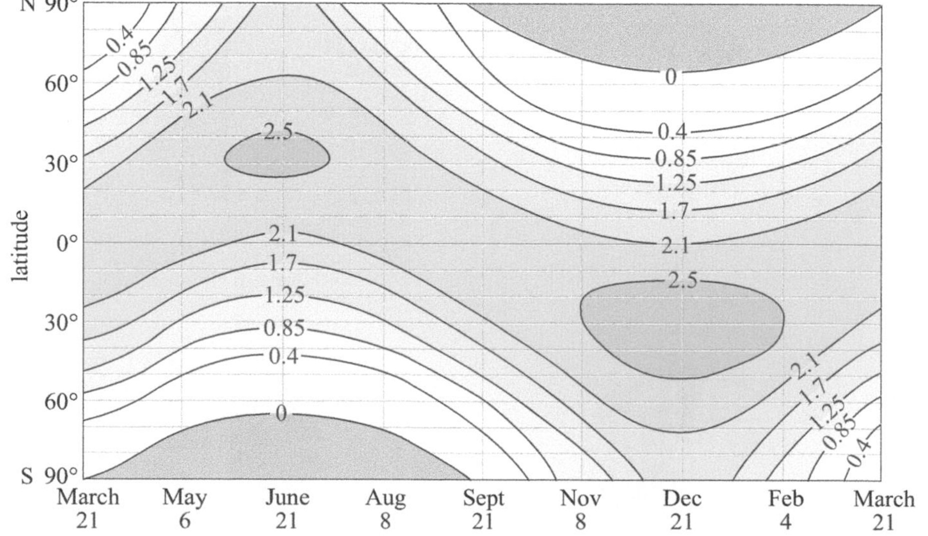

Figure 1.6 Seasonal variation of daily incoming solar radiation (in 10^7 J m^{-2}) at the Earth's surface, taking account of absorption by the atmosphere but ignoring the effect of topography. Note that this is not an ordinary spatial map, but a map, or plot, of incoming solar energy against latitude on the one axis and time of year on the other.

Question 1.1

(a) (i) With reference to Figure 1.6, describe briefly how the incoming solar radiation changes over the course of the year at 50° N.

(ii) The units used in Figure 1.6 are J m^{-2} because it shows values of incoming solar radiation *per day*. Convert the contour values so that they are for *average* incoming solar radiation in W m^{-2}.

(b) Over the year as a whole Figure 1.6 shows that, on average, the Equator receives the most solar radiation. Which latitudes receive the most solar radiation at any one time? Why is this?

One aspect of Figure 1.6 that may be initially puzzling is that the maximum amount of solar energy received by southern mid-latitudes in the southern summer is *greater* than the maximum amount received by northern mid-latitudes in the northern summer (e.g. the areas enclosed by the 2.5×10^7 J m^{-2} contour). Furthermore, careful study of Figure 1.6 indicates that in the southern summer *all* latitudes receive more energy than the corresponding latitudes in the other hemisphere in the northern summer. This is because the Earth's orbit is elliptical, and at the present time the Earth comes closest to the Sun (i.e. is at **perihelion**) during the southern summer (on 3 January), and is furthest from the Sun (i.e. is at **aphelion**) during the northern summer (on 4 July). It is because the Sun is at one of the two foci of the ellipse, rather than at its geometric centre, that perihelion and aphelion occur only once a year rather than twice a year.

Because of the varying gravitational attraction of the Sun and of the other planets (notably Jupiter and Saturn), the degree of ellipticity (*eccentricity*, or off-centredness) of the Earth's orbit varies with time and, over a period of about 110 000 years, changes from its most elliptical (maximum eccentricity) to nearly circular and back again (Figure 1.7a). This 110 000-year cycle is the longest of three astronomical cycles that affect the amount and distribution of solar radiation reaching the Earth's surface, and it is the only one that affects the *total* amount of solar radiation reaching the Earth. The two shorter cycles (tilt and precession: Figure 1.7b) involve the orientation of the Earth's spin axis, and so affect the *distribution* of solar radiation over the Earth's surface.

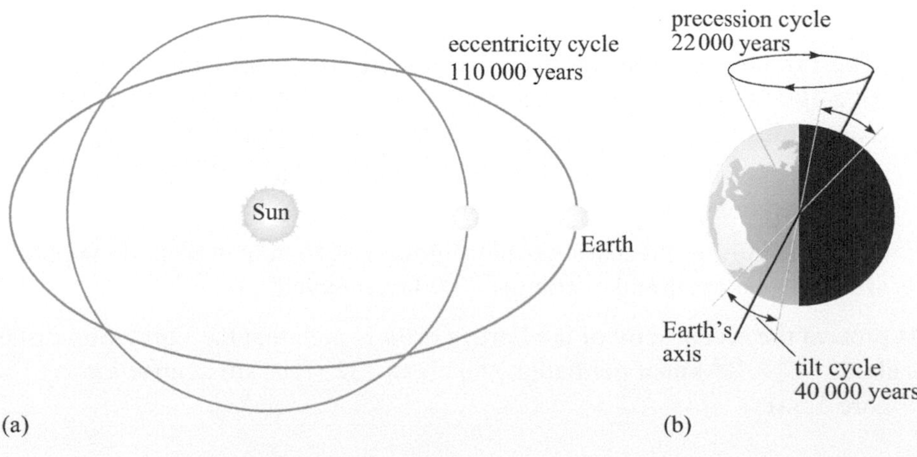

Figure 1.7 The component Milankovich cycles. (a) Plan view of the Earth's orbit to show how it changes shape from circular to almost elliptical and back again, over the course of the 110 000-year eccentricity cycle. (b) The Earth showing the 40 000-year tilt cycle and the 22 000-year precession cycle.

These three astronomical cycles (eccentricity, tilt and precession) are usually known as **Milankovich cycles**, after Milutin Milankovich, a Serbian astronomer. Milankovich's work in the 1930s and 1940s was an improvement and refinement of the work of Scotsman James Croll, for this reason the cycles are sometimes referred to as Milankovich–Croll cycles.

Over the course of about 22 000 years, the direction in which the Earth's spin axis points traces out a circle. Therefore, from about 11 000 years from now, the positions in the orbit of the northern and southern summer will be reversed as the seasons (i.e. the solstices and the equinoxes) will have moved clockwise around the orbit. In another 11 000 years, they will be back in their current positions. This phenomenon is often referred to as the *precession of the equinoxes*. (Of course, calendars will continually have to be adjusted to take account of this so that, for example, the northern summer solstice will remain in June and not gradually drift towards December.) Eccentricity will be roughly as it is now, thus the Earth will be at perihelion in the northern, rather than the southern, summer.

At the same time that the Earth's spin axis traces out a circle, the angle it makes with the normal to the orbital plane varies between about 21.8° and 24.4°, and back again, with a periodicity of about 40 000 years; at the moment, the angle of *tilt* is about 23.4°.

■ Bearing in mind Figure 1.5, what do you think the latitude of the tropics would be if the angle of tilt increased so that it was 24.4°, rather than 23.4°? What effect would this have on Figure 1.6?

■ If the angle of tilt increased to 24.4°, the tropics (over which the Sun would be directly overhead during the summer solstice) would be at 24.4° N and 24.4° S. This would mean that on a diagram like Figure 1.6, the areas of maximum incoming solar radiation corresponding to summer months would be shifted polewards slightly and, for the winter hemisphere, the zero contour (for example) would extend a little further towards the Equator.

So the greater the angle of tilt, the greater is the difference between winter and summer. At present the angle of tilt is in fact *decreasing*, so summers should be very gradually becoming cooler and winters should be very gradually becoming warmer.

The form of the three cycles over the past 800 000 years can be seen in Figure 1.8.

- The 110 000-year eccentricity cycle (Figure 1.8a) is actually a combination of a 100 000-year cycle with a much weaker 413 000-year cycle, and is sometimes referred to collectively as the 100 000-year cycle.
- The tilt cycle (Figure 1.8b) is sometimes referred to as the 40 000-year or 41 000-year cycle.
- The 22 000-year precession cycle (Figure 1.8c) can be broken down into a 19 000-year cycle and a stronger 23 000-year cycle.

At present the eccentricity of the Earth's orbit is such that the Earth–Sun distance is about 147×10^6 km at perihelion and about 152×10^6 km at aphelion (Figure 1.5a).

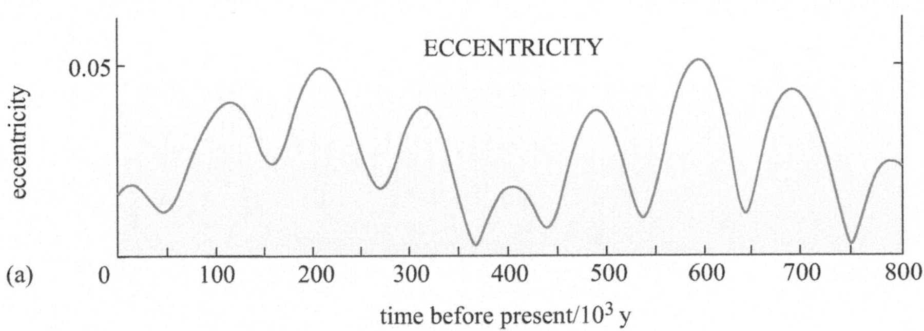

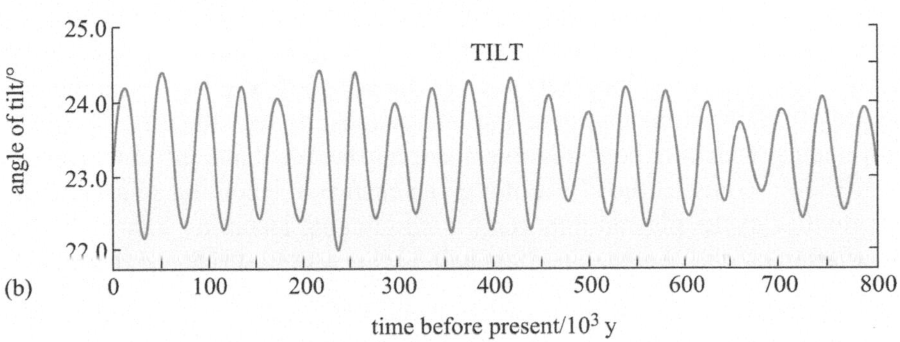

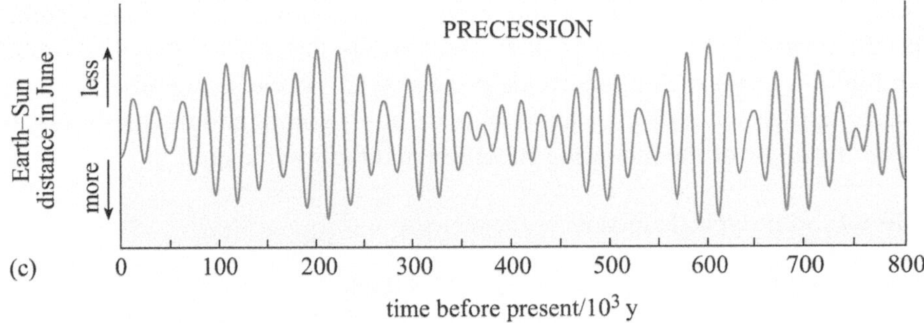

Figure 1.8 Milankovich orbital changes over the past 800 000 years, based on astronomical data. (a) Eccentricity (the higher the value, the more elliptical the orbit: an eccentricity of zero corresponds to circular). (b) Tilt. (c) Precession expressed in terms of the Earth–Sun distance in June. (Imbrie et al., 1984)

■ According to Figure 1.8, is the Earth's orbit currently becoming more or less elliptical? Approximately when was it last nearly circular?

□ It is currently becoming less elliptical. At present, the eccentricity curve (Figure 1.8a) is tending towards smaller values, and observation of the wave-like shape of the curve shows that it last passed through a minimum (i.e. a period when the orbit is almost circular) about 50 000 years ago.

Note that an elliptical orbit tends to exaggerate the seasons in one hemisphere (the one for which winter occurs during aphelion and summer occurs during perihelion) and to moderate them in the other; the more elliptical the orbit, the more extreme this effect is.

Figure 1.9 (overleaf) was computed using the information shown in Figure 1.8 and shows how the intensity of summer sunshine at high northern latitudes (65° N) has varied in response to Milankovich orbital changes over the past 625 000 years.

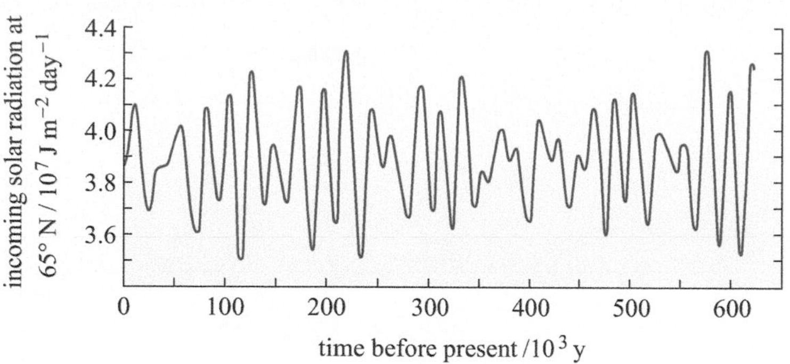

Figure 1.9 The variation in incoming solar radiation in summer at northern latitudes (65° N) over the past 625 000 years.

There is good geological evidence that Milankovich cycles have been important agents of climatic variation over much of the last ~500 Ma. However, the complexity of the climate system and interactions between the different components, including both **positive** and **negative feedback** effects, mean that it is difficult to predict the effect of a given change in incoming solar radiation. Despite this, cycles play an important role in linking incoming solar radiation and its variations to the climate, and thus habitability, of the surface of the Earth.

One way in which astronomical variations may affect the Earth's climate is through the influence they have on the growth and decay of the polar ice caps and, hence, the amount of solar radiation reflected by the Earth. As mentioned earlier, the planetary albedo (i.e. the fraction of incoming solar radiation reflected from the Earth) is about 30%. Albedos for various surfaces are given in Table 1.1. Note that it is not possible to provide precise values because the reflectivity of a surface depends not only on what it is made of, its colour and its roughness, but also on the angle of the incoming radiation and the wavelength.

Table 1.1 Some typical albedos.

Type of surface	Albedo
fresh snow and sea ice	80–90%
clouds	average ~55% (~15–80%, depending on type and thickness)
thawing snow	~45%
desert	35%
grassland	25–33%
forest, bare soil, rock, cities	~10–20%
water:	
moderate–high sun (elevation >40°)	<5%
low sun (elevation ~10°)	>50%

- Apart from changes in snow and ice cover, how might changes in climate cause changes in planetary albedo?

- Changes in cloud cover and vegetation type would also affect planetary albedo.

1.2.1 The atmosphere – a protective filter

So far, illumination and heating by solar radiation have been taken as being almost interchangeable, but visible and thermal radiation occupy different parts of the Sun's spectrum (i.e. they have different frequencies and wavelengths). As frequency and wavelength of radiation have very important climatic implications you should now read Box 1.3, which describes important facts about the **electromagnetic spectrum** and thermal energy.

Box 1.3 The electromagnetic spectrum and thermal energy

The Sun emits electromagnetic radiation over a wide range of the electromagnetic spectrum, from gamma-rays to radio waves (Figure 1.10), but ~99.9% of the energy is in the wavelength range 0.15–5 μm (Figure 1.11 overleaf), which includes ultraviolet radiation, visible radiation (violet to red), and infrared (heat) radiation. As with any propagating waves, the shorter the wavelength (λ), the higher the frequency (f), and multiplying the two together will give the speed (c):

$$c = f\lambda \qquad (1.3)$$

where the term '*ultra*violet' refers to radiation with a higher frequency than that of violet, and the term '*infra*red' refers to radiation with a lower frequency than that of red.

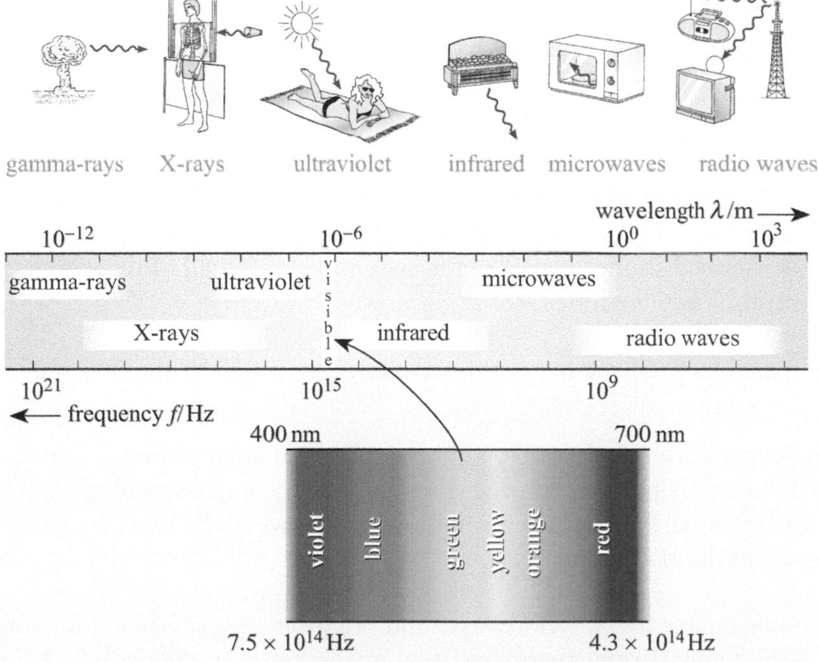

Figure 1.10 The electromagnetic spectrum. Wavelength is given in metres. For the expanded visible spectrum, wavelength is given in nanometres (1 nm = 10^{-9} m); frequency is given in hertz (Hz = cycles s^{-1}).

Note that the process of radiation is completely different from reflection, in which energy is not absorbed and reradiated, but 'bounced off', with its frequency and wavelength unaffected (think of light reflected by a mirror).

All molecules vibrate, and when a molecule absorbs electromagnetic energy, the amplitude of its vibration increases. Once 'excited' in this way, molecules can lose the energy again, either by re-emitting infrared radiation or by converting it into kinetic energy, by colliding with other molecules. The effect of such collisions is to raise the internal energy of the material, i.e. to raise its temperature. In the case of a solid or liquid, the vibrations of the excited atoms and/or molecules not only cause the surface to radiate longwave radiation, but the collisions of excited atoms (particularly electrons) with adjacent less energetic atoms result in the transfer of energy down the temperature gradient, by **conduction**.

It is often scientifically convenient to consider the Sun (and indeed the Earth) as a black body, i.e. a body that is radiating energy at the maximum possible rate for its temperature. If this assumption is made, then energy radiated per unit area per unit time is proportional to T^4, where T is the absolute temperature in kelvin (K). This expression is known as the *Stefan–Boltzmann law*, which states:

1 Any body above absolute zero emits radiant energy.

2 The energy emitted per unit area per unit time is proportional to the fourth power of the temperature in kelvin. Remember that the kelvin scale starts at absolute zero, or –273.15 °C, so 1 °C = 274.15 K.

Figure 1.11a shows the spectral curves for the Sun and the Earth, assuming that they radiate like black bodies of 6000 K and 255 K, respectively. These curves demonstrate a general principle: the higher the temperature of a surface, the more the maximum in the spectrum of energy it radiates is shifted towards shorter wavelengths and higher frequencies. As a practical example, think of a lump of coke in a furnace. As the coke starts to get hot it glows red; as it gets hotter and hotter, shorter and shorter wavelengths are emitted, until all of the visible part of the spectrum is being emitted, resulting in white heat.

■ To what extent do the ranges of wavelengths emitted by the Sun and by the Earth overlap?

□ The curves overlap only very slightly. Incoming solar radiation is in the ultraviolet, visible and infrared (thermal) bands; outgoing radiation is all in the infrared, and there is a small overlap between the two curves at wavelengths of 4–5 μm.

■ Is it reasonable to use 'shortwave' and 'longwave' as shorthand for solar radiation and thermal energy radiated by the Earth, respectively?

□ In the context of radiation incident on and emitted by the Earth, it is reasonable to think of solar radiation as shortwave and terrestrial radiation as longwave.

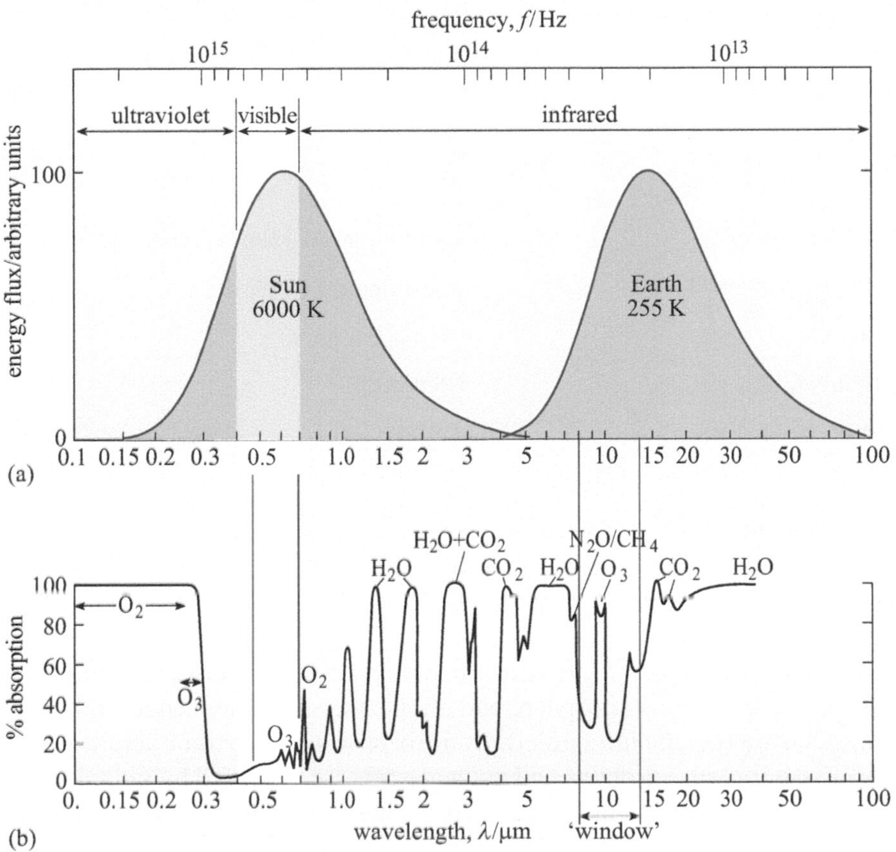

Figure 1.11 (a) The spectrum of electromagnetic radiation emitted by the Sun (assuming that it is a black body at 6000 K) and the Earth (assuming that it is a black body at 255 K). This latter value (255 K = −18 °C) is the Earth's *effective* planetary temperature, consistent with the radiation it emits to space. Because the Earth has an atmosphere, the average temperature *at its surface* is much higher, at about 288 K (= 15 °C). *Note*: both horizontal scales are logarithmic. Wavelength and frequency are inversely related, so while the wavelength increases from left to right, frequency increases from right to left (Equation 1.3). (b) The absorption spectrum of the Earth's atmosphere, showing the total percentage of radiation absorbed on passing through the atmosphere as a function of wavelength. This is a composite curve with the effects of the different atmospheric gases added together (CH_4 is methane and N_2O is nitrous oxide). Also shown within the visible band are the wavelengths of the radiation most used by photosynthesising organisms (~0.45 μm and ~0.65–0.7 μm). *Note*: these are the wavelengths *absorbed*, not the ones reflected (it is the latter that determine the colour of chlorophyll pigment). (Mitchell, 1989)

The curves in Figure 1.11a are emission curves. They show the spectra of radiation emitted by the Sun and the Earth. The spectrum of solar radiation reaching the Earth's surface, however, is strongly modified by absorption and reradiation within the atmosphere. The atmosphere acts as a filter, removing dangerous ultraviolet radiation and maintaining habitable conditions at the surface. To understand how it does so, you need to consider its gaseous composition.

Table 1.2 (overleaf) lists the gases in the atmosphere in order of abundance. By far the most abundant gases are nitrogen and oxygen, which together make up ~99% of the total. Apart from nitrogen and argon, all of these gases affect the Earth's climate through their interaction with incoming (shortwave and longwave) radiation, outgoing longwave radiation, or both. As shown in Figure 1.11b, each gas absorbs particular wavelengths, often in more than one part of the spectrum.

Figure 1.11a shows, in effect, the spectrum of wavelengths emitted by the Sun, and the spectrum of wavelengths reradiated by the Earth. The curve in Figure 1.11b shows the percentage of absorption of radiation passing through the atmosphere, as a function of wavelength. This is a composite curve with the effects of the different atmospheric gases added together – some peaks are entirely attributable to a particular gas, others result from the combined effect of two or more gases.

Table 1.2 Gases naturally present in the atmosphere and their current concentrations.

Gas	Approximate concentration by volume*
nitrogen, N_2	78%
oxygen, O_2	21%
argon, Ar (and other noble gases)	1%
water vapour, H_2O	~ 3000 ppm (0.3%), but very variable
carbon dioxide, CO_2	~ 380 ppm (0.038%)†
ozone, O_3	~ 0.01–0.1 ppm
methane, CH_4	1.8 ppm§
nitrous oxide, N_2O	0.3 ppm

*Measured as the proportion of the total number of molecules in the atmosphere contributed by each component, equivalent to its proportion by volume in the atmosphere (ppm = parts per million). *Note*: these values are averages.

†Value in 2006, increasing by ~1–3 ppm annually.

§Value in 2006, increasing by 0–0.01 ppm annually.

As far as oxygen and ozone are concerned, most absorption of incoming solar radiation occurs in the stratosphere and affects the short wavelength 'tail' of the Sun's spectrum (i.e. the ultraviolet). Its absorption has important implications for life on Earth, because proteins and nucleic acids are damaged by radiation with wavelengths less than about 0.29 μm. As Figure 1.11b shows, for most ultraviolet radiation, absorption by oxygen provides an effective filter but, for frequencies approaching 0.29 μm, only ozone has any impact.

Question 1.2

(a) To what extent are (i) visible radiation and (ii) incoming infrared radiation absorbed by atmospheric gases? Which gases are mainly responsible in each case?

(b) Which atmospheric gases are mainly responsible for the absorption of outgoing longwave radiation?

The gases that absorb most energy from incoming solar radiation are oxygen and ozone, water vapour and carbon dioxide. For outgoing longwave radiation, ozone, water vapour and carbon dioxide are again important, along with nitrous oxide (N_2O) and methane (CH_4). Much of the outgoing longwave radiation is in the range 8–13 μm, where there is relatively little absorption. This range is therefore known as the **atmospheric window**. Gases that absorb radiation reradiate energy *in all directions* and the absorption of outgoing longwave radiation and the subsequent reradiation, much of it back towards the Earth's surface, is what keeps the Earth's average surface temperature (~15 °C) so much higher than the effective planetary temperature (~ –18 °C) at which the system as a whole emits radiation to space. This is referred to as the **greenhouse effect**.

Figure 1.12 summarises the overall energy budget for the Earth and its atmosphere, including all the factors considered so far. The 100 'units' of

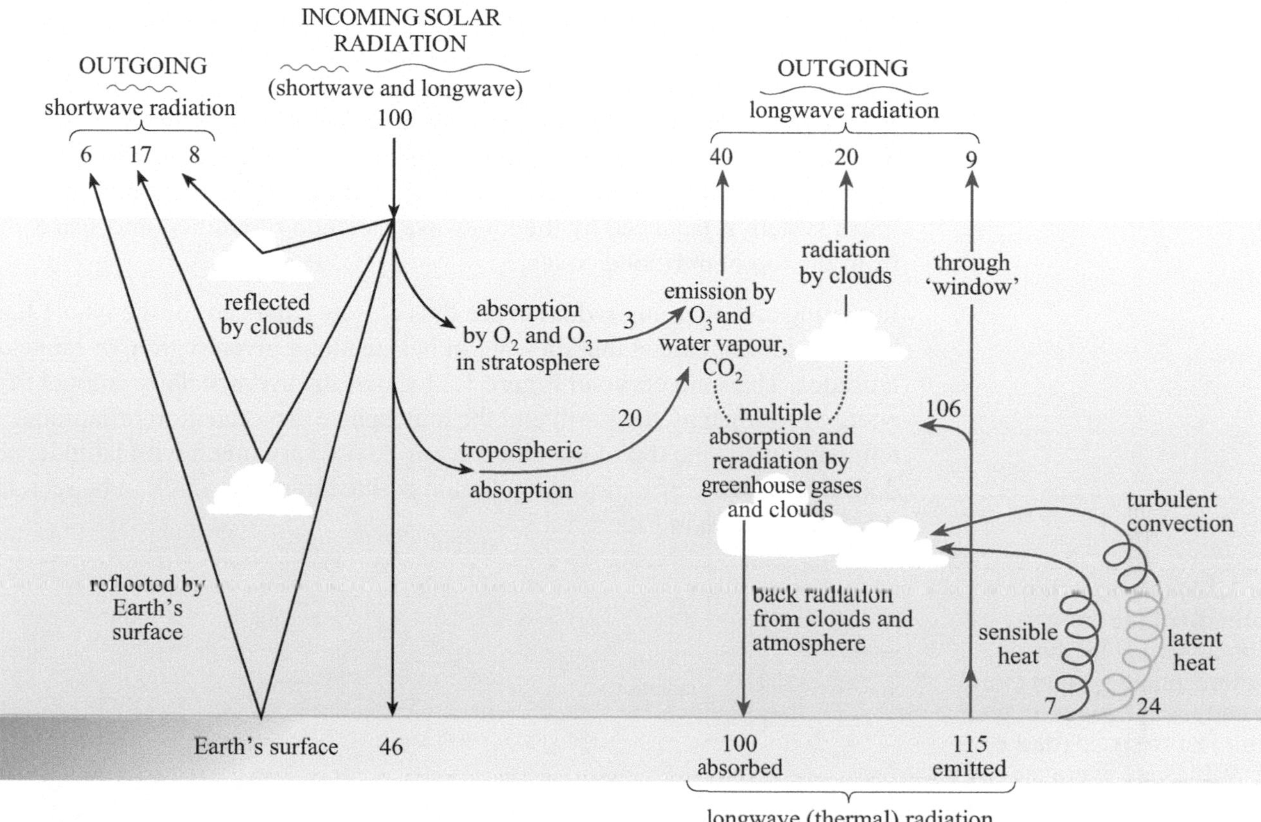

Figure 1.12 Schematic diagram of the overall energy budget of the atmosphere. Values for the outgoing radiation have been measured by satellite-borne radiometers; whereas the *reradiated* radiation (back radiation) has a longer wavelength than the incoming radiation, *reflected* radiation has the same wavelength after reflection as before. Other values are derived from model calculations or measurements, and you may find slightly different values given elsewhere. The effects of greenhouse gases and clouds result in the energy that the Earth's surface radiates (115 units) being greater than that originally absorbed as solar radiation (46 units).

incoming solar radiation represent the effective solar flux at the top of the atmosphere of ~ 343 W m^{-2}. Note that:

- *at the top of the atmosphere*, incoming and outgoing radiation are in balance (100 units = 6 + 17 + 8 units reflected + 40 + 20 + 9 units reradiated)
- *within the atmosphere*, total energy absorbed, from both the Sun and back radiation (20 + 3 + 106 + 24 + 7 = 160 units), balances the total energy emitted, both as back radiation and from the top of the atmosphere (100 + 40 + 20 = 160 units).

The percentage of incoming solar energy reflected by the Earth (i.e. 6 + 17 + 8 = 31%) is close to our earlier estimate for the Earth's albedo.

Note that liquid water in the form of clouds absorbs and reradiates longwave radiation in the same way as water vapour does. Clouds thus play a very important role in the greenhouse effect but, unlike gases, they also strongly affect the Earth's albedo. The net contribution of clouds to climate change is extremely difficult to gauge.

In summary, although around 30% of the shortwave energy input from the Sun is reflected back to space, while some of the remainder is absorbed directly by the atmosphere, most of it is absorbed at the Earth's surface, only to be re-emitted at longer wavelengths to the atmosphere. Little of this longwave radiation is radiated directly into space however, as most is absorbed in the atmosphere, particularly by carbon dioxide, water vapour and cloud droplets. The atmosphere is therefore heated primarily from below and the overall energy budget of the Earth system is balanced by the longwave radiation re-emitted into space, mostly from the top of the cloud cover.

Incoming and outgoing radiation are thus closely balanced for the whole Earth, but that does not mean that they are in balance for a given region, or range of latitudes. The solid curve in Figure 1.13 shows the average daily amount of solar energy absorbed by the Earth and the atmosphere, as a function of latitude. As temperatures at the top of the cloud cover do not vary much with latitude, neither does the intensity of longwave radiation emitted to space: this can be seen from the dashed curve in Figure 1.13.

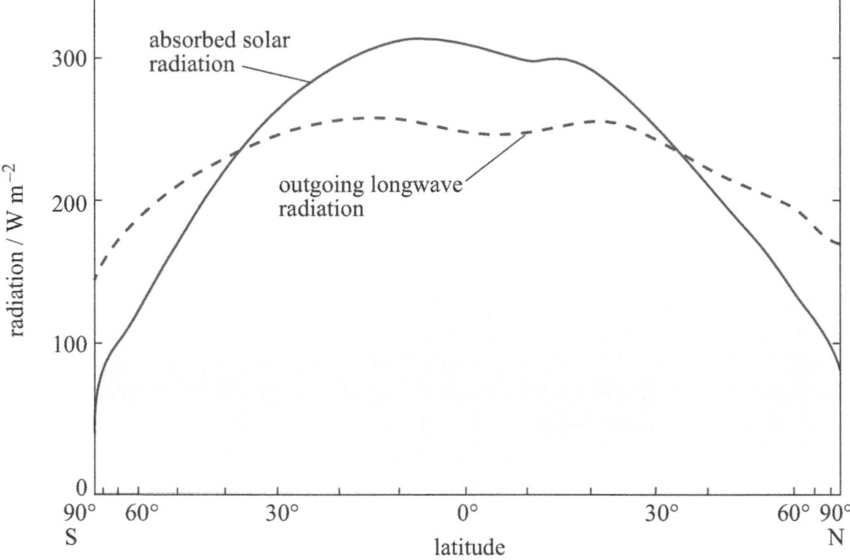

Figure 1.13 The variation with latitude of the solar radiation absorbed by the Earth–atmosphere system (solid curve) and the outgoing longwave radiation lost to space (dashed curve). Values are averaged over the year, and the latitude axis is scaled according to the area of the Earth's surface in different latitude bands so that the area under the graph is proportional to the net radiation of the planet in each band. (Vander Haar and Suomi, 1971)

Question 1.3

(a) Looking first at the full curve for the solar radiation absorbed by the Earth, give two reasons why it has the general shape that it does. *Hint*: refer to Figures 1.1 and 1.4, and Table 1.1.

(b) Over which latitudes does the Earth–atmosphere system have a net gain of heat, and over which latitudes does it have a net loss?

Note that together the two regions in Figure 1.45 (in the 'Answers to questions' section towards the end of this book) marked as net loss are approximately equal in area to the region marked as net gain, demonstrating that – at least over short timescales – the overall Earth–atmosphere radiation budget is in balance, so the Earth is neither cooling down nor heating up. Over timescales of decades or more, however, small imbalances may lead to net heating or cooling.

Despite the positive radiation balance at low latitudes and the negative one at high latitudes, there is no evidence that low-latitude regions are steadily heating up while high-latitude regions are steadily cooling as a result of natural processes. The reason for this is the continual redistribution of heat around the globe by winds in the atmosphere and currents in the ocean, which are the subject of the next section.

It may have struck you that in this text an implicit assumption has been made: namely, that the intensity of radiation emitted by the Sun remains constant. Is this a valid assumption? The answer is no, it isn't. For one thing, according to theories of stellar evolution, the amount of radiation emitted by the Sun at the formation of the Solar System 4600 Ma ago would have been only 70–75% of what it is now. Changes in solar intensity have also been shown to affect climate on timescales of decades to centuries; indeed, while the dominant cause of present global warming is human activity, climatic changes in the 20th to 21st centuries can only be fully explained when both natural and **anthropogenic** effects are taken into account.

Radiation from the Sun is an external input to the Earth's climate system that drives or forces much of the activity in the system. Such an external input is described as a **forcing function**, where the word 'function' refers to the distribution or map of the forcing in time or space, represented in this case by Figure 1.6. The human emission of pollutant gases can similarly be considered as being an external input (in this case external to the natural part of the Earth system), and would similarly give rise to a forcing function with a certain distribution in time and space.

The size of the response of a system to a forcing of given magnitude is referred to as its **sensitivity**, the response of the Earth's surface temperature to a change in solar forcing being one example. Unfortunately, the Earth's sensitivity to climate forcing is hard to measure, and (as implied in the discussion of the effect of Milankovich cycles) hard to predict. Observation and modelling suggest that an increase in the average amount of incoming solar radiation of 4.0–4.5 W m^{-2} could cause a change in temperature of anything from between 1.5 °C and 5.5 °C, which means that the climate's sensitivity is of the order of 0.5–1.0 °C for 1 W m^{-2} of radiative forcing (say 0.75 °C per W m^{-2} of forcing).

Note the uncertainty in this estimate. One reason for this is the time it takes for the Earth to respond to a change: as you will see shortly, its systems have huge inertia. If the Sun were to be completely extinguished for a few minutes (as it is locally during eclipses), the effects would be negligible but, if the solar flux were to decrease abruptly and permanently by 0.1%, this change would ultimately show up in all aspects of the climate system: for example, the temperature distribution within the atmosphere and the oceans, the extents of polar ice caps and of tropical rainforests, and so on. Each of these changes would feed back to other parts of the system. Thus, in trying to assess the sensitivity of the Earth's climate to changes in forcing, all of the various feedback processes that may come into play have to be taken into account, along with their characteristic timescales. An important implication of this is that the sensitivity of a system relates the change in a particular system property, or output, to the change in a particular input, over a particular timescale. The sensitivity for a different input, output or timescale will generally be different. Nevertheless, the term 'climate sensitivity' is often used as shorthand for the change in average surface temperature resulting from a given change in atmospheric CO_2 concentration or solar forcing over a timescale of up to a few millennia.

Question 1.4

During the Little Ice Age, which altogether lasted from the 15th century until well into the 19th century, different parts of the Earth experienced unusually cold periods, e.g. icebergs became common off Norway, glaciers advanced down valleys and the River Thames froze in winter. It has been suggested that these cold periods, which were most marked from 1640 to 1720, were caused by a prolonged period of low solar activity known as the 'Maunder minimum'. Given that average global temperatures were about 1 °C cooler than today, by how much (as a percentage) would incoming solar radiation (i.e. the effective solar flux) have been below its present value (~343 W m^{-2})?

Note: assume a climate sensitivity of 0.75 °C per W m^{-2} of solar forcing; both theoretical modelling and studies of the Earth's glacial history indicate that this is a reasonable average value.

In fact, various other factors have been proposed as causing (or contributing to) the Little Ice Age, including natural changes in concentrations of greenhouse gases in the atmosphere and changes in patterns of ocean circulation. Although internal to the natural Earth system, such factors can be considered as forcing agents for the atmospheric part of the climate system.

■ Can Milankovich variation in orbital parameters be described as a forcing function?

□ Yes, because their origin is external to the Earth system. In this case the variation is principally a function of time, as shown in Figure 1.8. This is often described as 'Milankovich', 'astronomical', or 'orbital' forcing.

Examples of other forcing functions will be discussed later, but first you will look at how the winds and ocean currents combined form the Earth's air-conditioning and/or central heating or cooling system, which in some cases moderates and in others mediates the effects of the various forcing factors and their influence on the habitability of the Earth.

1.3 The Earth's surface temperature pattern

The redistribution of solar energy from low to high latitudes is the principal driving force behind the motions of the atmosphere and the oceans and, as such, strongly controls the diversity and habitability of environments on the Earth's surface. Not surprisingly, given the complexity of the Earth's geography and the turbulent, chaotic motions of the atmosphere and the oceans, the smooth annual variations of solar forcing seen in Figure 1.6 transform into a much more complicated pattern of average climatic conditions. This section begins by investigating how and why the Earth's 'climate engine' carries out this transformation by considering the geographical variation of average surface temperature.

Figure 1.14 (overleaf) shows daytime surface temperatures as measured by a satellite-borne radiometer. In Figure 1.14a and b, temperatures below 0 °C are blue and dark blue (and occur mostly at high latitudes); the highest temperatures are shown by red and dark brown (and are at lower latitudes). In January (Figure 1.14a), temperatures at high northern latitudes are very low, having fallen below 0 °C in eastern Europe and northern USA, and are approaching –30 °C over Siberia and most of Canada. In the Southern Hemisphere it is summer, with mid-latitude temperatures of 20–30 °C. By July (Figure 1.14b), areas of the Northern Hemisphere have warmed by 10–20 °C. The Greenland ice cap remains frozen, but temperatures are considerably lower in the Antarctic, where there is now a large area of sea ice. Temperature differences between January and July are shown in Figure 1.14c. Areas of greatest increase in temperature are red and dark brown (between the Tropic of Cancer and the North Pole); areas of greatest decrease in temperature are bright blue and dark blue (ranging from the continental areas between the Tropic of Capricorn and the Antarctic Circle, and the Antarctic Circle to the South Pole). The greatest changes are over land in mid-latitudes, while low latitudes are more stable.

- ■ With what are the most extreme departures from a simple east–west trend associated?
- □ The distribution of continents and oceans. In the hemisphere experiencing summer, at a given latitude continental temperatures are generally *higher* than sea-surface temperatures; in the hemisphere experiencing winter, continental temperatures are generally *lower* than sea-surface temperatures.

This point is even more forcefully borne out in Figure 1.14c. The large areas of red and brown in the Northern Hemisphere and the even bigger areas of blue and green in the Southern Hemisphere show that the greatest warming and cooling occurs over the continents. Seasonal changes of up to 30 °C occur over land in both hemispheres; by contrast, seasonal changes in sea-surface temperature, which are greatest in mid-latitudes, rarely exceed 8–10 °C.

- ■ Read Box 1.4 (overleaf). Which one of the properties of water listed explains the contrast in seasonal temperature changes between land and sea?
- □ The high specific heat of water, which results in the thermal capacity of the oceans being much greater than that of the continents. In other words, a much greater heat input is needed to raise the temperature of a mass of ocean by 1 °C than is needed to raise the temperature of the same mass of continental rock by 1 °C. As a result, continental areas heat up and cool down more quickly than oceanic areas.

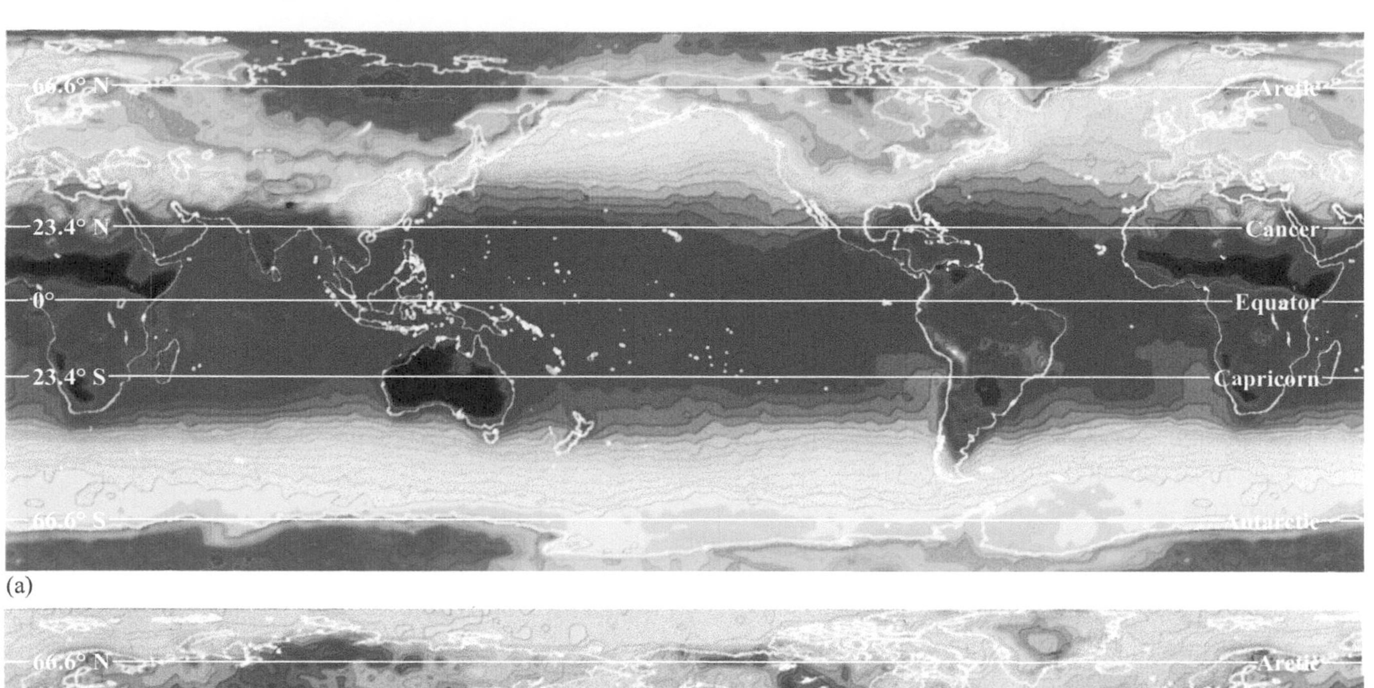
66.6° N
Arctic
23.4° N
Cancer
0°
Equator
23.4° S
Capricorn
66.6° S
Antarctic
(a)

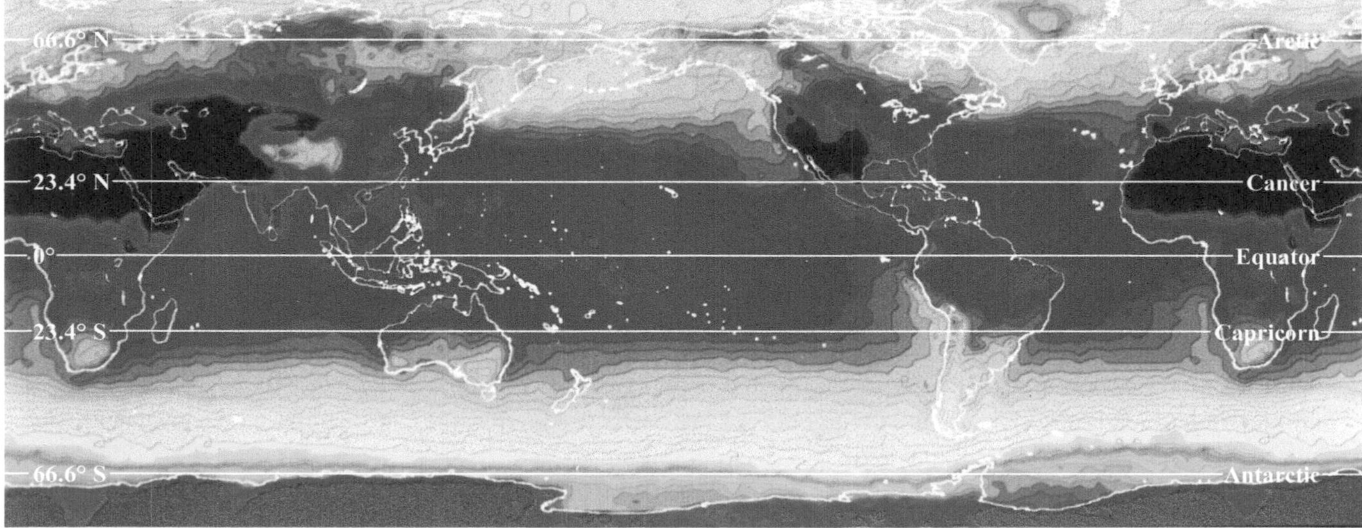
66.6° N
Arctic
23.4° N
Cancer
0°
Equator
23.4° S
Capricorn
66.6° S
Antarctic
(b)

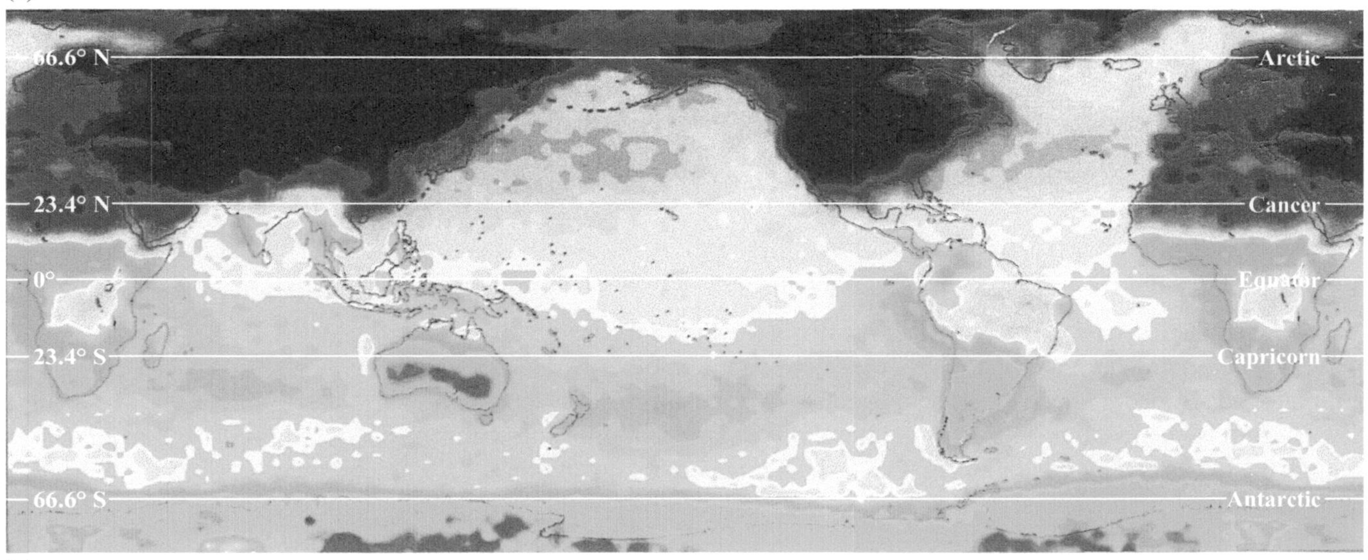
66.6° N
Arctic
23.4° N
Cancer
0°
Equator
23.4° S
Capricorn
66.6° S
Antarctic
(c)

◀ **Figure 1.14** (a) and (b) Daytime surface temperatures for January and July respectively, as measured by a satellite-borne radiometer (which measures thermal radiation emitted by the Earth). Surface temperatures generally decrease from low to high latitudes, but the **isotherms** (contours of equal temperature) do not run simply east–west (blue denotes low temperature). (c) Differences in surface temperature between January and July (large temperature differences are denoted by reds and browns). (See text for explanation.) (Internetwork Inc., NASA/JPL/GSFC)

Box 1.4 Properties of water that are of importance for the climate

The relevance of some of these properties of water to the Earth's climate may already be obvious; the importance of others will become clear as you read further.

- At the range of temperatures found at the surface of the Earth, water can exist as a gas (water vapour), as a liquid, and as a solid (ice).
- Its specific heat, i.e. the amount of heat needed to raise the temperature of 1 kg of water (in any state) by 1 °C (or 1 K) is 4.18×10^3 J kg^{-1} °C^{-1}. This is the highest specific heat of all solids and liquids except ammonia.
- Its latent heat of fusion, i.e. the amount of heat needed to convert 1 kg of ice to water at the same temperature (and the amount of heat *given up* to the surrounding environment when 1 kg of ice forms), is 3.3×10^5 J kg^{-1}. This is the highest latent heat of fusion (or freezing) of all solids and liquids except ammonia.
- Its latent heat of evaporation, i.e. the amount of heat needed to convert 1 kg of liquid water to water vapour at the same temperature (and the amount of heat *released* to the surrounding environment when 1 kg of water vapour condenses), is 2.25×10^6 J kg^{-1}. This is the highest latent heat of evaporation (condensation) of all substances.
- It dissolves more substances, and in greater quantities, than any other liquid.
- It conducts heat more efficiently than most other liquids naturally occurring on the Earth.
- Its temperature of maximum density decreases with increasing salt content; pure water has its maximum density at 4 °C, but the density of seawater increases down to its freezing point at about −1.9 °C.
- The density of ice is less than that of water, the result of which is that ice occupies more space than the water from which it formed, and ice floats on water. (For most substances, the solid phase is denser than the liquid phase.)
- Compared with other liquids, it is relatively transparent.

The property that the oceans have of heating up and cooling down slowly is sometimes referred to as their 'thermal inertia'. It is part of the reason why the range of temperatures found in the oceans is less than half that which occurs on land and one of the principal reasons for zonal (east–west) variations in surface temperatures. In the next section, you will look at another reason for this temperature difference: atmospheric and oceanic circulation patterns.

1.4 The Earth's air-conditioning and heating systems

In Section 1.2.1, you saw that the Earth's radiation budget has an excess at low latitudes and a deficit at high latitudes (Figure 1.13). Figure 1.15 shows, very schematically, how heat is redistributed over the surface of the Earth. Put simply, the three principal processes involved are:

1 **Wind-driven surface currents**

Under the influence of winds, surface ocean water warmed at low latitudes flows polewards in surface currents, while that cooled at high latitudes flows equatorwards.

2 **Winds in the atmosphere**

Moving air takes up heat from the surface of the oceans and continents. Warm air rising at low-pressure regions such as the Equator and moving polewards in the upper troposphere transports heat from low to high latitudes, as does any warm air moving polewards. Net polewards heat transport results

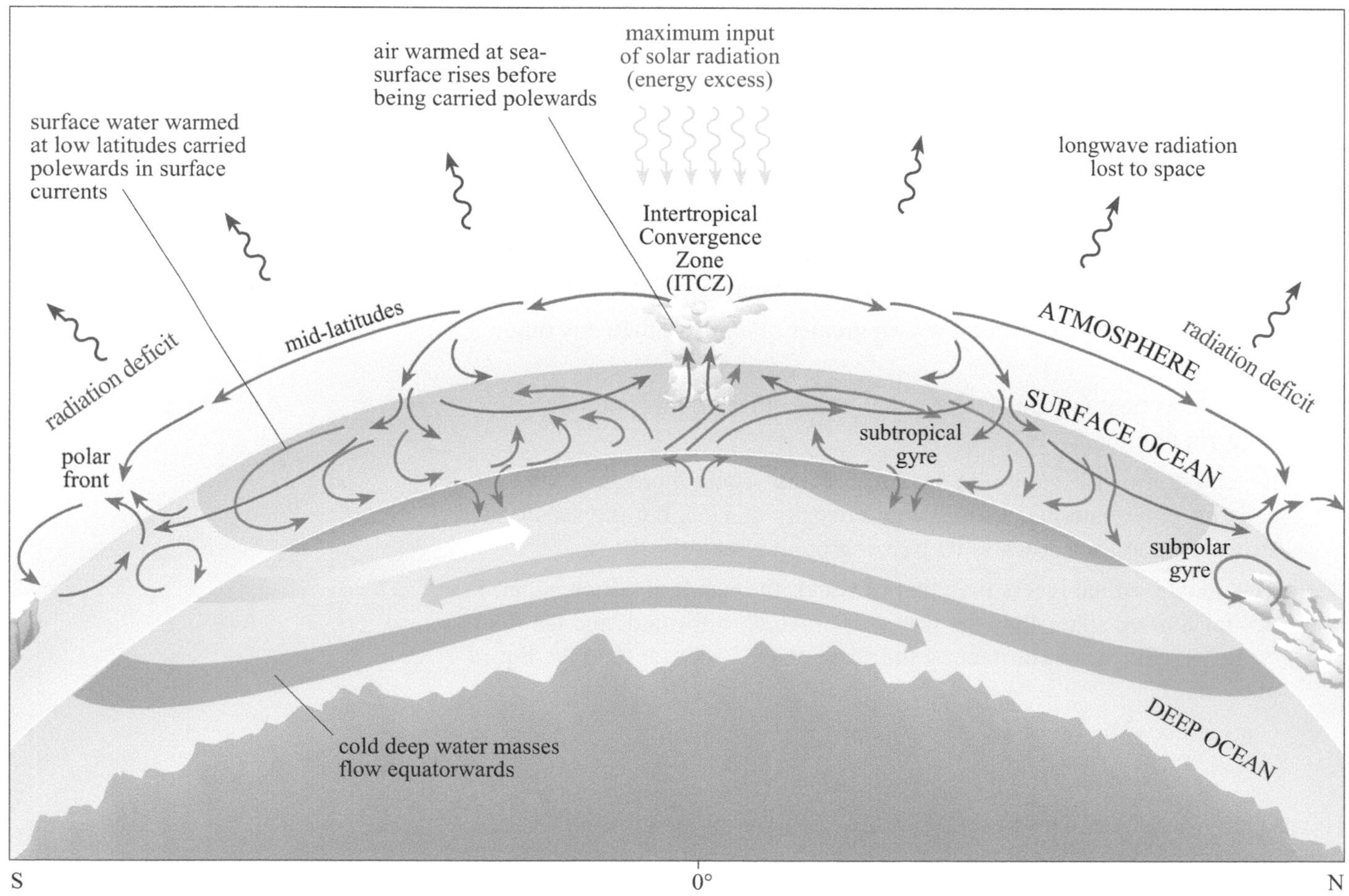

Figure 1.15 Schematic diagram of the Earth's heat-redistribution system (not to scale). The three interlinked circulatory systems are: wind-driven surface currents, winds in the atmosphere, and density-driven currents in the deep ocean. The **Intertropical Convergence Zone (ITCZ)** is the region where the wind systems of the two hemispheres meet. (See text for more details.)

because the air that moves towards the Equator at lower level to compensate is relatively cool from contact with ice and cold land and sea surfaces. Furthermore, warm, humid air from the Equator releases latent heat due to the condensation of water during its journey polewards.

3 **Density-driven currents in the deep ocean**

Surface ocean water cooled at high latitudes increases in density, sinks and flows equatorwards in the deep ocean.

Another way of looking at these three processes is as interrelated circulatory systems, largely driven by the transfer of heat and of momentum (i.e. mechanical or frictional forcing) at the ocean surface. An important feature of both oceanic and atmospheric circulation systems is **convection**. When a pan of water is heated from below, heat is transmitted through the pan to the water at the bottom of the pan by conduction. Being heated, this water expands, becomes less dense and rises. On reaching the surface, the warmed water begins to lose heat to the air; it cools, becomes denser and sinks, then is heated again and rises, and so on (Figure 1.16), forming convection 'cells'. (You can sometimes see such cells when cooking spaghetti, because the spaghetti strands tend to become aligned with the flow.)

- Does the convection occurring (i) in the atmosphere, and (ii) in the ocean (see Figure 1.15) resemble that in a pan of water being heated (Figure 1.16)?

- Convection occurring in the atmosphere does resemble that occurring in a pan of water because it is driven by heating from below. Convection in the ocean is apparently driven by cooling from above, but technically the process is identical, as it is only the temperature difference between the top and the bottom that matters.

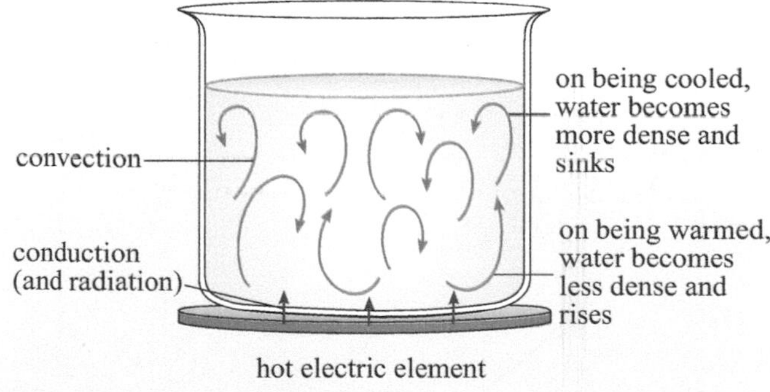

Figure 1.16 The circulatory pattern in a pan of water heated on an electric element.

1.4.1 Transport of heat and water by the atmosphere

Warm air rises or, to be more precise, air that is warmer than its surroundings (and is therefore less dense) rises. As the analogy of the pan of water demonstrates, it is the convective bulk mixing of water that distributes the heat supplied at the bottom of the pan so that eventually all of the water becomes warm. So it is for the atmosphere: when air is warmed by contact with a warm sea or land surface and rises, it is replaced by cooler air, which is warmed in turn. By contrast, the transfer of energy by conduction occurs at the molecular level (see Box 1.3) – if you had to rely on conduction to heat a pan of water, you would have to wait a very long time indeed.

What happens in practice, however, is complicated by two factors:

- air, like all fluids, is compressible
- air contains variable amounts of water vapour.

First, consider compression: when a fluid is compressed, the internal energy that it possesses per unit volume by virtue of the motions of its constituent atoms, and which determines its temperature, is increased. Thus, a fluid heats up when compressed (a well-known example is the compression of air in a bicycle pump), and cools (i.e. undergoes a decrease in energy per unit volume) when it expands (this is what occurs in the cooling system of a refrigerator). Changes in temperature that occur in this way, and not as a result of a gain or loss of heat from the surroundings, are described as **adiabatic**. When warmed air rises, the atmospheric pressure it is subjected to decreases (see the solid curve in Figure 1.17), and so it expands and becomes less dense; it therefore cools by adiabatic expansion. It will only continue to rise as long as it continues to be warmer than its surroundings.

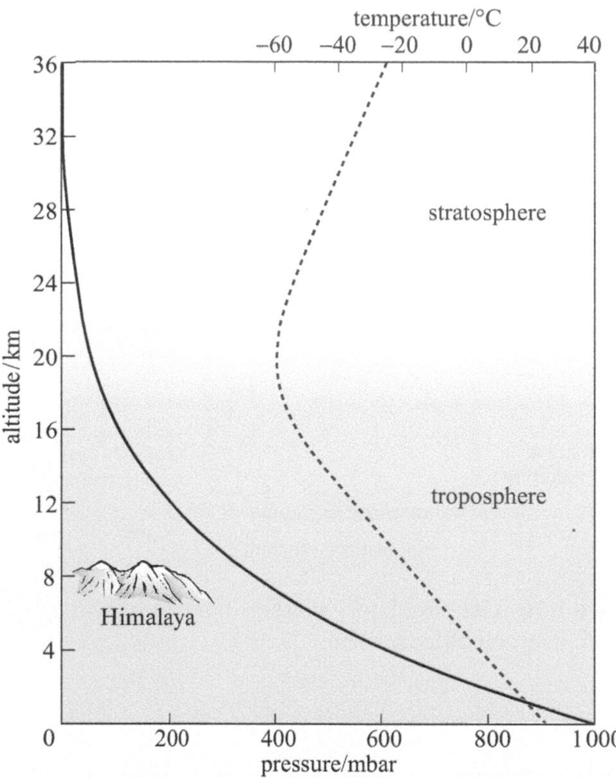

Figure 1.17 Schematic diagram to show how both pressure (solid curve) and temperature (dashed curve) decrease with increasing height in the lowermost atmosphere, or troposphere. Above that, in the stratosphere, temperature increases again. The curves are generalised, and intended to illustrate the general principle only. The thickness of the troposphere and the actual values of temperature and pressure near the Earth's surface vary with latitude and location; the temperature curve shown is appropriate for a latitude of about 30°. One millibar (mbar) is one-thousandth of a bar, where 1 bar is approximately 1 atmosphere (atm).

Imagine a parcel of air heated by contact with the ground and beginning to move upwards. Temperature decreases with height in the lower atmosphere (see the dashed curve in Figure 1.17) but, as long as the adiabatic decrease in temperature of a rising parcel of air is *less* than the decrease of temperature with height in the lower atmosphere, the rising parcel of air will be warmer and less dense than its surroundings and will continue to rise: the situation will be *unstable*, in the sense that convection, once initiated, is reinforced. Furthermore, any small random motions (which always occur) will initiate convection. Such conditions are said to be 'conducive to convection'. On the other hand, if the adiabatic cooling of the rising parcel of air is sufficient to reduce its temperature to below that of the surrounding air, conditions are said to be *stable*, in that the air will sink back to its original level and convection will be inhibited.

So far, the assumption is that the rising parcel of air is dry: that is, it has no gaseous water vapour. Rising air, particularly over the ocean, may be saturated with water vapour or become saturated as a result of adiabatic cooling (warm air can hold more water vapour than cool air). Continued rise and associated adiabatic cooling result in cloud formation (Box 1.5 overleaf). This condensation releases latent heat to the rising air (compare with Box 1.4), offsetting the effect of adiabatic cooling. In other words, humid air convects much more easily than dry air because the condensation of water vapour releases additional heat energy, keeping the rising air less dense than the surrounding air for longer than would be the case for dry air.

Rising air warmed locally by conduction and/or convection becomes part of the global-scale atmospheric circulatory system shown schematically in Figure 1.15 (and Figure 1.20), whose horizontal flows over the surface of the Earth are the surface winds. Wind systems redistribute heat partly by the **advection** (bulk transport) of warm air masses into cooler regions (and vice versa), and partly by the transfer of latent heat bound up in water vapour, which is released when the water vapour condenses to form cloud in a cooler environment, perhaps thousands of kilometres from the site of evaporation. Most of this moisture comes from the surface of the ocean; indeed, at any one time a large proportion of the water in the atmosphere (water vapour and clouds) has only recently evaporated from the tropical ocean, and polewards transport of warm, humid air is the most important way in which heat from the ocean at low latitudes is transferred to higher latitudes.

The amount of evaporation from land depends on the moisture content of the exposed soil or rock. Vegetation is also a source of atmospheric moisture, both through simple evaporation from surfaces and through **transpiration**, whereby water drawn up from the soil by roots is lost to the atmosphere through pores in leaves; together, these two processes are known as **evapotranspiration**.

■ From which type of land areas would you expect transfer of latent heat to the atmosphere to be greatest?

□ From tropical rainforests, where evapotranspiration releases large amounts of water vapour to the atmosphere.

Indeed, being both warm and a good source of atmospheric moisture, rainforests behave climatically rather like the tropical ocean.

As mentioned earlier, heat is also transferred from the Earth's surface to the atmosphere by conduction and convection. This is referred to as the transfer of **sensible heat** (heat which can be felt, or sensed) and it increases as the temperature difference between the Earth's surface and the overlying atmosphere increases.

■ By reference to Figure 1.14a–c, would you expect the loss of sensible heat to the atmosphere to be greatest from land surfaces or from the ocean?

□ From land surfaces, especially at mid and low latitudes in summer when the land masses have heated up.

At low latitudes, sensible heat loss from land surfaces to the atmosphere is an order of magnitude greater than heat loss by evaporation. It is true that there is also a loss of sensible heat from the sea to the air, if the sea-surface is warmer than the air above it (which is the case more often than not) but, as far as the ocean is concerned, *an order of magnitude more* heat is transferred to the atmosphere by evaporation and subsequent condensation than is transferred by conduction, even when enhanced by convection.

Cumulus and cumulonimbus clouds (over both the land and the sea), are visible evidence of convection in the atmosphere (see Box 1.5). Cloud formation, the turbulence of atmospheric convection and the winds that redistribute heat over the Earth's surface are mostly confined to the lower atmosphere. This is known as the **troposphere** (from the Greek word *tropos*, meaning 'turn', i.e. causing mixing), which makes up ~80% of the total mass of the atmosphere. Temperature decreases with height within the troposphere (Figure 1.17) up to its upper boundary (the *tropopause*), then begins to increase again in the overlying stratosphere because of absorption of radiation by ozone there. It is this temperature *inversion* which generally prevents convection in the lower atmosphere from reaching any higher, since rising air from below is almost always denser than warmer air above. There is some interchange of air between the troposphere and the stratosphere, particularly in mid-latitudes, and in certain circumstances rapidly rising air masses can overshoot the tropopause. In general, however, stratospheric winds do not interact strongly with the winds of the lower atmosphere and so do not directly affect conditions at the Earth's surface. It is within the troposphere that the Earth's weather occurs.

Box 1.5 Clouds

Clouds form when water vapour in the atmosphere condenses around solid particles or nuclei (e.g. pollen grains, dust or salt from sea spray) to form tiny water droplets or, at greater heights and lower temperatures, ice crystals. Condensation can also occur on **aerosols**, which are minute droplets, often of sulfate compounds, notably sulfuric acid formed from water vapour and sulfur dioxide (SO_2) emitted from volcanoes or produced by industrial processes.

As pressure decreases with height in the atmosphere (solid curve in Figure 1.17), water droplets can exist in clouds at temperatures down to −12 °C, as a mixture of water droplets and ice crystals at temperatures between −12 °C and −30 °C, and predominately as ice crystals at temperatures less than −30 °C. Clouds consist entirely of ice crystals at temperatures below −40 °C.

As shown in Figure 1.18, clouds can have a variety of forms: for example, cumulus and cumulonimbus are associated with convection and have flat bottoms (marking the height at which condensation began to occur) and bubbly tops; stratus clouds form by more gentle uplift of air at a front or as a result of the presence of mountains and are layered. Wispy cirrus-type clouds are made of ice and are usually associated with flow in jet streams (see Figure 1.27), cumulus-type clouds consist of water droplets, although the tops of towering cumulonimbus clouds consist of ice crystals. Stratus-type clouds, which at ground level are called fog, can be made of either.

All clouds are highly reflective, being white when illuminated (they only appear grey if their undersides are in shadow). However, the veil-like structure of cirrus clouds means that they have a fairly low albedo (<20%), while stratocumulus and cumulonimbus, which are thicker and consist of more densely packed droplets and/or ice crystals, have fairly high albedos of about 50% and 70%, respectively.

The role of clouds in the climate system is still unclear. Because of their high albedo, they may greatly reduce the amount of solar radiation reaching the Earth's surface, particularly at high and low latitudes (see the discussion in connection with Table 1.1). As you will see shortly, however, they absorb radiation as well as reflect it. Last but not least, their complex three-dimensional shapes make their effect on the radiation budget much more difficult to quantify than, say, land ice. The role of clouds in the climate system is discussed in more detail later in this chapter.

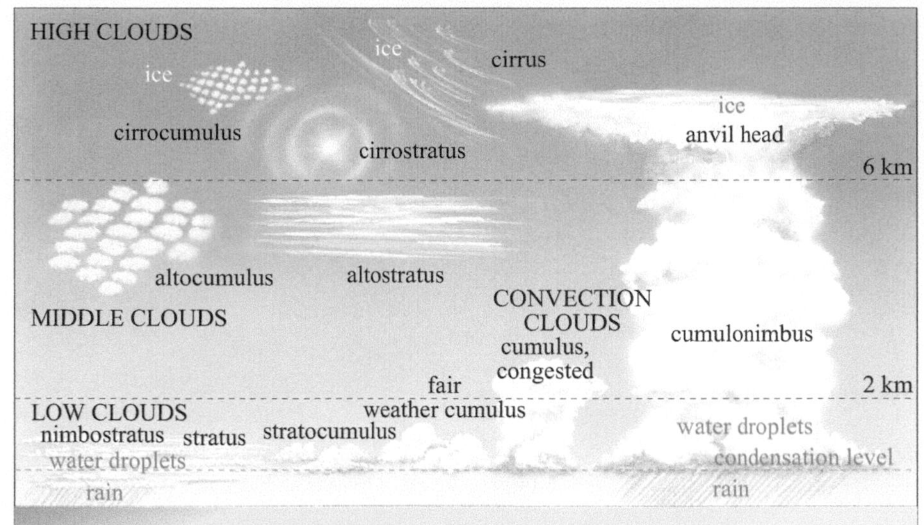

Figure 1.18 Various types of cloud and their characteristic levels and/or extents in the atmosphere.

Figure 1.19 shows how temperature varies with height in the atmosphere as a whole (compare with Figure 1.17 for the troposphere), and that above the troposphere and the stratosphere there are two more changes in temperature gradient within two more 'spheres', although it is generally thought that these outer 'spheres' are not important influences on the Earth's climate system. Part of the reason for this is that the atmosphere is so rarefied at these altitudes that the heat content of the air is very low, even when the temperature (determined by the vibrations of individual molecules; Box 1.3) is high. The increase of temperature in the stratosphere is caused by absorption of ultraviolet radiation there by oxygen and ozone. This absorption of energy results in that layer heating up directly, rather than by absorption of longwave radiation from below, as occurs in the troposphere.

Figure 1.20 (overelaf) shows what the wind system and surface atmospheric pressure pattern would be if the Earth were completely covered by ocean.

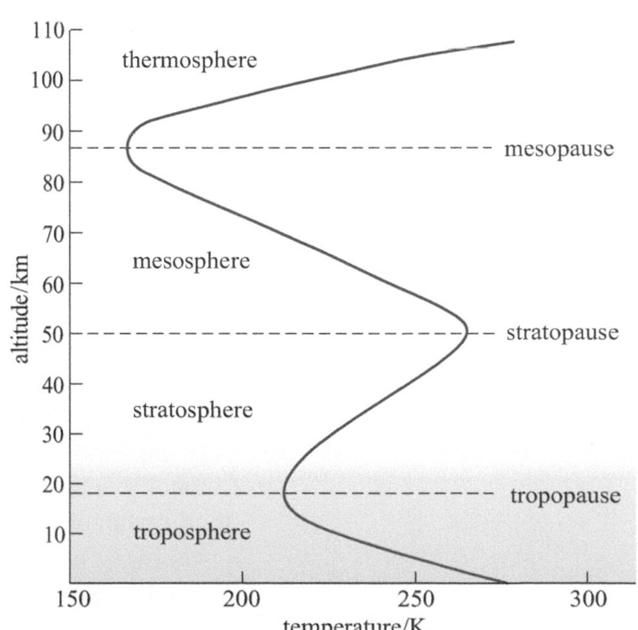

Figure 1.19 The vertical temperature 'structure' of the atmosphere, defined by the way in which temperature varies with altitude. In each successive 'sphere', the temperature gradient is reversed: temperature decreases with height in the troposphere, so that conditions are conducive to convection; by contrast, temperature increases with height in the stratosphere, and vertical motions are suppressed. Note that, as indicated by the vertical air movements in Figure 1.20, the tropopause is higher at low latitudes (where it is at ~17 km) than at high latitudes (where it is at ~8–10 km).

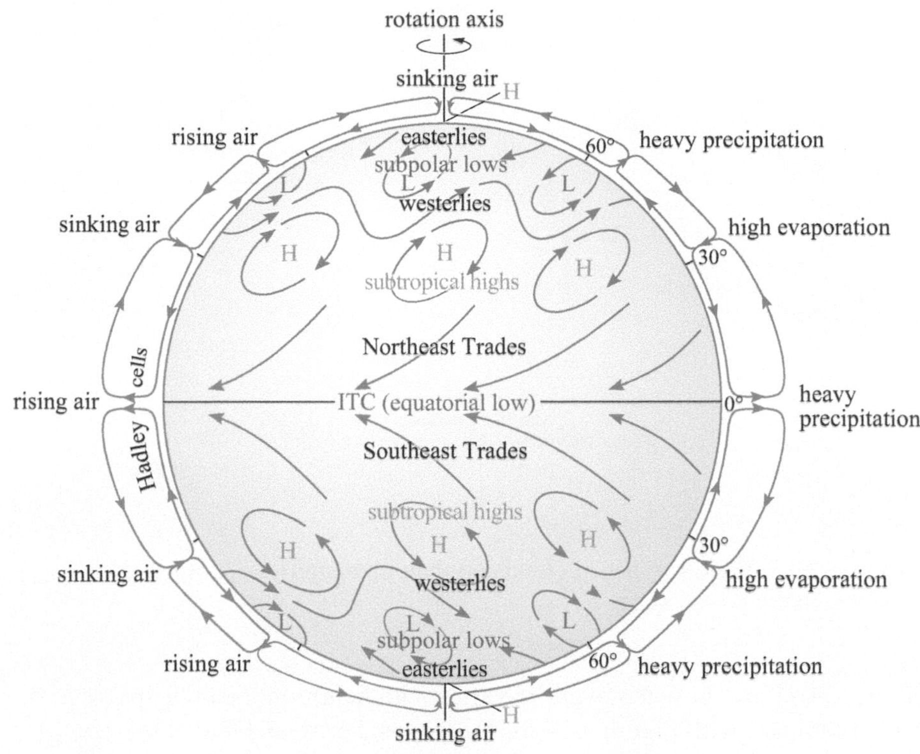

Figure 1.20 Wind system for a hypothetical water-covered Earth, showing the major surface winds and the zones of low and high pressure. Vertical air movements are indicated on the left-hand side of the diagram; characteristic surface conditions are given on the right-hand side.

Question 1.5

(a) By reference to Figure 1.20, describe how *surface* wind directions relate to regions of high and low surface pressure.

(b) By reference to the vertical air motions shown in Figure 1.20, describe how vertical air movements relate to regions of high and low surface pressure.

Air sinks and flows anticyclonically *outwards* from regions of high surface pressure, and flows cyclonically *inwards* and rises at regions of low surface pressure (Figure 1.21).

But why are the paths of winds curved? And why isn't the global wind system a simple convection system in which air sinks only at the poles, and surface winds blow directly from the poles to the Equator, as illustrated in Figure 1.22? The answer lies in the rotation of the Earth. Air masses moving above the surface of the Earth are bound extremely weakly to it by friction and, as they move, the Earth turns beneath them. The resulting deflection of winds relative to the surface of the Earth is known as the **Coriolis effect** and it applies equally to ocean currents as to the atmosphere. The effect increases with latitude, from no deflection at the Equator to a maximum at the poles. The Coriolis effect can be understood in terms of an imaginary Coriolis *force*, acting at right angles to the direction of flow:

- in the Northern Hemisphere, the Coriolis force acts to the right of a flow and, in the absence of balancing forces, turns it clockwise
- in the Southern Hemisphere, the Coriolis force acts to the left of the flow, and in the absence of balancing forces, turns it anticlockwise.

The origin of the Coriolis effect is explained in more detail in Box 1.6.

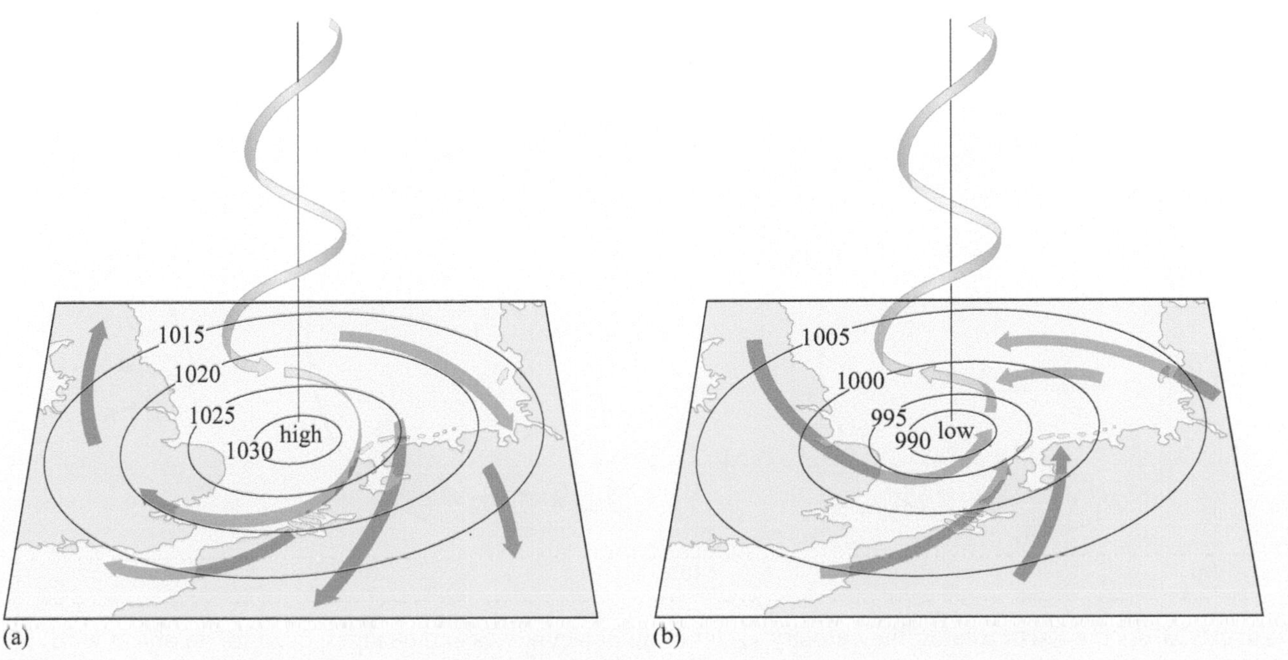

Figure 1.21 (a) Air spiralling downwards and outwards from an atmospheric high (an anticyclone). (b) Air spiralling inwards and upwards towards an atmospheric low (a cyclone or depression). Note that the flow directions shown are for the Northern Hemisphere; in the Southern Hemisphere, flow around anticyclones is anticlockwise, and flow around cyclones is clockwise (Figure 1.20). Contour values are typical atmospheric pressures at sea level, expressed in millibars (10^{-3} bar). Surface wind speeds typically reach a few tens of metres per second.

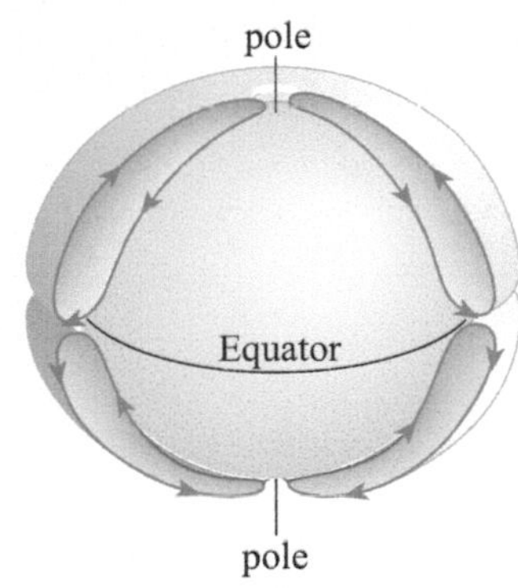

Figure 1.22 Simple hypothetical wind system for a non-rotating Earth.

Box 1.6 The Coriolis effect

An object which is stationary relative to the Earth is, in fact, moving in a circle as the Earth rotates. If it starts to move relative to the Earth, its total velocity is the sum of two parts:

- its velocity relative to the Earth
- the circular velocity due to the Earth's rotation.

According to Newton's laws, a change of velocity is by definition an acceleration and therefore requires a force.

Now consider the situation illustrated in Figure 1.23 (overleaf). If the velocity of the object is constant relative to the Earth, the velocity vector itself is also moving in a circle because of the Earth's rotation, but this constitutes a change of absolute velocity, and hence an acceleration. This is the Coriolis effect. Technically, the effect is doubled because the velocity of the object itself relative to the Earth changes the speed and radius of the circular motion and, again, this amounts to a change of velocity and hence an acceleration.

In the ocean and the atmosphere, there is no force to supply the acceleration (which is to the left of the motion in the Northern Hemisphere), and hence water and air do not generally move with constant velocity relative to the Earth, but show a deflection to the right in the Northern Hemisphere and to the left in the Southern Hemisphere. Thus, in the absence of

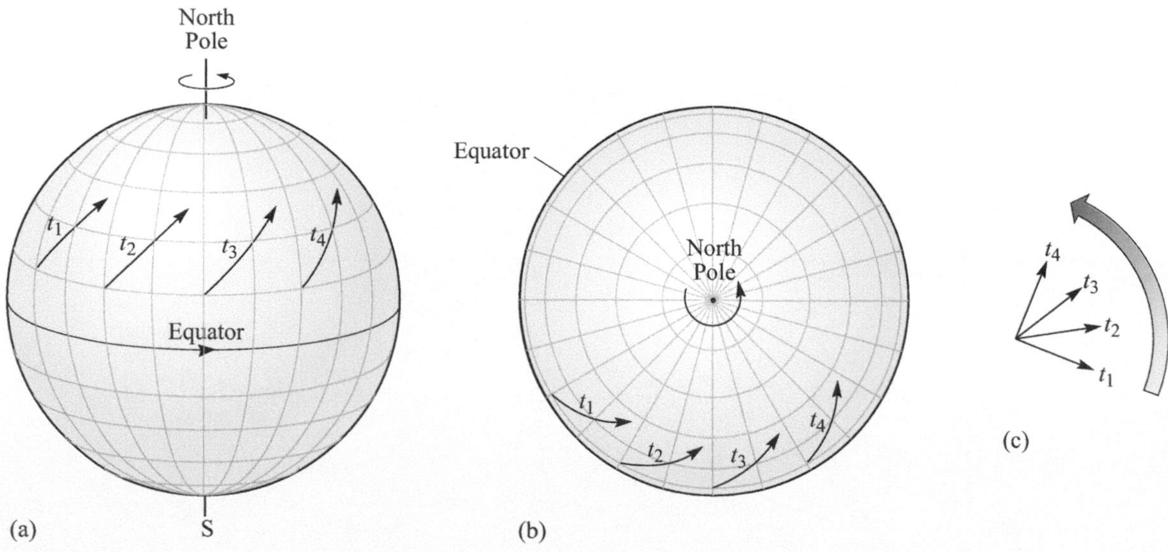

Figure 1.23 As the Earth rotates, the velocity vector of an object moves and changes, even for an object with constant velocity relative to the Earth: (a) shows the velocity of such an object at successive times, assuming, for convenience, it moves a small distance relative to the Earth; (b) shows the same scene looking down on the North Pole, illustrating how the velocity rotates towards the axis; (c) again shows the change in velocity vector, but relative to a single point. This change amounts to the Coriolis effect.

a *real* force to supply the acceleration, there is a deflection (as if by an *imaginary* force) in the opposite direction. Note that it is only the horizontal component of the acceleration that is relevant because any vertical part is swamped by other forces.

The change in velocity is always towards the axis of rotation; thus, on the Equator it has only a vertical component and is thus irrelevant.

Note, too, that the Coriolis effect is only felt for large horizontal scales or rapid rotations. The familiar centripetal effect, which you feel when driving round a bend, for instance, is not important for understanding global-scale flows because you simply experience it as a modification to gravity.

Figure 1.24 Hypothetical cylindrical Earth, spinning about its axis.

- So, what can be said about the Coriolis effect for motion over a hypothetical *cylindrical* Earth, rotating on its axis as shown in Figure 1.24?
- For a rotating cylindrical Earth, there would be no visible Coriolis effect, except at the flat ends of the cylinder.

Since the acceleration is always towards the axis, the horizontal deflection on a cylinder would be zero, as it is on the Equator, and both Coriolis and centripetal accelerations would simply amount to changes in the effective force of gravity. The ends, however, would still experience deflections, as on a sphere.

The characteristic conditions indicated on the right-hand side of Figure 1.20 are a result of the fact that while sinking air is generally dry, rising air often has a high moisture content. The most dramatic common manifestations of rising (i.e. convecting) moist air are cumulonimbus clouds, the tallest of which form at low latitudes in the Intertropical Convergence Zone (ITCZ), where the wind systems of the two hemispheres meet (Figures 1.15 and 1.20). Here, moist air carried in the Trade Winds converges and rises, resulting in the formation of towering thunderheads.

As a result of the Coriolis effect, the Trade Winds blow towards the Equator not from the north and the south, but from the northeast and the southeast. The convention for naming winds is to use the direction *from* which they blow, so these winds are known as the Northeast Trades and the Southeast Trades. As mentioned above, in the lower atmosphere, pressure is low along the Equator (or, more accurately, along the ITCZ, which corresponds more or less to the region of highest surface temperature), and here air converges and rises. It then moves polewards in the upper troposphere, with much of it sinking in mid-latitudes where, as a result, the pressure at the Earth's surface is high. Thus, the Trade Winds are the surface expression of a helical circulatory system known as the **Hadley circulation** or Hadley cells (Figure 1.25). Strictly speaking, only the latitudinal–vertical component of the circulation, as shown in Figure 1.20, is named the Hadley circulation.

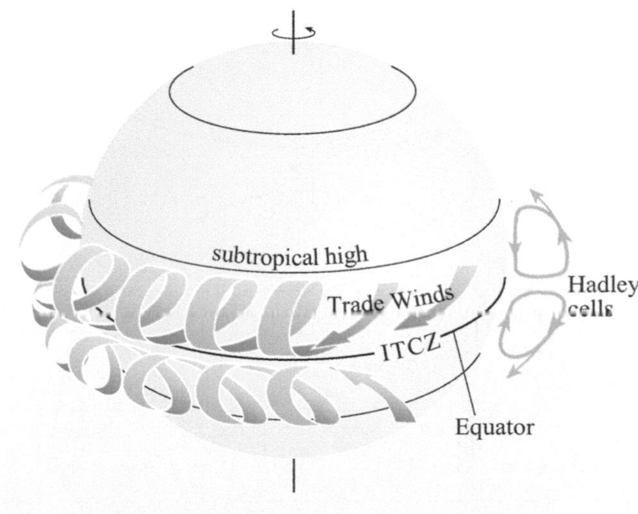

Figure 1.25 The helical circulation patterns of which the Trade Winds form the surface expression; the north–south component of this helical circulation is known as the Hadley circulation; the two 'Hadley cells' can be seen on either side of the Equator (see also the left-hand side of Figure 1.20).

As a result of the Coriolis effect increasing with increasing latitude, there is a change in the *style* of atmospheric circulation with latitude. As the Trade Winds blow in tropical latitudes where the Coriolis effect is relatively small, they are deflected laterally only slightly. At higher latitudes, where the degree of deflection is much greater, vortices tend to form in the lower atmosphere with a predominantly horizontal, or slantwise, circulation. As shown in Figure 1.21, these are the anticyclonic and cyclonic winds familiar from weather charts.

- How can mid-latitude cyclones and anticyclones contribute to the polewards transport of heat?

- Moving air masses mix with adjacent air masses along *atmospheric fronts*, and heat is exchanged between them. Air moving northwards in a Northern Hemisphere cyclone or anticyclone will transport relatively warm air polewards, while air returning equatorwards will have been cooled. This is, in effect, a kind of large-scale stirring, and is shown schematically in Figure 1.26.

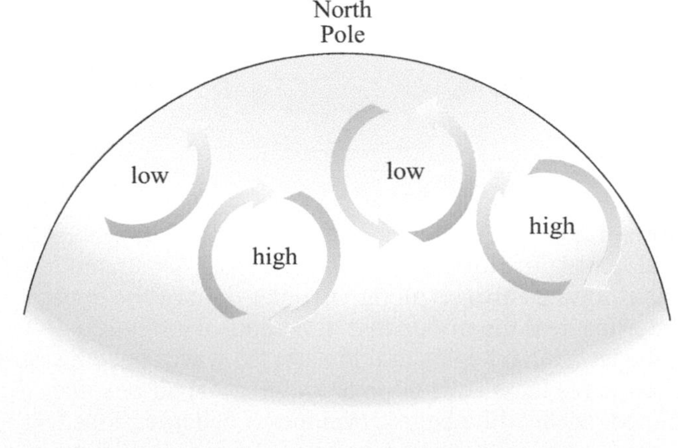

Figure 1.26 Highly schematic diagram to show the polewards transport of heat through the action of mid-latitude cyclones and anticyclones.

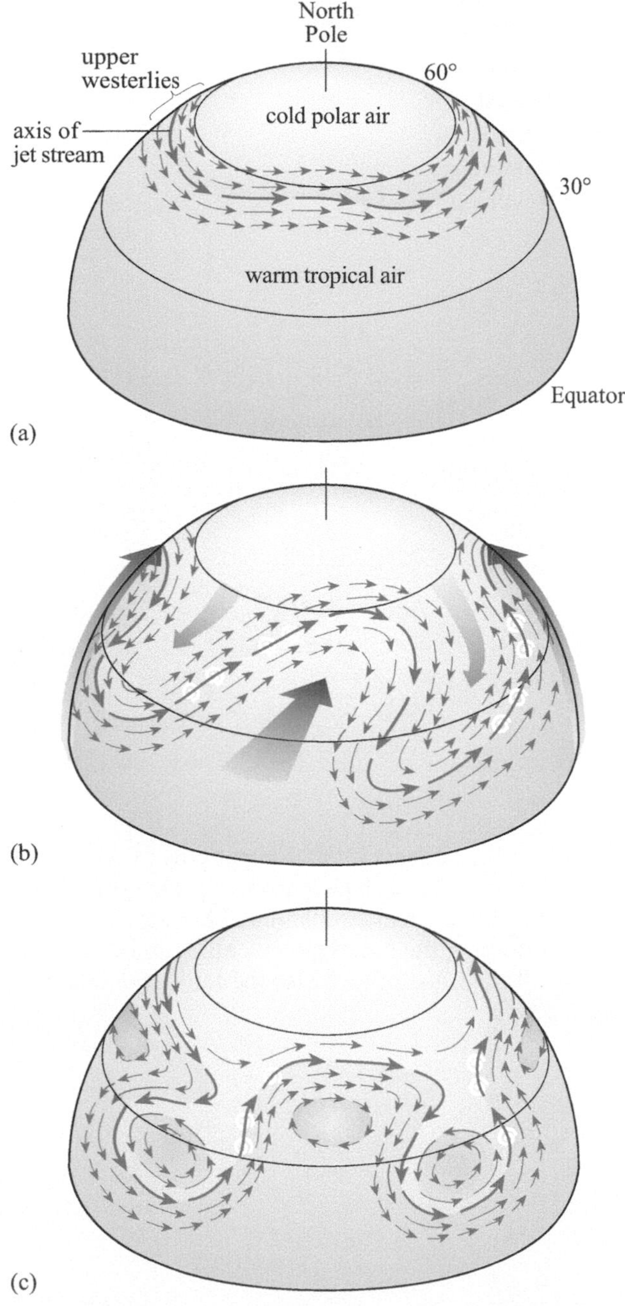

Figure 1.27 Schematic diagram showing stages in the development of waves in the northern polar jet stream, which flows eastwards along the polar limit of the upper westerlies near the tropopause at heights of ~10 km: (a) the jet stream begins to undulate; (b) waves become more extreme; (c) large cells of polar and tropical air become isolated. Meanwhile below, cyclones (i.e. depressions, shown here as white 'swirls' of cloud) form along the poleward-trending parts of the front, and anticyclones (not shown) form along the equatorward-trending parts.

The paths taken by mid-latitude depressions and anticyclones are determined by the behaviour of the **polar jet stream** (Figure 1.27). This is a high-level, fast air current that in each hemisphere flows around the Earth above the boundary between warm tropical air and the underlying cold polar air, known as the **polar front**. The jet stream tends to develop large undulations, typically three to six in number, which eventually become so extreme that cells of tropical air become isolated at relatively high latitudes, and cells of polar air become isolated at relatively low latitudes (Figure 1.27c). This mechanism results in the horizontal transport of enormous amounts of warm air polewards and cold air equatorwards. Meanwhile, the vertical air movements that of necessity accompany these large-scale horizontal wave motions (known, incidentally, as *Rossby waves*) lead to the development of cyclonic flow and low-pressure centres, and anticyclonic flow and high-pressure centres near the Earth's surface.

■ What would be the effect of the undulations in the jet stream becoming more pronounced?

□ Cold polar air would flow to lower latitudes than normal, while warm tropical air would flow to higher latitudes than normal.

The areas of the globe that are affected by pronounced undulations in the jet stream depend on the configuration of the jet stream at the time; the undulations themselves travel westwards around the Earth, relative to the air flow. In recent years, the unusual behaviour of the jet stream (perhaps related to global warming) has resulted in droughts and crop failures in the USA. The effect of droughts on agriculture, whether in the USA or the African Sahel, is a reminder that whether a particular type of vegetation flourishes or dies is determined by the prevailing climatic conditions. Natural ecosystems are more robust than modern agricultural monocultures, but they are attuned to the prevailing climate and, if that changes, then so will they.

Now look at Figure 1.28 (overelaf), which shows the prevailing winds at the Earth's surface. Note the zones of westerly winds between the subpolar low-pressure regions and the subtropical highs. These are the parts of the globe that, depending on the configuration of the polar jet stream (Figure 1.27), may be affected by cold polar air or warm tropical air. These regions are often referred to as temperate.

■ Compare the wind patterns in Figure 1.28 with those shown in Figure 1.20. Over which parts of the globe are the subtropical highs and subpolar lows most evident? What features of the real world cause the actual wind pattern to differ from the hypothetical one shown in Figure 1.20?

□ Subpolar lows and subtropical highs are most evident over the oceans; this is not surprising given that Figure 1.20 shows the wind pattern appropriate to an Earth completely covered in water. It is the presence of the continental land masses that makes the real wind pattern so much more complicated.

The presence of the continents 'distorts' the global wind system from the hypothetical east–west zones of polar high/subpolar low/subtropical high/ equatorial low. Question 1.6 investigates the effects of the distribution of land and sea on the global wind system in more detail.

Question 1.6

The ITCZ, where the wind systems of the two hemispheres meet (Figure 1.28), generally follows the zone of highest temperature at the Earth's surface.

(a) In general terms, what are the main differences between the position and/or shape of the ITCZ in July (Figure 1.28a) and January (Figure 1.28b)? How do Figure 1.14 and the related discussion help to explain these differences?

(b) The ITCZ moves seasonally northwards and southwards between the extreme positions shown in Figure 1.28. What implications does the changing position of the ITCZ have for the prevailing wind direction at the Earth's surface? To answer this, describe how the wind direction changes over tropical West Africa (Mauritania, Senegal and Guinea) between July and January.

(c) Over what region is the north–south seasonal shift in the position of the ITCZ greatest?

Many low-latitude regions experience such seasonal changes in wind direction. Air masses that travel over the oceans pick up moisture, while those that move over arid regions are dry, so seasonally reversing winds often bring with them contrasting climatic conditions (Figure 1.29 overleaf).

■ Bearing this in mind, how might the conditions in the south of the Eurasian continent differ between July and January?

□ Conditions will be wet in July when the prevailing winds are southwesterlies blowing off the Indian Ocean and dry in January when the prevailing winds are northeasterlies blowing off the Eurasian continent.

The rains that fall over southern Asia in the summer are often referred to as the monsoon rains. However, as the word 'monsoon' derives from the Arabic for 'winds that change seasonally', there are really two Asian monsoons:

- the warm, moist South–West Monsoon
- the relatively cool, dry North–East Monsoon.

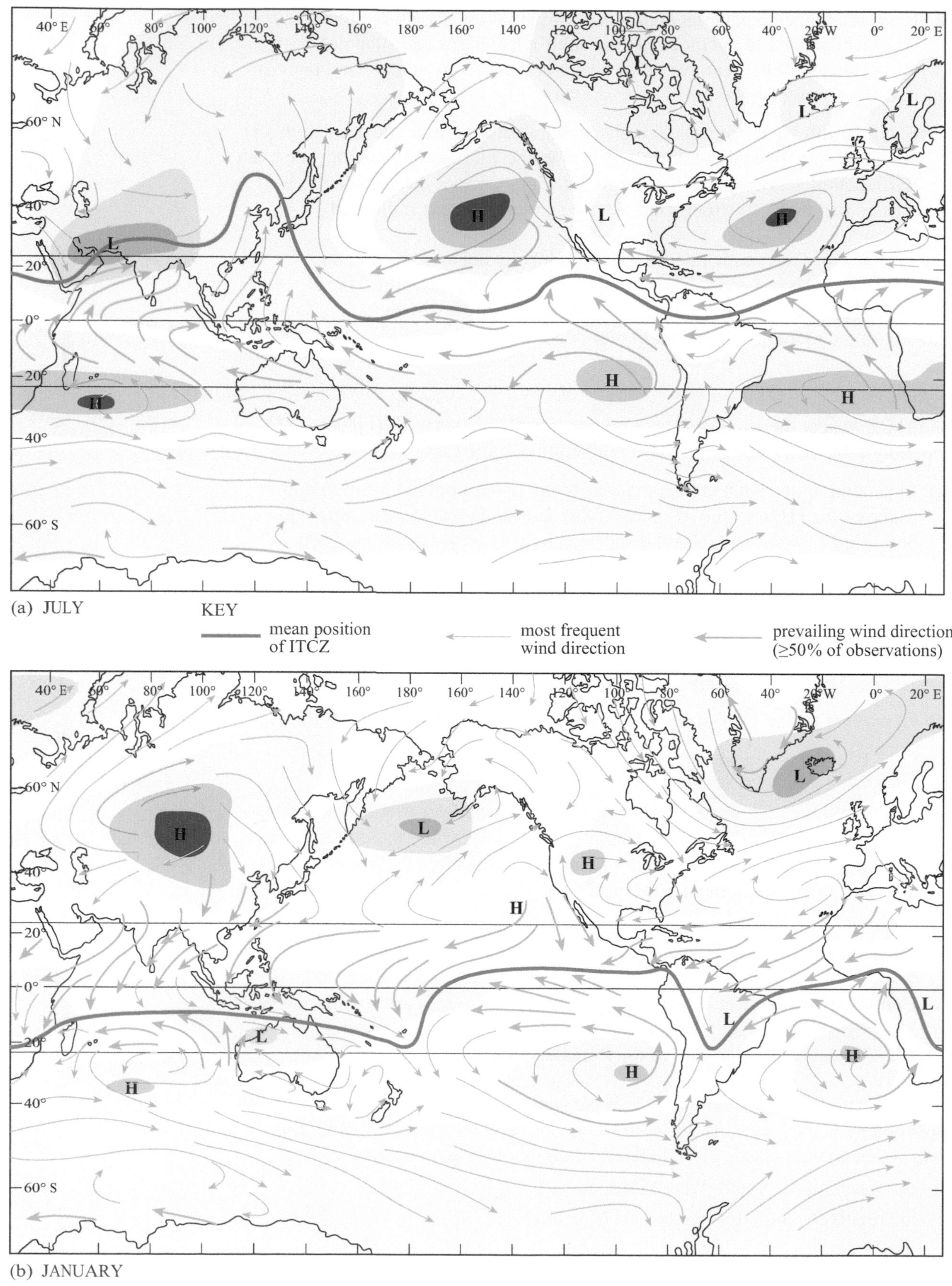
40° E
60°
80°
100°
120°
140°
160°
180°
160°
140°
120°
100°
80°
60°
40°
20° W
0°
20° E
60° N
40°
20°
0°
20°
40°
60° S
H
L
(a) JULY
KEY
mean position of ITCZ
most frequent wind direction
prevailing wind direction (≥50% of observations)
(b) JANUARY

◀ **Figure 1.28** The prevailing winds at the Earth's surface, and the position of the Intertropical Convergence Zone (ITCZ) where the wind systems of the two hemispheres meet, in (a) July (northern summer/southern winter) and (b) January (southern summer/northern winter). Also shown are the positions of the main regions of high (H) and low (L) surface pressure for these seasons of the year.

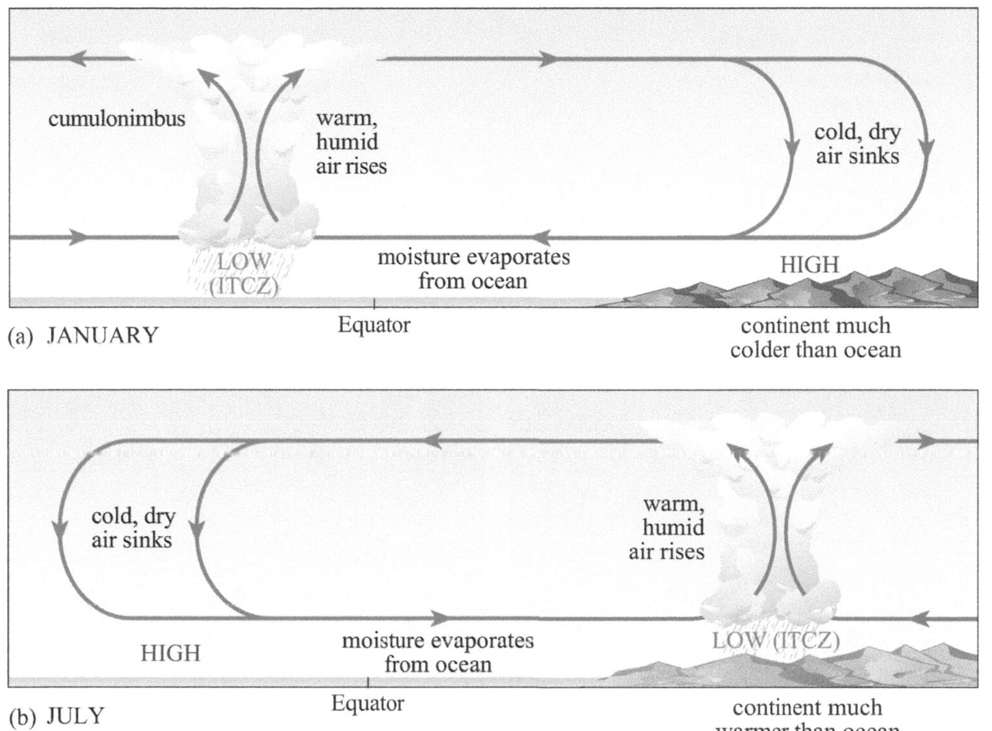

Figure 1.29 Schematic diagrams of the changing conditions over continent and ocean in a monsoon climate, drawn assuming that the continent is in the Northern Hemisphere. (a) Northern winter: dry, cooled air subsides over the continent, which is a region of high surface pressure; winds blow off the continent picking up moisture from the much warmer ocean and eventually moist air rises at the Intertropical Convergence Zone, causing abundant rain. (b) Northern summer: land is now much warmer than the ocean, so the region of low surface pressure corresponding to the ITCZ and its zone of rain has moved northwards towards the interior of the continent. (*Note*: the latitudes shown are notional, as the northernmost and southernmost positions of the ITCZ vary from place to place, as can be verified from Figure 1.28, which should be studied in conjunction with this figure.)

It has been worth considering monsoonal reversals in some detail because, as you will see later, the interaction between winds and continents may also have a more subtle and longer term influence on global climate.

The large-scale reversals in atmospheric circulation over southern Asia and the Indian Ocean are particularly marked because the continental land mass involved is not only large but also in places very high. In general, it is important to remember that climatic conditions are affected not only by the distribution of land and sea but also by the shape of the land masses in the vertical dimension (i.e. their topography). Air that is forced to rise over high ground may be triggered to convect, leading to rainstorms. Furthermore, moisture-laden air that is forced to rise by the presence of a topographic high will cool adiabatically and, being cooler, will no longer be able to hold as much moisture. If cooling is sufficient, condensation will occur, clouds will form and rain or snow will result (Figure 1.30a overleaf). By contrast, air subsiding over the leeward slope will warm adiabatically and, having no moisture source, may become very dry, resulting in the formation of a *rain shadow*. Rain and snowfall triggered by mountains (known as *orographic* precipitation) occurs on all scales – from small volcanic islands in mid-ocean to mountain chains on land. In southwestern USA, the western slopes of the Coast Range and the Sierra Nevada receive ample precipitation from the moisture-laden northwesterlies blowing off the

Figure 1.30 (a) Orographic precipitation: mountains cause moisture-laden air to rise, with the result that the windward slopes are wet and well vegetated, while to leeward there is an area of rain shadow. (b) Two locations on the Hawaiian island of Kauai: (i) the verdant eastern side of the island, thought to be one of the wettest places on Earth; (ii) Waimea Canyon on the western side, which has a much drier climate and desert-like vegetation. (Tony Waltham/ Geophotos)

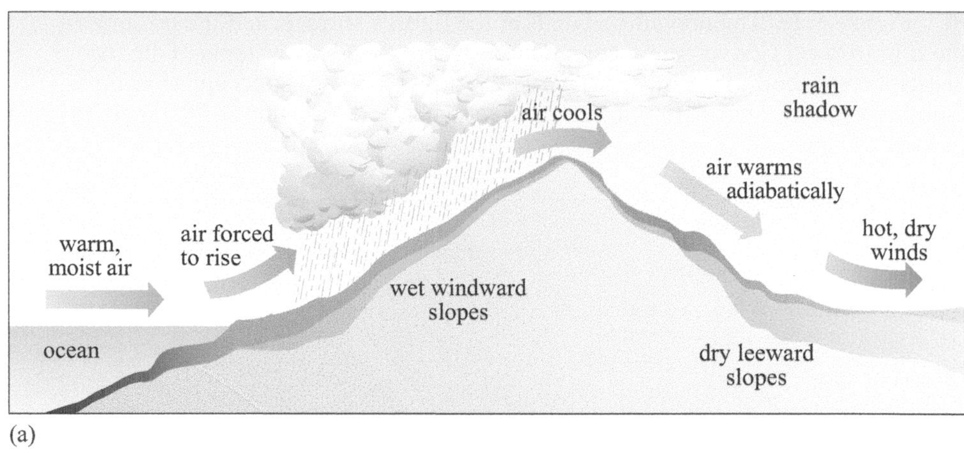

(a)

(b)(ii)

(b)(i)

Pacific Ocean (Figure 1.28); to the southeast of these mountains, in their lee, are the desert regions of Nevada and eastern California (also visible in Figure 1.1). Figure 1.30b shows the dramatic effect of orographic precipitation and rain shadow on the vegetation of the mountainous island of Kauai in Hawaii.

Before moving on to look more closely at the role of the ocean in the climate system, it is important to remember the influence that living terrestrial organisms have on their environment. This is most dramatically illustrated by the effects of deforestation, so often shown on the television. The removal of large areas of rainforest means that rainwater is no longer trapped but runs away, carrying with it the topsoil (itself largely a product of the forest). As a result, productive ecosystems that trapped and recycled water, and provided moisture and heat to the overlying atmosphere, are all too often converted to arid wastes.

1.4.2 Heat transport by the upper ocean

As shown by the summary diagram in Figure 1.15, the ocean has a surface current system that is primarily wind-driven and a deep current system driven primarily by heating and cooling at the surface. The surface current system is confined to the uppermost layers of the ocean, which are separated from deeper, colder water by a zone of marked decrease in temperature known as the **thermocline**.

Comparisons between Figure 1.31 (overleaf), which shows the average surface current pattern, and the winds in Figure 1.28 (particularly (b)), reveal a fairly close correspondence between the two, although current patterns are of necessity modified by the presence of land masses, which cause flow within ocean basins to form more-or-less closed circulatory systems, or **gyres**. In the Atlantic, anticyclonic wind systems blowing around the high-pressure regions in mid-latitudes give rise to anticyclonic gyres (often referred to as subtropical gyres), in both hemispheres; similar subtropical gyres can be seen in the Pacific.

- To what extent can cyclonic gyres be identified that correspond to the cyclonic winds associated with subpolar low-pressure regions (Figure 1.20)?

- Cyclonic gyres can be seen in the subpolar regions of both the North Atlantic and the North Pacific, but *not* in similar latitudes in the Southern Hemisphere. In the Southern Ocean, where there are no land barriers, the westerly winds drive the Antarctic Circumpolar Current eastwards around the globe.

To see how surface currents influence the distribution of heat, compare Figure 1.31 with Figure 1.14a and b, showing surface temperature. Although it is quite hard to see from Figure 1.14, sea-surface temperatures in low to mid-latitudes are generally higher on the western sides of the Atlantic and the Pacific than on the eastern sides.

- Why is this?

- Water warmed at low latitudes is carried polewards in the western 'limbs' of the subtropical gyres. Although some water is carried to high latitudes, some must also circulate in the subtropical gyres for a considerable time. However, because flow is anticyclonic, the western sides of the gyres will have been warmed more recently than the eastern sides.

One of the clearest manifestations of surface current flow in the temperature distributions in Figure 1.14 is the shape of the isotherms in the North Atlantic, which follow the northeastwards flow of the Gulf Stream (Figure 1.31). By contrast, in the South Pacific, the finger of low temperatures extending northwards up the west coast of South America (particularly clear in Figure 1.14b) may be correlated with flow in the cold Peru or Humboldt Current, carrying water cooled at high latitudes equatorwards; a similar although less-marked influence on the temperature distribution may be seen in the South Atlantic associated with the Benguela Current (Figure 1.31).

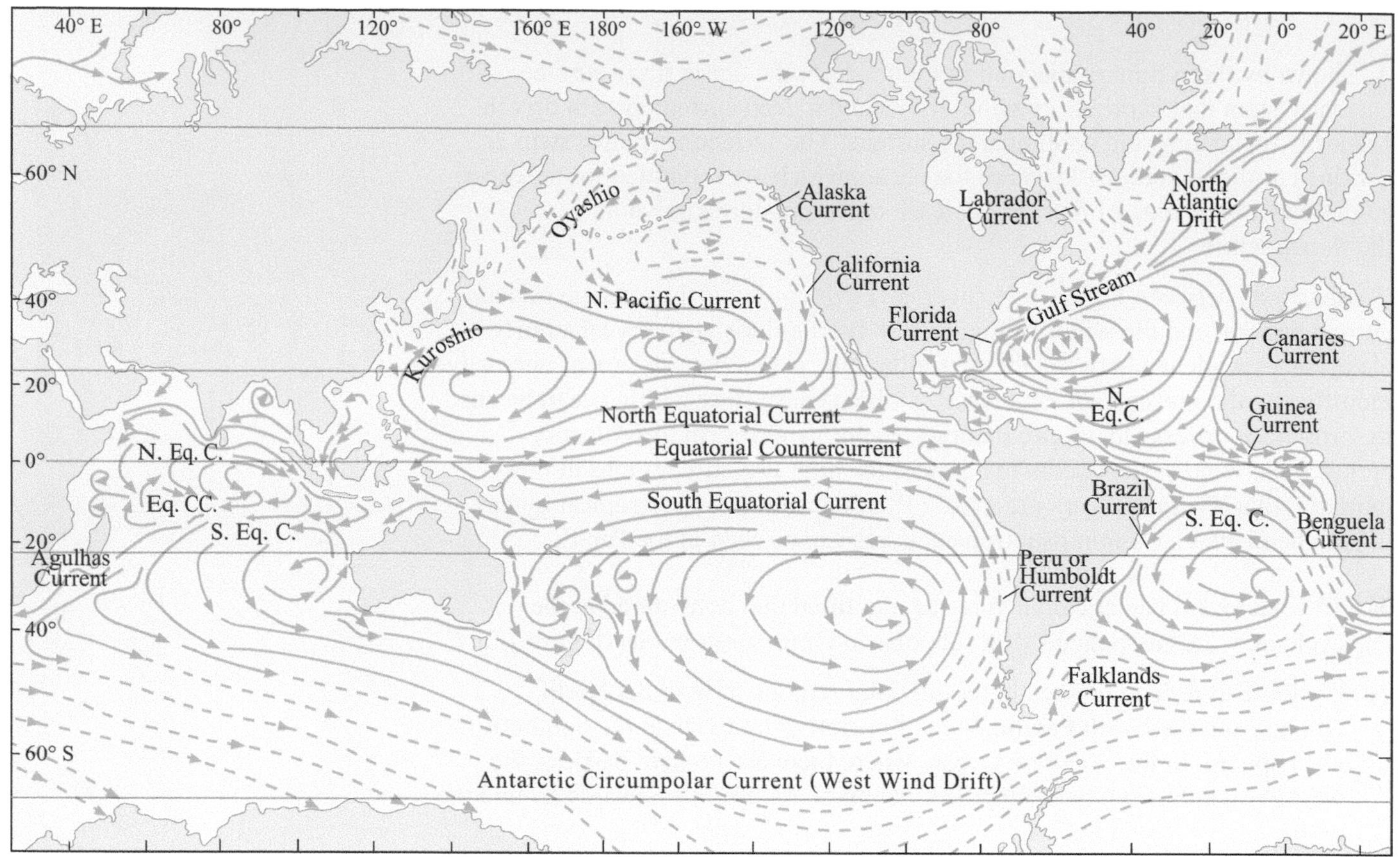

Figure 1.31 The global surface current system in the northern winter; this is the long-term average pattern – at any one time, the pattern will differ in detail. There are local differences in the northern summer, particularly in regions affected by monsoonal reversals; see also Figure 1.14. Dashed lines are cold currents. Note that even in strong currents, such as the Gulf Stream, current speeds rarely exceed a few m s^{-1}. (Strahler, 1973)

The flow of warm water northeastwards across the North Atlantic in the Gulf Stream moderates the climate of Britain and northwest Europe, making the region much warmer than it would otherwise be. The Gulf Stream is the narrow, fast-flowing western side of the North Atlantic subtropical gyre; similar western boundary currents are found in all the ocean basins. They owe their intensity to the fact that the Coriolis effect increases with latitude. Were it not for the speed at which the Gulf Stream flows polewards, much more of its heat would be lost to the atmosphere and adjacent ocean en route, and so be unavailable to warm higher latitudes.

The gyres and the Gulf Stream, and other current patterns on maps like Figure 1.31, are examples of large-scale, long-term features of the oceanic circulation. They represent the *average* situation in the ocean (at any one time the current in a particular place could be flowing in the opposite direction to that shown on the current map), and are often thought of as the ocean's climate. The ocean also has 'weather' (i.e. short-term, smaller-scale features). Most significant in this context are eddies about 50–250 km across (i.e. about a quarter of the size of mid-latitude atmospheric cyclones and anticyclones), which because of their intermediate size are known as **mesoscale eddies**.

Such eddies are frequently generated by meanders in strong currents like the Gulf Stream or the Antarctic Circumpolar Current. As these currents flow along boundaries between bodies of water with contrasting temperatures, they are known as **frontal currents**; the formation of eddies from meanders in frontal currents is similar to the formation of weather systems along atmospheric fronts (Figure 1.32). Mesoscale eddies clearly play an important role in transporting heat around the oceans, particularly across oceanic frontal regions but, until recently, computer simulations of the ocean have not been of sufficiently fine resolution to include realistic representation of the eddies. Indeed, computer modelling of global oceanic flows poses particular problems because the high density of water compared with air means that the length scales of current patterns are generally shorter and the timescales longer than is the case for winds.

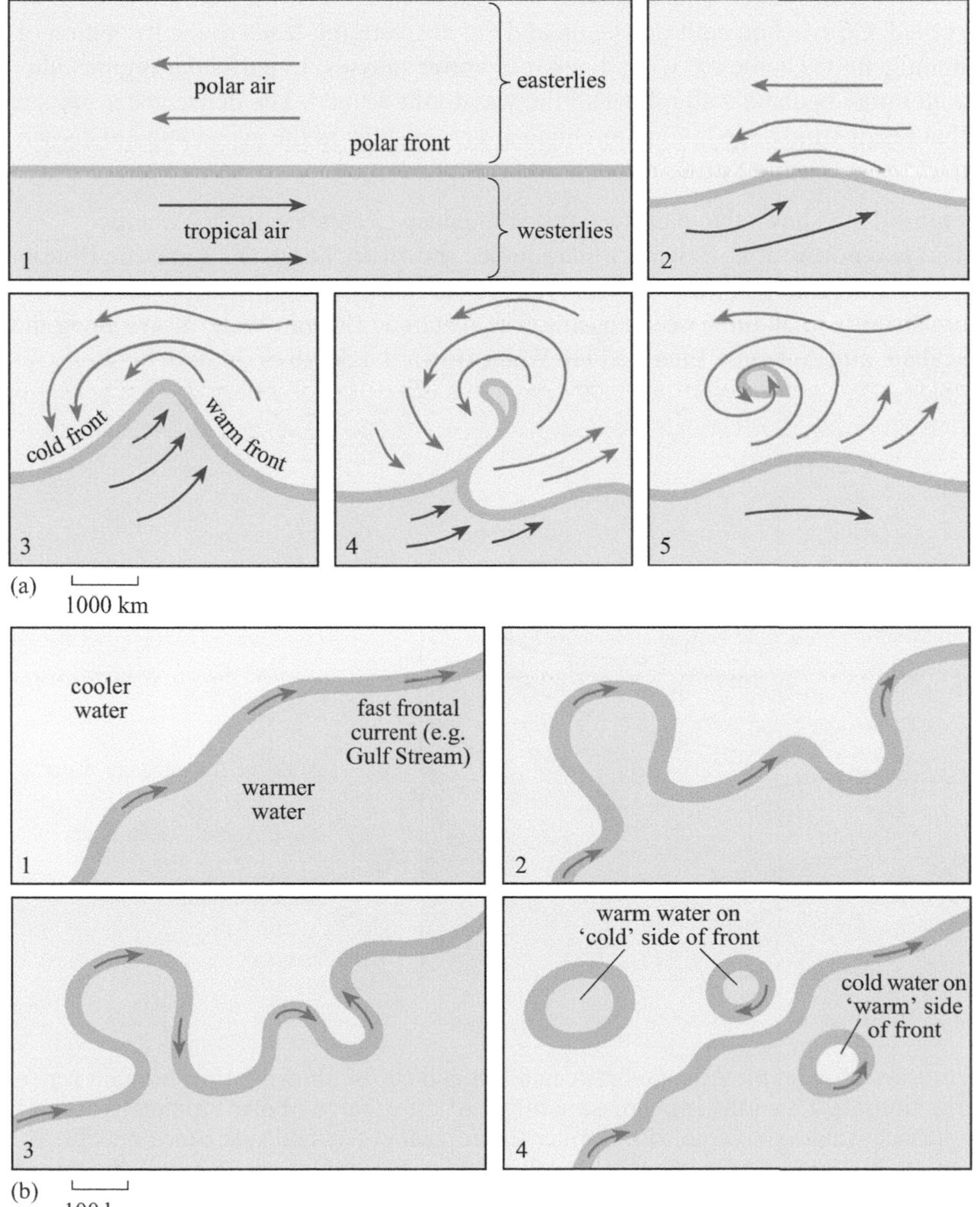

Figure 1.32 Plan view diagrams showing weather in the atmosphere and in the ocean. (a) Stages in the development of a mid-latitude cyclone (depression) in the Northern Hemisphere, showing how it contributes to the polewards transport of heat. (b) The formation of mesoscale eddies at a strong frontal current like the Gulf Stream, showing how they act to transport heat from one side of an oceanic front to the other.

1.4.3 Thermohaline circulation

The previous section considered how the ocean can transport heat by horizontal recirculations in the upper ocean, but the ocean also has vertically overturning circulations similar to the Hadley cells of the atmosphere. Where these overturning cells consist of relatively warm upper-layer water moving polewards above relatively cool deep water moving equatorwards, as is most often the case, the net effect is a polewards heat transport. Although the flows are far slower than in the atmosphere, the much greater heat capacity of water means that enormous quantities of heat can be transported. The properties of the slowly moving deep ocean currents are therefore surprisingly important for climate. These properties are mostly set in high latitudes in winter where, in certain locations, cold winds cool the surface water to such an extent that it becomes denser than the water beneath it and sinks. This displaces the water beneath it, which in turn rises to be cooled. Convection cells are set up and the deep mixing leads to the formation of homogeneous bodies of water known as **water masses**. In particular regions, the water may become well mixed all the way to the seabed. The dense water masses that result from deep mixing circulate at great depths in the ocean and are known as *deep* (or *bottom*) water masses.

Figure 1.33 shows the main features of the deep circulation in the Atlantic. The two northwards-flowing water masses shown are known as Antarctic Bottom Water and Antarctic Intermediate Water; both of these water masses extend northwards in all three ocean basins. The Antarctic Bottom Water flows along the seabed, and Antarctic Intermediate Water flows at a depth of 1000 m or so.

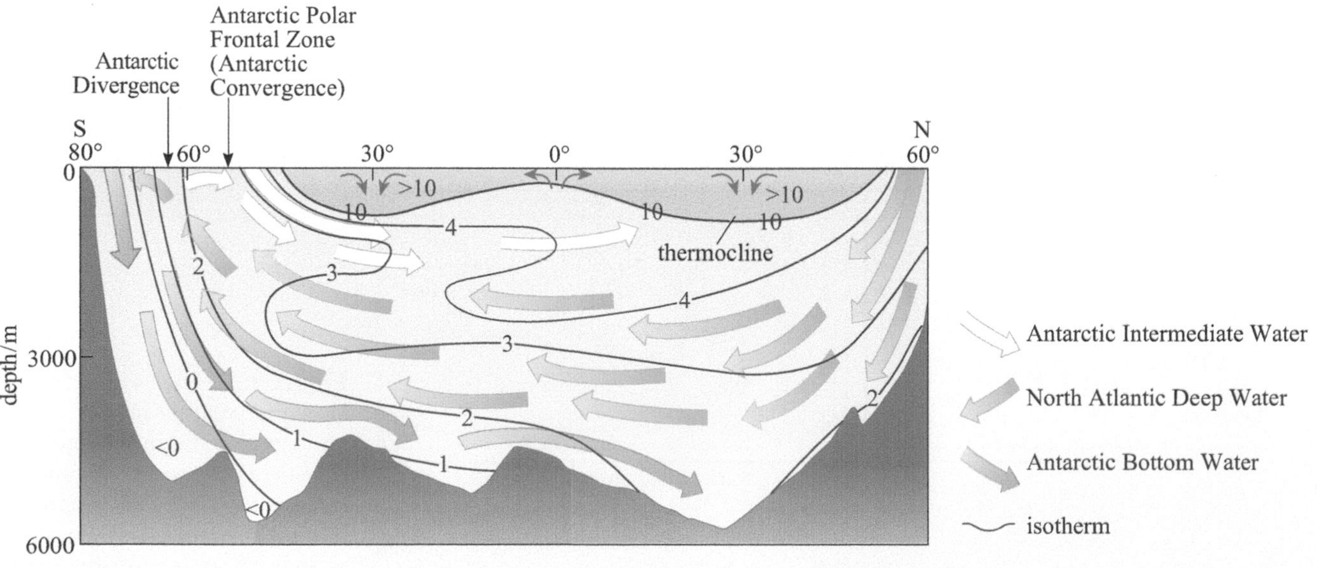

Figure 1.33 Generalised north–south cross-section of the Atlantic between 80° S and 60° N, showing flow of main water masses and resulting temperature distribution; values on the isotherms are in °C. At the Antarctic Polar Frontal Zone (or Antarctic Convergence), Antarctic surface waters converge with warmer subtropical waters and sink beneath them: at the Antarctic Divergence, water that sank in high northern latitudes eventually reaches the surface again. Small downward arrows near the surface at about 30° of latitude represent convergence in the subtropical gyres, and small upward arrows near the surface at the Equator represent upwelling at the Equatorial Divergence (see Figures 1.36 and 1.34). *Note*: vertical scale is greatly exaggerated.

Observation of the temperature contours between about 40° S and the Equator reveals that the Antarctic Intermediate Water may be distinguished as a 'tongue' of water that is cooler than both the overlying surface water and the underlying deep water mass – North Atlantic Deep Water.

■ Why, at first sight, would this seem to be an unlikely situation?

□ Cooler water is normally denser and would therefore be expected to mix very rapidly with the warmer water by convection.

The solution to this conundrum is that the Antarctic Intermediate Water is relatively fresh (for seawater), both because it forms in the subpolar regions where precipitation is high (Figure 1.20), and because it contains ice meltwater. By contrast, North Atlantic Deep Water is relatively saline; that is to say, its concentration of dissolved salts (its **salinity**) is relatively high. So, despite being cooler than North Atlantic Deep Water, Antarctic Intermediate Water is less dense and flows above it. Deep circulations in the ocean, in particular the overturning cells, are driven essentially by small variations in density. Since the density of seawater is determined by its temperature and its salinity, the ocean's deep circulation system is known as the **thermohaline** circulation, where 'thermo-' refers to its dependence on temperature, and '-haline' to its dependence on salinity.

The salinity of seawater may be altered in two ways:

- by addition of fresh water by precipitation (rain or snow) or its removal by evaporation
- by addition of meltwater from ice, or removal of freshwater by freezing.

When sea-ice forms (Figure 1.34), the ice itself is freshwater, so droplets of brine-rich water collect beneath it. This *brine rejection* in cold polar waters may produce water sufficiently dense to sink to the seabed, and occurs extensively around Antarctica, contributing to the formation of Antarctic Bottom Water. It also

Figure 1.34 Newly forming sea ice (also known as 'grease ice' or 'frazil ice'). Note that this ice, forming from seawater, should not be confused with the polar and Greenland ice caps, which form from snowfall. (Mark Brandon/Open University)

occurs around the Arctic Ocean, where ice forms over the continental shelf, and here it leads to the formation of a cold, saline water mass that sinks to fill all the deep basins of the Arctic. Eventually, it escapes into the North Atlantic either by flowing through the Canadian Arctic islands to the west of Greenland and out into the Labrador Sea, or by flowing out into the deep basins to the east of Greenland, and then into the northeastern Atlantic.

Either way, this cold, saline Arctic water flows below regions in the northernmost Atlantic where surface water may be cooled by winter winds to such an extent that deep convection is triggered. Surface water may sink down to mix with the Arctic water at several thousand metres depth, and the resulting mixture is North Atlantic Deep Water. As the water sinking from the surface has been carried north in the Gulf Stream, it is initially fairly warm, at about 12–15 °C, and it has been estimated that in winter about 1300 km^3 of water sinks from the surface of the northern North Atlantic each day, having been cooled by winds from 12–15 °C down to 1–4 °C.

Question 1.7

Assuming that the warm surface water cools by 11 °C, how much heat is given up each day to the atmosphere over the northern North Atlantic? (*Hint*: use information from Box 1.4 along with the fact that 1 m^3 of water has a mass of 10^3 kg and there are 10^9 m^3 in 1 km^3.)

This is an enormous amount of heat – four orders of magnitude more than that supplied by solar radiation at ~55° N in winter (Figure 1.6).

Having sunk, the waters flow equatorwards as North Atlantic Deep Water still carries heat, but much less than it carried polewards. Meanwhile, much of the heat given up to the atmosphere is carried eastwards in the prevailing winds, so the formation of North Atlantic Deep Water has important implications for the climate of Europe.

Furthermore, it is thought that, as water sinking from the surface draws in yet more water to fill its place, North Atlantic Deep Water formation effectively increases flow in the Gulf Stream. This idea, combined with what is known about the paths of water through the ocean, has led to the concept of the thermohaline conveyor belt, illustrated in Figure 1.35. Overall, heat is carried northwards in both the North and the South Atlantic, and this is compensated for by a net transfer of heat into the South Atlantic from the Pacific and Indian Oceans. Of course, the flow pattern shown is a generalisation (e.g. a particular parcel of water would be very unlikely to follow this particular path). To give some idea of the timescales involved, however, a parcel of water sinking in the northern North Atlantic would take in the order of a thousand years to come to the surface again in the northern Indian Ocean or the *North Pacific*.

Interestingly, the Pacific has no water mass analogous to North Atlantic Deep Water. The reason for this seems to be that, because of high precipitation, the salinity of surface seawater in the *North Pacific* is so low that cooling and evaporation can never increase its density sufficiently for sinking to occur. So does this mean that the formation of North Atlantic Deep Water can be 'turned off' by a decrease in the salinity of Atlantic surface waters? Perhaps so. Indeed,

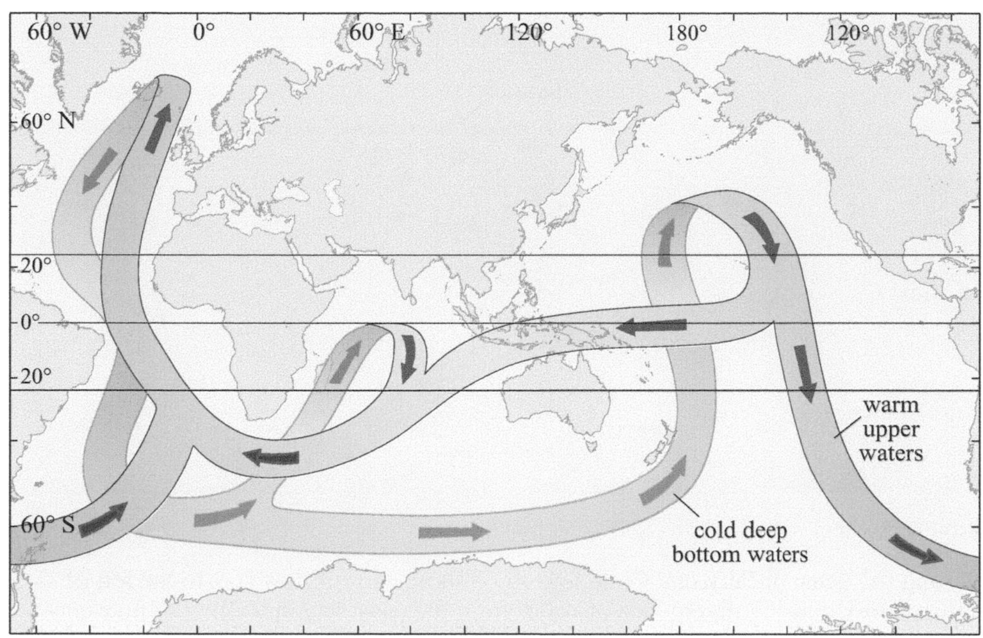

Figure 1.35 Schematic diagram of the global thermohaline conveyor, driven by the sinking of cold water at high latitudes. Warm water in the upper 1000 m or so of the ocean generally follows a pathway towards the northern North Atlantic. After sinking, cold water generally follows a path towards the North Pacific and the northern Indian Ocean. (As the current flows in the Pacific and Indian Oceans – and the seas between – are not as well known as those in the Atlantic, different versions of this diagram are used.) (Broecker, 1991)

many scientists believe that North Atlantic Deep Water formation was turned off by a lid of low-salinity meltwater between 12 900 and 11 600 years ago, causing the period of cold climate known as the Younger Dryas.

As they flow through the oceans, water masses gradually mix with adjacent waters, but they may nevertheless still be identified many thousands of kilometres from their sites of formation. Since they acquire their characteristic temperatures and salinities when they are in contact with the atmosphere at the surface and maintain these characteristics over long periods of time, water masses – or at least imprints of them – are invaluable for studying past climate.

Upwelling, downwelling and convection

The global-scale patterns of overturning seen in Figures 1.33 and 1.35 are driven by heating, cooling and freshwater forcing at the surface and only involve sinking or downwelling in very restricted, high-latitude regions with widespread upwelling elsewhere. This asymmetry, like the narrowness of the western boundary currents, is a consequence of the change in Coriolis effect with latitude. Localised downwelling is closely related to localised high-latitude convection, but there is a subtle difference between the two. Convection involves an exchange or mixing of denser surface water with lighter water from below and there is no net transfer of water downwards. Convection on its own, therefore, cannot supply the net downwelling needed to complete the overturning loop of the thermohaline circulation. It is, however, still closely linked to thermohaline downwelling because it is responsible for creating dense water masses. One reason for the strong connection is that convection results in a powerful *positive* feedback: when surface cooling causes convection, warmer water is lifted to the surface from below, thus releasing yet more heat to the atmosphere.

Downwelling occurs when surface waters move together, or converge. Similarly, diverging surface waters result in upwelling. Such convergence and divergence

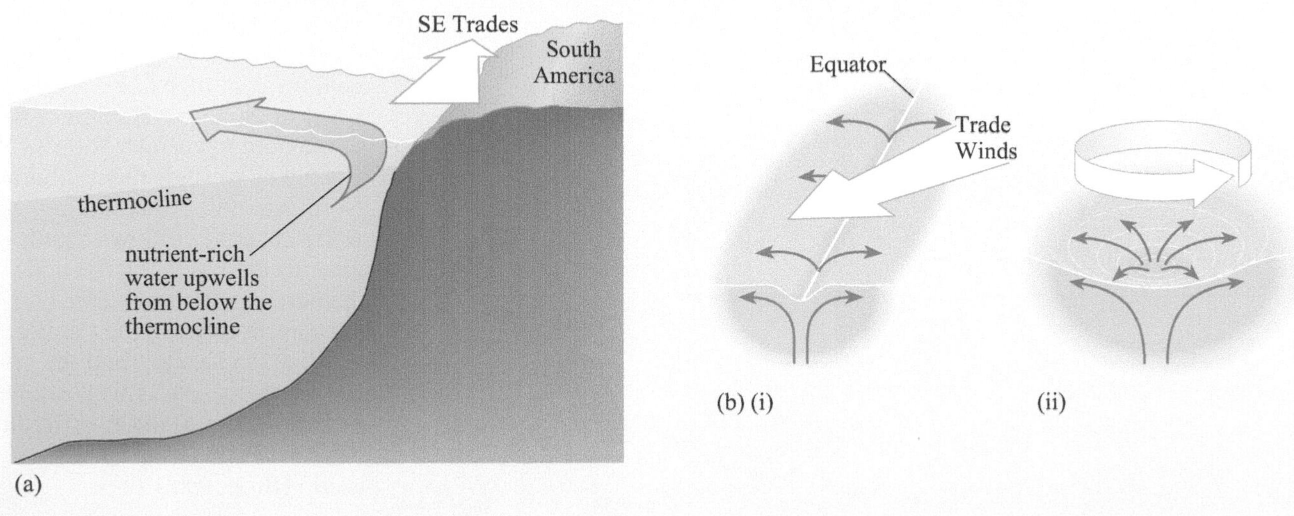

Figure 1.36 (a) Equatorwards winds along the coast of Peru and Chile lead to offshore current flow (i.e. to the left of the wind, as this is in the Southern Hemisphere), causing divergence of water from the coast and upwelling of nutrient-rich water from below the thermocline. (b) Schematic diagrams to show other types of wind fields that lead to upwelling: (i) Trade Winds crossing the Equator (where the Coriolis force is zero) lead to a zone of upwelling along the Equatorial Divergence. (ii) Cyclonic winds lead to divergence of surface water and mid-ocean upwelling (here shown for the Northern Hemisphere); by contrast, anticyclonic winds lead to convergence of surface water and downwelling.

can be driven by density variations, as in the thermohaline circulation, or by wind forcing. Wind-driven upwelling often occurs at coasts, where land–sea temperature contrasts lead to winds blowing roughly parallel to the shore. As a result of the Coriolis force, water acted on by the wind moves at an angle to it. So, for example, the Southeast Trades blowing *along* the western side of South America cause surface water to diverge from the coast, leading to upwelling of water from below. You do not need to worry about the details of exactly *how* coastal upwelling occurs, but instead look at the various wind patterns that lead to upwelling along coasts in Figure 1.36a and in the open ocean (Figure 1.36b). As you will see in later chapters, patterns of wind-driven upwelling and downwelling have important consequences for life in the oceans, and indirectly affect the climate.

Figure 1.37, which shows the upwelling and downwelling flow in July, was computed using the known wind field. It can clearly be seen that the Equatorial Divergence (the band of upwelling that occurs where the Southeast Trades cross the Equator; Figure 1.36bi), and the areas of coastal upwelling along the western coasts of North and South America and of Africa, is caused by the Trade Winds (Figure 1.36a).

- By reference to Figures 1.20 and 1.28, suggest why there is also some open ocean upwelling at latitudes above about 45°.

- Divergence and upwelling will occur in the subpolar gyres, which are driven by the cyclonic winds of the subpolar low-pressure systems (Figure 1.35bii).

By contrast, the centres of the subtropical gyres beneath the anticyclonic winds of the subtropical highs are sites of convergence of surface water and sinking.

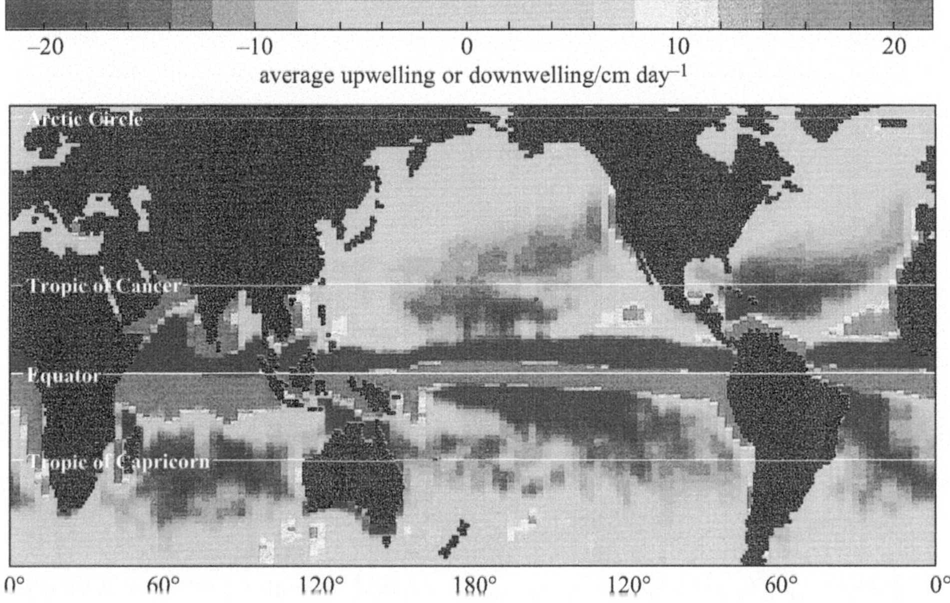

Figure 1.37 Upwelling and downwelling (vertical current flow in cm day^{-1}) for July, computed from the wind field. Areas of strongest upwelling (large positive values in the key, most notably along the West African, South Indian and Sri Lankan coasts and close to the Equator) are pink; areas of strongest downwelling (large negative values – notably the tropical oceans away from the Equator) are dark blue. (Xie and Hseih, 1995)

1.5 Earth–ocean–atmosphere: the support system for life

In Section 1.4, you considered the Earth's heating and air-conditioning systems in terms of two connected fluid envelopes – the atmosphere and the ocean – meeting and interacting at the sea surface by transfer of heat and momentum. You have already seen that the transfer of fresh water is important too, but atmosphere, ocean, and indeed the solid Earth and living organisms, are interlinked by the movement between them of many other materials. Although the timescale of movement can be months, millennia or even millions of years, depending on the circumstances and the elements involved, this continual redistribution may be described in terms of **biogeochemical cycles** – a concept that will be fundamental to the discussions regarding carbon in Chapter 2.

The transport of water (and hence heat) by means of evaporation (particularly from the sea-surface at low latitudes), followed by condensation in cooler environments, forms just one part of the **hydrological cycle** (Figure 1.38 overleaf). It is the movement of water (which is a powerful solvent) that underpins the cycling of many other constituents through the Earth–ocean–atmosphere system. In the atmosphere, gases, including CO_2 and SO_2, dissolve in rainwater, resulting in a weak acid which, when it falls onto land, is neutralised by reaction with minerals in soil and rocks, releasing their constituent ions into solution in a process known as **chemical weathering**. The hydrological cycle can thus be viewed as a simple but special example of a biogeochemical cycle: another indication of the central role of water in Earth's unique environment.

Question 1.8

(a) According to Figure 1.38, what percentage of the Earth's water inventory is in the oceans at the present time? What percentage is in ice caps and glaciers?

(b) There is currently much concern about the rise in sea level as a result of global warming. Assuming one-quarter of the ice caps eventually melted, what percentage increase in the volume of the oceans would result?

So, while the ice caps and glaciers contain most of the water that is on the land surface, they contain only a few per cent of the global water inventory: all but 3% of the Earth's water is in the oceans.

A concept that can usefully be applied to many different biogeochemical cycles is that of average **residence time** which, in the case of the hydrological cycle depicted in Figure 1.38, is a measure of the average length of time an individual water molecule spends in any particular stage or **reservoir** of the cycle. It is calculated by dividing the amount of water in that particular reservoir by the amount entering (or leaving) in unit time. It should be pointed out that, although

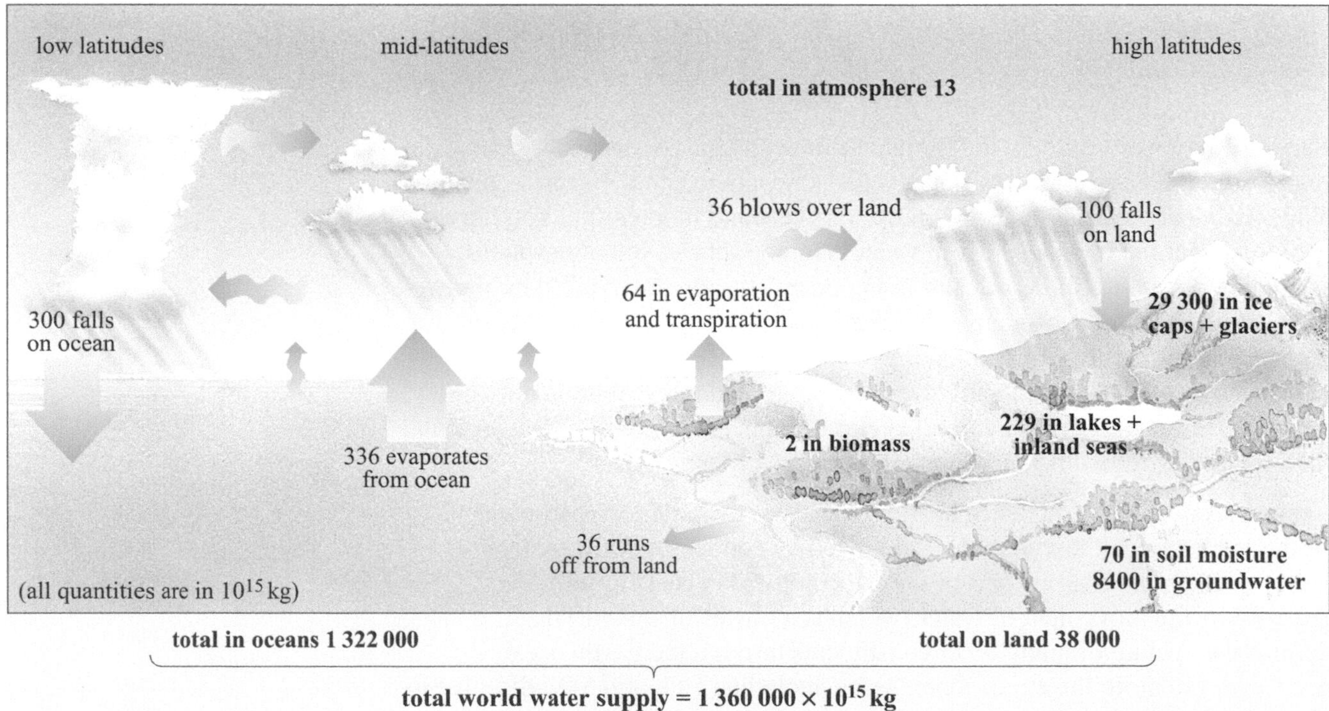

Figure 1.38 The hydrological cycle, showing annual movements of water through the cycle and amounts of water stored in different parts of the cycle (often referred to as reservoirs and here shown in **bold** type). Note the role played by vegetation in storing and cycling water. All quantities are × 10^{15} kg; note that 10^{15} kg liquid water occupies 10^3 km^3. The values given are estimates, and you may see different values elsewhere. For example, some authorities consider that groundwater storage accounts for a much larger proportion of the total water on land than shown here. The labels 'low latitudes', 'mid-latitudes' and 'high latitudes' are intended to remind you of the high precipitation at the ITCZ and in subpolar regions, and the high net evaporation in mid-latitudes; they are not supposed to imply that the processes shown are unique to those areas.

this is a useful concept, it is only an indication of average behaviour; individual water molecules may spend longer or shorter times in any given reservoir.

■ How much water (in 10^{15} kg) moves through the atmosphere annually? What is the average residence time of water in the atmosphere?

□ According to Figure 1.38, the amount of water entering the atmosphere yearly is 336×10^{15} kg + 64×10^{15} kg = 400×10^{15} kg (or the amount leaving it per year is 100×10^{15} kg + 300×10^{15} kg = 400×10^{15} kg); in other words, 400×10^{15} kg of water moves through the atmosphere annually. On average, 13×10^{15} kg of water resides in the atmosphere at any one time, so the residence time of water in the atmosphere is:

$$\frac{13 \times 10^{15}\ \text{kg}}{400 \times 10^{15}\ \text{kg y}^{-1}} = 0.033\ \text{y or about 12 days.}$$

Figure 1.38 and the above calculations are appropriate for the present time on Earth. In the past, and presumably in the future, under radically different climatic conditions, the relative magnitudes of the different processes and fluxes could be very different.

Looking at Figure 1.38, it would be tempting to conclude that, because rivers flow into the ocean, seawater is simply a concentrated form of river water. But is it? Figure 1.39 (overleaf) shows the average chemical composition of rainwater, river water and seawater. Study these histograms carefully, taking note of the different vertical axes (and for the moment ignoring the fact that some columns of the river water histogram consist of two parts).

Comparison of the top and bottom histograms in Figure 1.39 shows that the relative concentration of seawater compared with rainwater is:

$$\frac{34.4\ \text{g l}^{-1}}{7.1 \times 10^{-3}\ \text{g l}^{-1}} = 4800\ \text{(to 2 sig. figs).}$$

■ Comparing the histograms by eye (and ignoring the differences in concentration), would you say that seawater is closer in composition to rainwater or to river water?

□ Rainwater: the rainwater and seawater histograms bear a striking resemblance to one another.

One of the main differences between rainwater and seawater on the one hand, and river water on the other, is the relatively high concentrations of dissolved Ca^{2+}, HCO_3^- and SiO_2 in the latter. These are supplied to rivers as a result of chemical weathering of carbonate and silicate minerals, both common constituents of the rocks making up the Earth's crust. What happens to remove these constituents from solution in seawater will become clear in Chapter 2.

Another marked difference is the relatively large amount of chloride (Cl^-) in rainwater and seawater and the fairly small amount in river water. Only a tiny proportion of the chloride in the oceans comes from weathering. It originated as HCl gas released in volcanic eruptions (especially early in Earth's history when volcanism was more widespread than it is now) and is continually recycled by oceanic aerosols. These originate as droplets of seawater ejected into the atmosphere

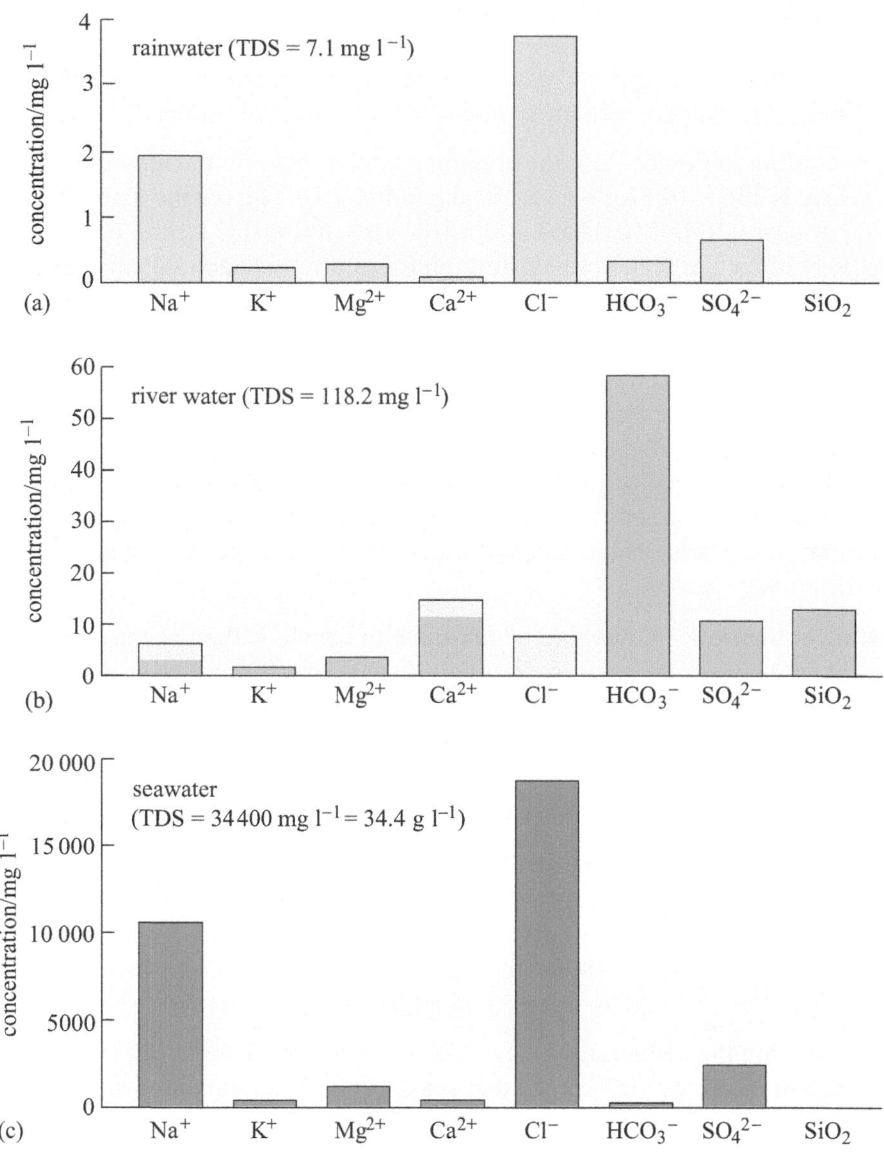

Figure 1.39 The average chemical composition of (a) rainwater, (b) river water and (c) seawater, shown in terms of the eight most abundant dissolved constituents, some at concentrations too low to appear. TDS = total dissolved salts. In (b), the darker lower parts of the columns are the contributions from weathering; the lighter upper parts correspond to salts supplied by aerosols (see text). Note the changes in scale of ×15 from (a) to (b), and ×400 from (b) to (c). Concentrations increase partly because of the addition of new dissolved constituents during weathering, etc., and partly because of loss of water through evaporation and transpiration.

when conditions are rough, along with any dissolved constituents, gases and particulate matter that they may contain. Particles of sea salt that end up in the atmosphere in this way are among those acting as nuclei for cloud condensation (Box 1.5), and their constituents are eventually rained out. As Figure 1.39 shows, while most dissolved ions in river water result from weathering (the darker parts of the histogram columns), a significant proportion of both sodium (Na^+) and calcium (Ca^{2+}), in addition to virtually all the chloride (Cl^-), comes from the ocean via rainwater (the lighter parts of the columns).

In total, all of the 92 naturally occurring elements found on Earth have been detected in seawater, in widely varying concentrations. Figure 1.40 summarises the various ways in which dissolved constituents enter and leave the oceans. Note that for some elements, hydrothermal vents at spreading axes are an important source. (Sedimentation, biological cycling and volcanism are discussed in Chapters 2 and 3.)

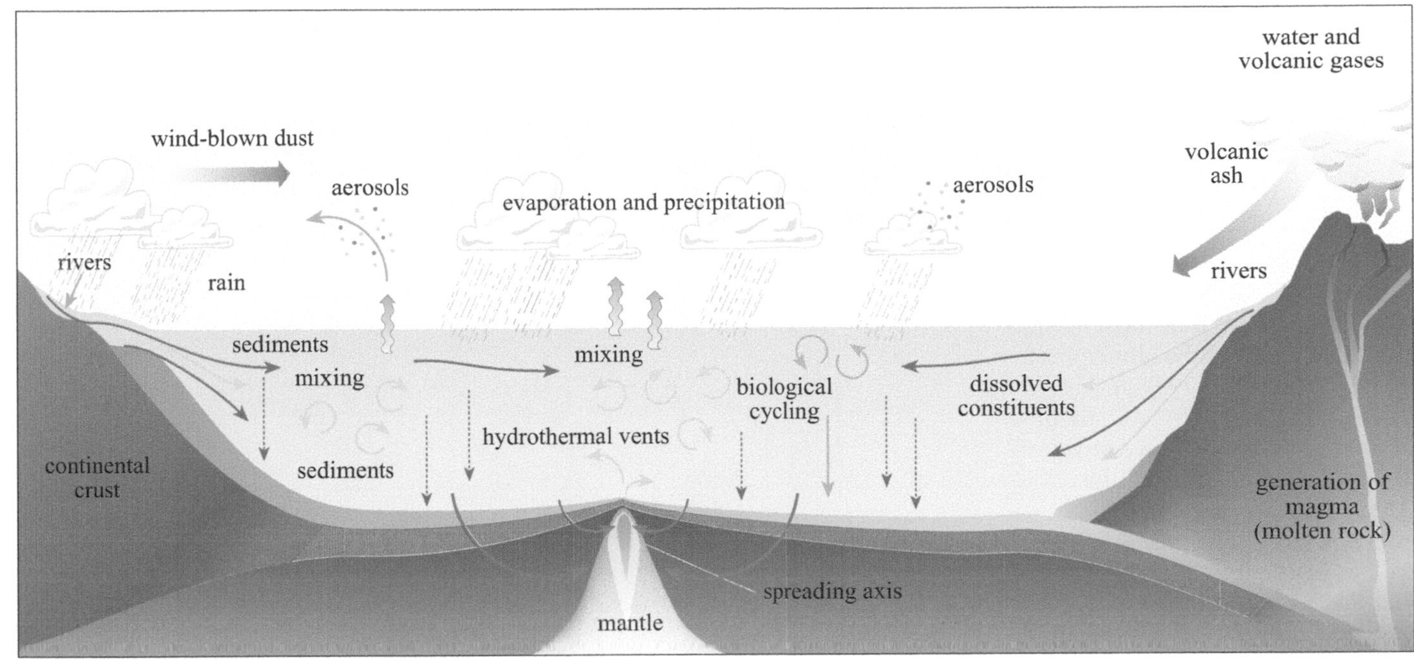

Figure 1.40 Diagrammatic cross-section illustrating the important role of the ocean in the global cycling of elements.

So, if water and all the elements are continuously recycled through the Earth system, what are the implications for life? The term biogeochemical cycles suggests that life plays a leading role in the story. Earlier, you saw how the tilt of the Earth's axis and the shape of its orbital path affect the distribution of incoming solar energy over the surface of the globe, and how the resulting pattern of surface temperature is modified by winds and currents. Intimately linked to the transport of heat around the globe is the transport of water.

■ To what extent is the following statement true?

Together, energy from the Sun and the water supply on Earth determine the distribution of primary production (mainly of plants) over the surface of the globe (Figure 1.1).

□ It is only partly true. As any gardener or farmer knows, some areas of land are more fertile (i.e. will support more plant growth) than others.

A fertile soil is one that contains an abundance of elements necessary for plant growth (i.e. **nutrients**) in a form that can be used by plants. Some landscapes will never be very green, irrespective of how much rain falls on them, because the minerals in the bedrock cannot be made *available* for plant growth (i.e. as dissolved ions) in sufficient amounts to support large stands of vegetation. Examples of such barren regions are steep rocky surfaces where no soil can accumulate, sand dunes and limestone pavements. To a large extent, however, the distribution of land plants is limited by the availability of light for photosynthesis, temperature and the availability of water. In the ocean, water is never in short supply; furthermore, temperatures are never too low to prevent plant growth (see Figure 1.2). Nutrients, on the other hand, are not so plentiful. A crucial role of biogeochemical cycling for life, particularly in the ocean, is thus to provide the nutrients essential for growth.

Biogeochemical cycles also apply to elements that are normally gaseous, and indeed seawater contains dissolved gases as well as dissolved solids, particles and ions. In some cases, these too have important implications for the distribution of life. Figure 1.41 shows the concentration in seawater of the four most abundant gases in the atmosphere. The data in (b) are given for a particular temperature and pressure because the solubility of gases generally *decreases* with increasing temperature and salinity, and *increases* with increasing pressure.

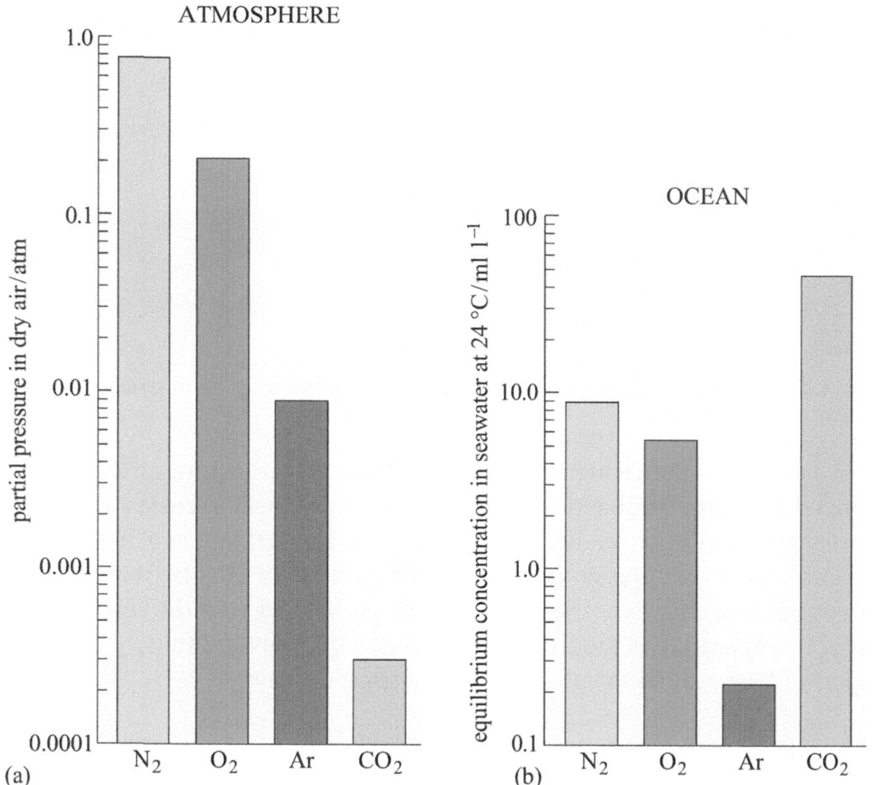

Figure 1.41 (a) Proportions by volume of the four most abundant gases in the atmosphere (together totalling 99.9%). (b) Concentrations by volume of these gases dissolved in seawater at 24 °C as controlled by the solubilities of the gases at their atmospheric concentrations. (b) has been plotted assuming that there is equilibrium between atmosphere and ocean across the air–sea interface; for the four gases shown, this is probably valid, but only to a first approximation, for reasons that will become clear later on. Note the different logarithmic scales on the vertical axes.

- Which one of the gases shown in Figure 1.41 is the most soluble in seawater? Which one is essential for the respiration of marine organisms?

- CO_2 is the most soluble, by a long way (the reason for this will become clear in Chapter 2). O_2 is essential for respiration.

Atmospheric gases dissolve into or escape from the ocean surface, depending on the relative concentrations either side of the air–sea interface (this is discussed further in Chapter 3). If it is very windy, and the sea-surface is rough, air and water mix together and the rate of gas transfer between them is increased. Once a gas is dissolved in the upper ocean, its transport throughout the ocean is the result of current flow and turbulent mixing.

- Given what you know about ocean currents (Figures 1.15 and 1.33), would you expect to find living organisms at great depths in the ocean?

- Yes, because dissolved oxygen is carried down into the deep ocean in the cold water masses sinking in polar regions.

So the ocean does not just contain dissolved constituents, particulate material and dissolved gases – it is also the largest 'living space' on Earth, a soup of living organisms, large and small. If asked to name a marine organism, many people might say 'fish' or 'whale', but the most important organisms are too small to be seen with the naked eye. These are:

- the *primary producers*, which include the minute floating algae or **phytoplankton** (Figure 1.42a), sometimes called 'the grass of the sea', and
- the *decomposers*, which are the **bacteria**.

Zooplankton are minute animals that feed on these smaller organisms and on each other (Figure 1.42b).

Not surprisingly, biological activity plays an important role in changing the concentrations of oxygen and carbon dioxide below the sea surface. Respiration by marine organisms, including bacteria, takes place throughout the ocean at all depths.

■ Is this also true of photosynthesis?

□ No. Photosynthesis uses light, and can therefore only occur in near-surface waters.

The oceans are pitch-black below about 1000 m; light is too weak for photosynthesis below ~250 m even in the clearest water. Since phytoplankton are plants and need to photosynthesise to grow (Equations 1.1 and 1.2), they (and the zooplankton that feed on them) live in the sunlit surface waters of the **photic zone**. As far as phytoplankton are concerned, the most important of the non-gaseous constituents dissolved in seawater are nitrate (NO_3^-), phosphate (PO_4^{3-}) and silica (SiO_2), i.e. the nutrients. Mysteriously, N and P occur in seawater in the same molar ratio (15 : 1) that they occur in the soft tissues of marine organisms. This suggests a certain level of feedback operates in the marine biosphere (probably on very long timescales) to maintain the dynamic balance between the compositions of the ocean and biosphere.

Other constituents are also used by living organisms, for example calcium, which, like silica, is used to build shells and skeletons, and sodium, which is one of a number of elements used in soft tissues. Calcium and sodium are available in great abundance in seawater and so their precise concentrations are not important for living organisms. Nitrate, phosphate and silica, on the other hand, may be completely used up by phytoplankton in surface waters and so their concentration may be the limiting factor determining whether or not phytoplankton can grow. For this reason, these three constituents are described as **biolimiting**.

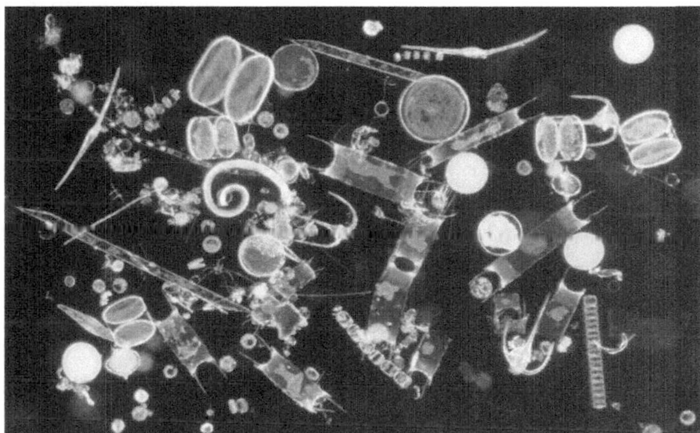

(a)

(b)

Figure 1.42 (a) Living phytoplankton: the needle-shaped, cylindrical and chain-like organisms are diatoms, which are encased in silica; the irregular ones with 'horns' or spines are dinoflagellates. The field of view is about 1.75 mm across. (b) Living zooplankton, including copepods (planktonic crustaceans) and the planktonic larvae of various animals. The field of view is about 1.75 cm across, so is ×10 that in (a). (These photographs show samples collected by special plankton nets that concentrate the organisms; they are not naturally found so closely packed in the open sea.) (N.T. Nicholl/Natural Visions)

Most of the organic matter forming the soft tissue of phytoplankton is recycled in near-surface waters through consumption by zooplankton (and other pelagic animals) and bacterial breakdown of detritus and excretion products. A small amount of debris does, however, escape from surface waters and sinks towards the deep ocean. As a result of decomposition of organic remains and dissolution of shells and skeletons, their constituents, including the nutrient elements, are gradually returned to seawater at greater depths.

■ How might these constituents be returned to surface waters, where they can again be used by planktonic organisms?

■ By upwelling of subsurface water in response to the divergence of water nearer the surface (Figure 1.36).

Question 1.9

Figure 1.43 shows the concentration of chlorophyll pigment in the surface waters of the North Atlantic in May and December. Bearing in mind Figures 1.6 and 1.28:

(a) Explain why there is such a difference in the primary productivity in the northern North Atlantic at the two seasons of the year.

(b) Explain why, even in May, there is very little chlorophyll pigment in the central North Atlantic.

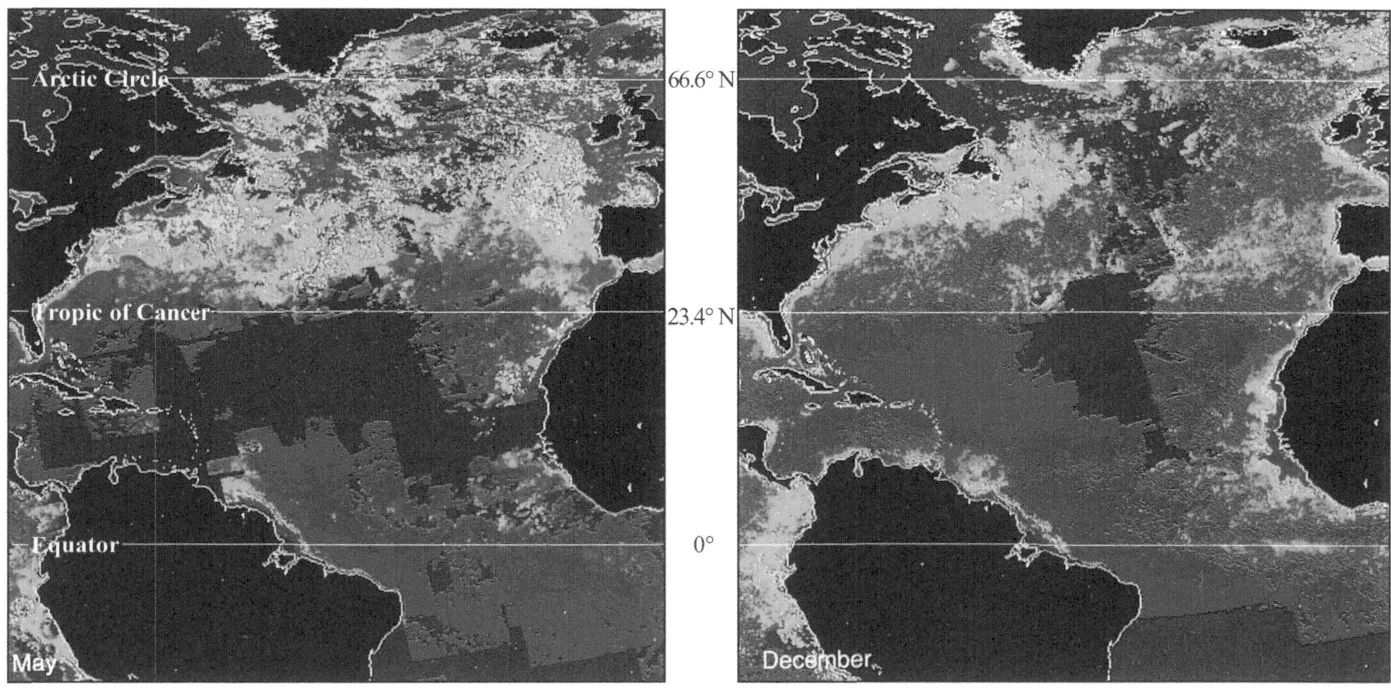

Figure 1.43 Seasonal changes in primary productivity in the North Atlantic, as indicated by concentration of chlorophyll pigment in May and December, measured by the satellite-borne Coastal Zone Color Scanner. Pink corresponds to lowest chlorophyll concentrations as in most open-ocean areas in December, and low-latitude regions in May (e.g. off the coast of eastern Brazil); bright red corresponds to highest chlorophyll concentrations widespread in mid- to high latitudes in May and in many coastal areas (e.g. off the coast of West Africa in May). Note that the areas of bright colour around the British Isles may be due partly to large amounts of suspended sediment and partly to real high primary productivity in response to nutrients (including agricultural fertiliser) contained in river runoff.

Therefore, as mentioned earlier, vertical current flows in the ocean have important implications for phytoplankton populations. Incidentally, it is important to bear in mind that the only reason changes in marine primary productivity can be so easily observed using satellite images is that phytoplankton increase their populations very fast but live only a short time, i.e. they have a very fast turnover rate. It would not be so easy to observe the changes in productivity of a forest of long-lived trees.

Recently, there has been much speculation as to whether, in certain circumstances and certain locations (e.g. the Southern Ocean around Antarctica), phytoplankton growth might be limited by lack of dissolved iron. Iron is found in seawater and in organisms in concentrations several orders of magnitude less than those of N, P and Si; it is therefore sometimes referred to as a *micronutrient.*

The clear blue waters of the tropical oceans are often described as barren because of their small phytoplankton populations. To test the hypothesis that tropical waters are iron-limited, one thousand square kilometres of the tropical Pacific was 'fertilised' by the addition of dissolved iron in 1995. The area concerned was southeast of the Galápagos Islands, at about 4° S, 104° W and the experiment was named IRONEX II. The large phytoplankton bloom that resulted from the addition of iron showed conclusively that primary production in tropical waters *is* iron-limited. Since then, a multitude of iron-fertilisation experiments have extended this result to other areas, in particular the Southern Ocean. Such controls on the amount of marine primary production may have important implications for climate.

As more is discovered about planet Earth, more links are discovered between the biosphere and the other components of the climate system – the atmosphere, the hydrosphere and the geosphere. In the late 1980s, a link was proposed between phytoplankton in the ocean and cloud formation. Many phytoplankton have population explosions or blooms during spring and summer (Figure 1.43), and blooms of certain species produce the volatile sulfur compound dimethyl sulfide (DMS). Once in the atmosphere, DMS undergoes a series of photochemical oxidation reactions, to produce sulfur dioxide (SO_2) and then sulfuric acid aerosols, which act as cloud condensation nuclei. This discovery excited many researchers, including environmental scientist James Lovelock (see Box 2.4), because increased concentrations of 'greenhouse' CO_2 in the atmosphere might also lead to increased marine primary productivity, i.e. more phytoplankton and bigger plankton blooms.

Question 1.10

How and why might such increased primary productivity affect the Earth's albedo?

By producing bigger blooms and hence more clouds, phytoplankton might counteract, or at least moderate, the effect of increased amounts of CO_2 in the atmosphere. This is an example of *negative* feedback, stabilising the climate system. This idea was favourably received when it was first proposed in the 1980s but, despite two decades of energetic research, conclusive evidence for or against the proposed feedback has remained elusive.

Finally, it should be noted that primary production can and does occur in the dark of the deep sea, by bacteria using chemical energy rather than the energy of sunlight. Like their counterparts photosynthesising in surface waters, these primary producers form the bases of whole ecosystems, of which the best known are the communities of animals found at hydrothermal vents. Unlike phytoplankton, however, **chemosynthetic bacteria** do not interact with the climate system in any significant way (at least as far as is known).

Figure 1.44 summarises much of what has been discussed about conditions at the surface of the Earth at the present time. As you will see later in this book, in the past, the Earth has been very different. This map is not intended to be comprehensive, but to remind you of some of the main points described in this chapter.

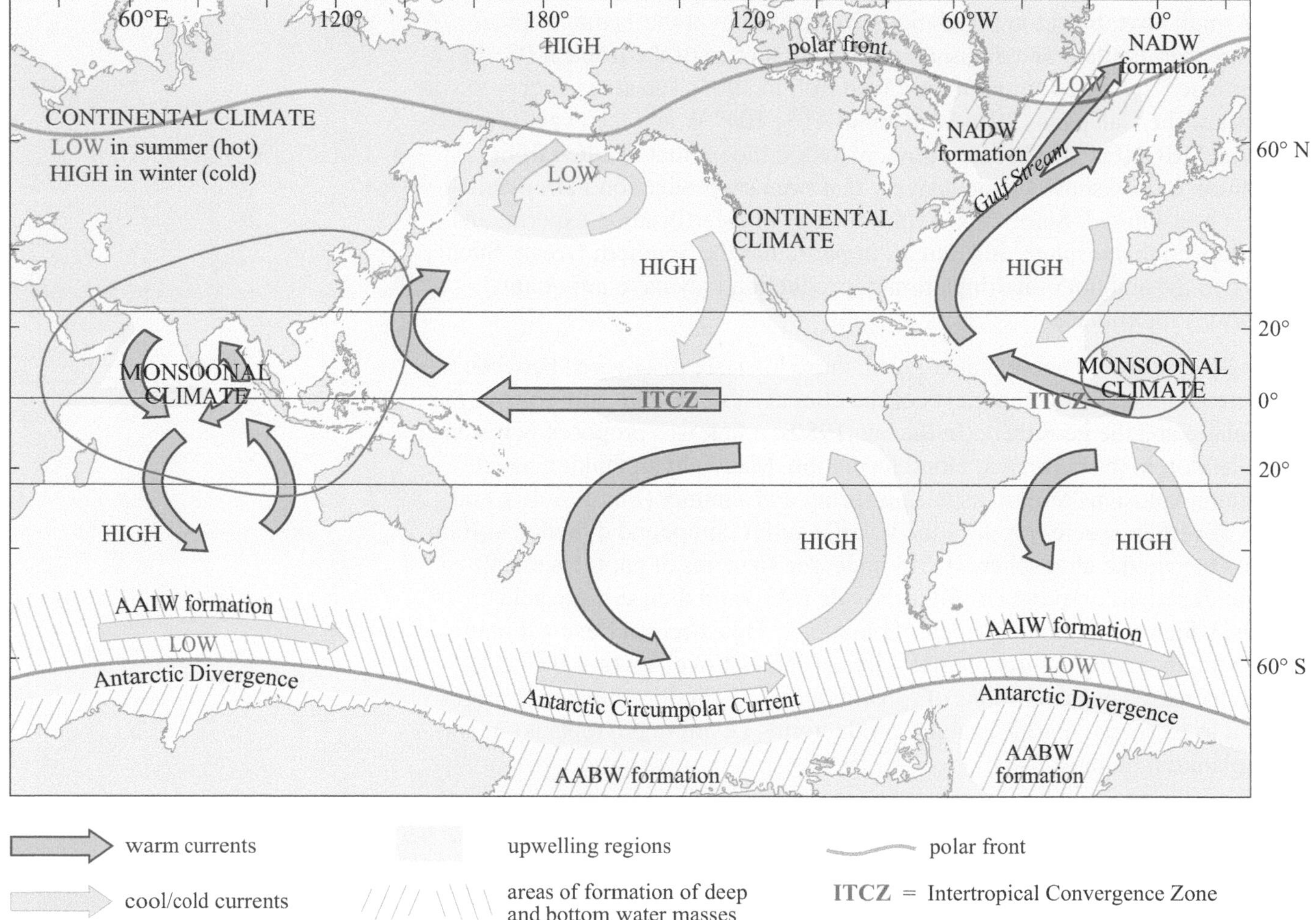

Figure 1.44 Highly schematic map showing general climatic features for land and sea, and the general position of the polar fronts in the atmosphere. HIGH and LOW refer to atmospheric pressure at the Earth's surface. Regions of deep and bottom water-mass formation are shown by blue hatching. Antarctic Bottom Water (AABW) formation, in particular, is localised, occurring only at very high southern latitudes, while North Atlantic Deep Water (NADW) formation occurs only in the northern North Atlantic. Antarctic Intermediate Water (AAIW) is formed throughout the Southern Ocean north of the Antarctic Divergence. Regions of upwelling, principally in equatorial and subpolar gyre regions and along the Antarctic Divergence, are shown in green.

Summary of Chapter 1

1 The Earth is hospitable to life because of its particular orbit around the Sun, which determines the amount of solar radiation that reaches it. The amount of solar radiation actually available to warm the Earth's surface is determined by how much is reflected rather than absorbed, i.e. by its albedo. At the temperatures obtained at the Earth's surface, water can exist as solid, liquid or gas; and, were it not for the presence of liquid water, life could not exist on Earth.

2 The fixation of carbon by primary producers (mainly plants) is the basis of all life on Earth.

3 Seasonal changes in incoming solar radiation are a result of the tilt of the Earth's axis in relation to the orbital plane. Over timescales of tens of thousands to hundreds of thousands of years, there are cyclical variations in incoming solar radiation caused by changes in the degree of *eccentricity* of the Earth's orbit, in the *angle* of tilt of the Earth's axis, and in the *direction* in which the axis points. These cycles have periodicities of ~110 000 years, ~40 000 years and ~22 000 years respectively, and are known as the Milankovich cycles.

4 Most of the radiative energy emitted by the Sun is in the wavelength range 0.15–5 μm, which includes ultraviolet radiation, visible radiation and infrared radiation. Solar radiation is often referred to as shortwave radiation, to distinguish it from longer-wavelength (thermal) radiation emitted by the Earth (and any other body) as a consequence of having been warmed. Certain gases in the atmosphere, notably carbon dioxide and water vapour, absorb outgoing longwave radiation and reradiate energy, much of it back to the Earth's surface. This trapping of radiant energy (which also involves clouds) is known as the greenhouse effect.

5 Although the Earth's surface is heated unevenly by the Sun, the redistribution of heat by winds and currents ensures that low latitudes do not continually heat up and high latitudes do not continually cool down. The continents have a much lower thermal capacity than the oceans, and so heat up and cool down much faster. This strongly affects the Earth's surface temperature distribution, particularly in the land-dominated Northern Hemisphere; in the oceanic areas of the Southern Hemisphere, temperature decreases more smoothly from Equator to poles, the temperature zones simply shifting northwards and southwards with the seasons.

6 Heat is redistributed over the surface of the Earth by winds and ocean currents, and by the evaporation, transport and condensation of fresh water. Rising air generally has a high moisture content (as at the Intertropical Convergence Zone); sinking air (as at the subtropical highs) is usually dry.

7 Clouds both reflect incoming solar radiation and absorb longwave radiation. Although they are a very important component of the climate system, and potentially the main driving mechanism for changes in the Earth's radiation budget, their role is by no means fully understood. Cloud condensation nuclei include sulfate aerosols, produced by volcanoes and industry, as well as forming from dimethyl sulfide, a waste product of phytoplankton populations.

8 The Coriolis effect acts to deflect flows to the right in the Northern Hemisphere and to the left in the Southern Hemisphere. The Coriolis effect increases with increasing latitude, leading to more circulatory atmospheric flow in mid-latitudes than nearer the Equator. The increase in Coriolis effect is also responsible for the narrowness of the western boundary currents found in all ocean basins. The formation of mid-latitude weather systems is determined by the behaviour of the polar jet stream, whose undulations at the top of the troposphere lead to the transport of enormous amounts of warm air polewards and cold air equatorwards. Extreme undulations of the polar jet stream may lead to severe droughts and unusual weather patterns. The ocean also has weather in the form of mesoscale eddies, which are an order of magnitude smaller than atmospheric cyclones and anticyclones.

9 Tropical regions where the ITCZ moves seasonally over land are subject to seasonally reversing winds (monsoons) and hence dry seasons alternating with seasons of heavy rainfall. The atmosphere and the ocean are tightly coupled by positive feedback loops, especially in tropical areas. Positive feedback loops cause instability in systems; negative feedback loops are stabilising.

10 While light, suitable temperatures and availability of water are the principal factors limiting primary production (and hence most other life) on land, in the oceans the limiting factors are light and the availability of nutrients. Nutrients depleted from the sunlit surface layer may be returned to it by upwelling. Upwelling occurs along the Equator, along western coastlines under the Trade Winds, and as a result of cyclonic winds.

11 While phytoplankton live in near-surface waters, in the photic zone, animals and bacteria live throughout the ocean, at all depths. This is possible because the cold water masses sinking at high latitudes carry dissolved oxygen down into the deep ocean. Some organisms, using chemical sources of energy, thrive in deep-sea hydrothermal vents.

Learning outcomes for Chapter 1

You should now be able to demonstrate a knowledge and understanding of:

1.1 How interactions between physical, chemical, biological and geological processes on Earth, along with its position relative to the Sun, make it a habitable planet for life to develop, exist and evolve.

1.2 The range of factors influencing the amount of solar radiation that reaches the Earth over both short and geological timescales, and difficulties associated with calculating such values.

1.3 The role of wind and oceanic currents in maintaining the Earth's surface temperature conditions and how perturbations in these circulatory systems can result in extreme short (and longer-term) changes in weather patterns.

1.4 The different circulatory systems that operate within the distinct oceanic and atmospheric layers (or spheres), and the physical and chemical processes that result in interaction and mixing between these layers.

1.5 How the concept of biogeochemical cycling and residence time can be used to investigate the extent and rate of redistribution of materials between the different biological, chemical, geological and physical spheres on Earth.

The carbon cycle

Chapter 2

All life, from bread mould to beetles, and from streptococci to professors, is composed of similar combinations of carbon-containing molecules and water. Life on this planet is based on the chemistry of carbon, which is why carbon-based compounds are referred to as 'organic'.

2.1 Carbon and life

There are good reasons for carbon's unique role in the living world. Here are the main ones:

1 *Carbon can form both soluble and other insoluble compounds*. Life is thought to have originated in water, and water is essential to life. Elements fundamental to the development of living organisms must be able to interact freely with one another, and that occurs most readily in the presence of water. Carbon dioxide in the atmosphere readily dissolves in water, as do many organic molecules that make up living organisms.

2 *Carbon can combine with itself and other elements to make more compounds than any other element*, and these compounds may be solid, liquid or gaseous. It is continually cycled through living organisms, and between living, dead and inorganic components of the Earth.

3 *Carbon permits the storing and passing on of information*. The chemical properties of carbon allow the construction of large, complex three-dimensional molecules (RNA and DNA) that are unique in the extent to which they can store, replicate and convey large amounts of information.

4 *Carbon compounds can store and release energy*. Carbon exists in both oxidised and reduced compounds. The energy released when carbon is transformed from a reduced to an oxidised state can be used for biochemical reactions in cells.

The underlying reason for all of these characteristics is the unique chemistry of the carbon atom (Box 2.1 overleaf).

Chemical reactions can be of two types:

- **endothermic** reactions, which *require* energy to proceed
- **exothermic** reactions, which *release* energy as they proceed.

Reactions involving carbon are no exception. For instance, burning coal is an *exothermic* reaction in which the carbon in coal is combined with oxygen in air, releasing energy in the form of heat and light. For life, the most important *endothermic* reaction of carbon is photosynthesis, whereby plants make organic compounds and oxygen.

■ The outside energy source is light from the Sun, but what are the raw materials used in photosynthesis?

□ Carbon dioxide in the air and hydrogen in water.

The carbon dioxide diffuses into the plant through **stomata**, which are special cells (mainly on the underside of leaves) that can open and close to the air and, along with the roots, regulate the gas and water balance of the plant.

Box 2.1 Carbon compounds

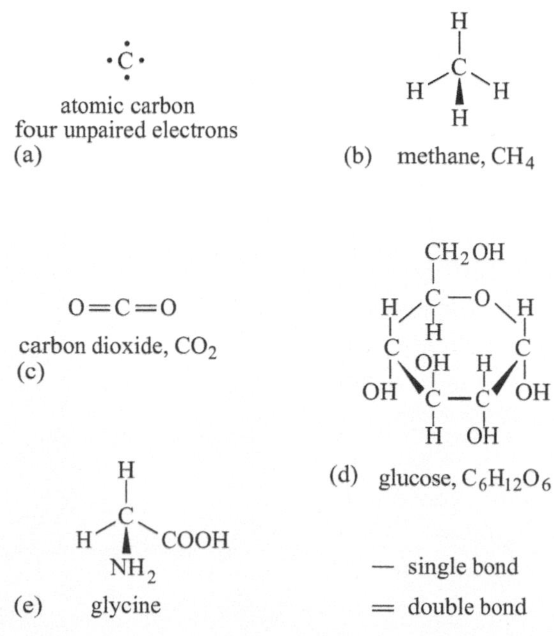

Figure 2.1 The carbon atom and some common carbon compounds: (a) atomic carbon, with its four unpaired electrons that enable each carbon atom to form up to four bonds; (b) methane, which is a tetrahedral molecule; (c) carbon dioxide; (d) glucose, which is a sugar; (e) glycine, which is an amino acid. The structures in b, d and e use a solid wedge symbol for groups and atoms projecting forwards (e.g. $-NH_2$ and –H).

The carbon atom has four unpaired electrons in its outer shell and forms covalent bonds by sharing these four electrons with other atoms (Figure 2.1a). As a result, carbon atoms may form up to four separate bonds with other atoms and so build complex three-dimensional structures. For example, carbon can share four electrons with four hydrogen atoms to form the simplest organic carbon compound, methane (CH_4), or share two pairs of electrons with two oxygen atoms to form carbon dioxide (CO_2) (Figure 2.1b and c). Moreover, carbon readily forms similar shared-electron bonds with oxygen, nitrogen, phosphorus and sulfur, which are all key elements in living organisms. Most important of all, carbon can share electrons with other carbon atoms. This enables it to form large molecules such as sugars (e.g. glucose, Figure 2.1d), chain-like molecules such as **carbohydrates** (which are sugars strung together), and cyclic molecules. Figure 2.1e shows the structure of glycine, which is an amino acid and one of the building blocks of the large and complex protein molecules.

Photosynthesis is a complex process involving many chemical reactions but, as mentioned in Section 1.1, it may be expressed as:

$$\underset{\text{carbon dioxide from atmosphere}}{6CO_2(g)} + 6H_2O \longrightarrow \underset{\text{organic matter}}{C_6H_{12}O_6} + \underset{\text{oxygen}}{6O_2(g)} \qquad (1.2)$$

where organic matter is represented by the carbohydrate molecule, $C_6H_{12}O_6$ (see Figure 2.1d). The *total* amount of carbon fixed in this way by plants is called **gross primary production**. Primary production (and by extension, organic matter) is usually expressed in terms of grams of carbon (gC) or kilograms of carbon (kgC).

Some of the CO_2 taken up by plants is returned to the atmosphere through the plant **respiration**, which is essentially the reverse reaction of photosynthesis:

$$C_6H_{12}O_6 + 6O_2 \longrightarrow 6CO_2 + 6H_2O + \text{energy} \qquad (2.1)$$

Respiration releases the energy stored in organic matter and provides energy for the plant to sustain itself by, for example, synthesising tissues and reproduction. The carbon that is *not* released back into the atmosphere through plant respiration is called the **net primary production** (or productivity). In other words, net primary production is the carbon that becomes incorporated into plant tissue. (Note that the term respiration encompasses what is commonly referred to as breathing *and* eating. Eating supplies organic matter to be oxidised with oxygen, releasing energy and CO_2 as a by-product. Breathing brings in the oxygen and removes the carbon dioxide. Both processes are combined in Equation 2.1.)

Plants are autotrophic (i.e. they manufacture their own food). Plant tissue may be consumed by organisms that cannot photosynthesise and depend ultimately on plants, the primary producers, for energy. These include carnivores which consume animals which consume plants. Such consuming organisms (e.g. fungi, cows and humans) are heterotrophic. Like plant respiration, aerobic (oxygen-using) heterotrophic respiration uses oxygen to convert organic carbon back into CO_2 and water, releasing stored energy in the process (Equation 2.1). So plants cannot live without sunlight and humans cannot live without plants.

Question 2.1

How would you classify the chemical reactions involved in (a) photosynthesis (Equation 1.2) and (b) respiration (Equation 2.1) in terms of oxidation or reduction?

The oxygen liberated in photosynthesis comes from the breakdown of the *water* molecules. The hydrogen so produced reduces the CO_2 to form carbohydrate molecules. In addition to being a reducing reaction, the conversion of CO_2 into organic matter during photosynthesis is an endothermic reaction (as seen above), taking in energy in the form of sunlight. Taking in energy is a primary characteristic of living things; on death (and during respiration), the energy is released to their surroundings.

Incidentally, you may wonder why, if it releases energy, Equation 2.1 does not simply occur spontaneously. The answer is that it does, but very slowly. Both plants and animals secrete enzymes (biological catalysts) that allow them to regulate the rate of this reaction.

2.2 Carbon and climate

As mentioned in Section 1.2, CO_2 and CH_4 are greenhouse gases that absorb longwave (infrared) radiation emitted by the Earth's surface and reradiate energy at longer wavelengths, much of it back towards the Earth (Figure 1.12). The warming caused by these and other natural greenhouse gases (e.g. water vapour, ozone and nitrous oxide) means that, rather than being about −18 °C, the average surface temperature of our planet is about +15 °C, which is favourable for the maintenance of liquid water and life. Indeed, without the 'blanket' of greenhouse gases, the Earth's surface would be completely covered in ice.

The current interest in atmospheric CO_2 in the context of global warming is heightened because there is good evidence that its concentration has varied over the history of the Earth and that these variations are at least partially linked to changes in the Earth's climate. But what are the factors that control atmospheric CO_2? For reasons that will become clear later, the best way to begin to solve this problem is to study the fluxes of carbon through the natural carbon cycle.

2.3 The natural carbon cycle: a question of timescale

Largely because of its involvement in living processes, carbon continually cycles through the different spheres of the Earth, i.e. the biosphere, the atmosphere, the hydrosphere and even the outer part of the solid Earth – the lithosphere. This global cycling of an element used by organisms (e.g. carbon, oxygen, nitrogen, phosphorus and sulfur) in and out of living, dead and inorganic reservoirs (Figure 1.40) is a biogeochemical cycle (Section 1.5). Such elements periodically move between reservoirs and rearrange themselves into different compounds and may be thought of as the currency exchanged between Earth and life.

There are six main carbon reservoirs, as shown in Figure 2.2.

■ What are these reservoirs in order of decreasing size?

□ They are: the deep ocean; deep-ocean sediments; soil, including organic debris; the surface ocean (including its **biota**); the atmosphere; and terrestrial organisms (the land biota).

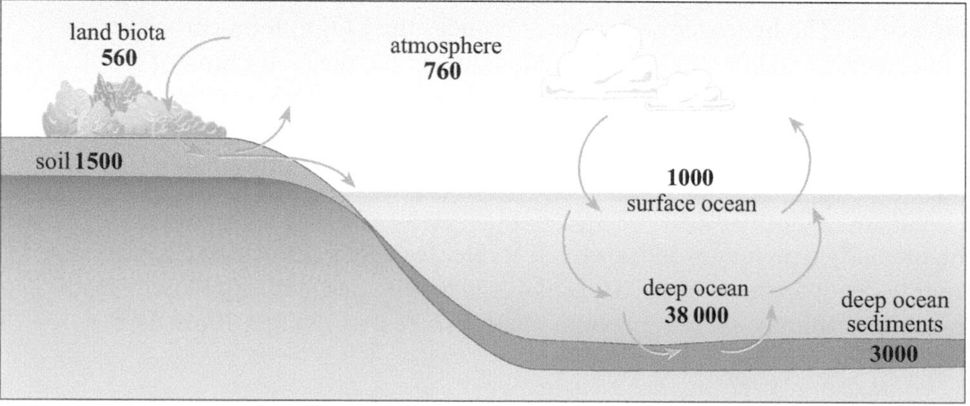

Figure 2.2 Schematic representation of the natural carbon cycle, with six main reservoirs as described in the text. 'Biota' means living organisms – in this context mainly plants. Bold numbers give the sizes of the reservoirs in 10^{12} kgC. In the rest of this chapter, estimated values will be added to the fluxes indicated by the arrows.

The different amounts of time that carbon atoms spend on average in each of these reservoirs have enormous implications for their potential to affect climate.

The overall natural carbon cycle is actually a number of cycles that link together and occur on several different timescales (Figure 2.2):

- a **short biological timescale** of months or years to decades
- an **intermediate geological timescale** of up to hundreds of thousands of years, involving chemical, biological and physical components
- a **long geological timescale** of up to hundreds of millions of years, involving rocks and sediments (not shown in Figure 2.2).

In the next section, you will look at the carbon cycle driven by land-based organisms, which is often referred to as the *terrestrial carbon cycle*.

2.3.1 Short timescales: the terrestrial carbon cycle

The terrestrial carbon cycle is dominated by the processes of photosynthesis and respiration and involves three major reservoirs:

- the atmosphere
- plant biomass
- soil organic matter (including detritus).

The reservoir sizes of plant biomass and the atmosphere are similar at approximately 560×10^{12} kgC and 760×10^{12} kgC respectively (Figure 2.3). Soils store about three times as much carbon as plants (about 1500×10^{12} kgC).

■ There is no box in Figure 2.3 for terrestrial animal biomass. Why do you think this has been omitted?

□ Animals are an insignificant carbon reservoir when compared with plant biomass (in fact, they comprise only about 0.01% of the carbon resevoir).

The land surface of the Earth is about 150×10^6 km^2, of which a large proportion is covered by plants (Figure 1.2).

■ If plant biomass were spread evenly over the land surface, how much (in kilograms of carbon, kgC) would there be per square metre?

□ $$\frac{560 \times 10^{12}\ \text{kgC}}{150 \times 10^{6}\ \text{km}^2} = 3.7 \times 10^{6}\ \text{kgC km}^{-2}$$

or 3.7 kgC m^{-2} (to 2 sig. figs).

Clearly, this average conceals huge variations – think of the amount of plant biomass in 1 m^2 of lawn compared with the average in 1 m^2 of forest.

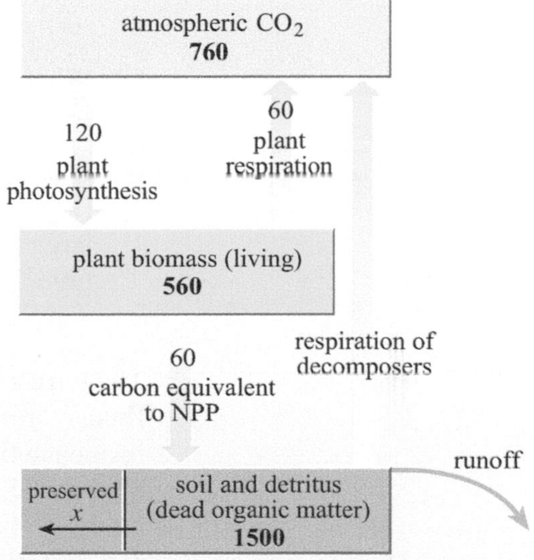

Figure 2.3 The biospheric carbon cycle on land. The bold numbers are the sizes of the reservoirs and the numbers on the arrows are the sizes of the annual fluxes (both in 10^{12} kgC). x denotes the carbon that becomes sequestered in forms that are not readily returned to the system, e.g. plant residues. NPP is net primary production.

Plants convert about 120×10^{12} kg of carbon from atmospheric CO_2 into organic carbon each year. This is the annual terrestrial gross primary production. About half of this is returned to the atmosphere and the soil as CO_2 through plant respiration, so that *net* primary production (i.e. the annual *net* fixation of carbon from the atmosphere into terrestrial biomass) is about 60×10^{12} kg y^{-1}. Most of this carbon goes into soil organic matter when vegetation dies or trees shed leaves or needles, so the amount of carbon deposited in the soil is also about 60×10^{12} kg y^{-1}. This is balanced globally by an approximately equal amount of carbon released back into the atmosphere when plant debris decomposes or vegetation burns in natural fires (ignoring the effect of human activities, such as deforestation, on this flux). Remember, the amount of carbon that goes into animal biomass is trivial compared with plant biomass; respiration fluxes from animals – particularly the less numerous 'larger' animals – are also small. The return of CO_2 to the atmosphere from recycled plant material is therefore almost entirely accounted for by the respiration of decomposers (bacteria, etc.) in the soil.

As with the hydrological cycle (Figure 1.38), if the size of a reservoir and the flux of a substance into and out of it are known, it is a simple matter to calculate the residence time of the substance in the reservoir:

$$\text{residence time} = \frac{\text{mass of substance in reservoir}}{\text{flux into (or out of) it}}. \qquad (2.2)$$

In doing such a calculation, it is assumed that an equilibrium has been reached, i.e. that the fluxes in and out are equal. Imagine a plant that has been uprooted. Within a short time it wilts: the flux of water molecules into the roots has been cut off, but the flux out of the leaves has not. The fluxes of water are no longer in balance and so the size of the reservoir (the amount of water in the plant) decreases.

Question 2.2

(a) What is the average residence time of carbon in plant biomass? What is it in soil organic matter?

(b) In global terms, 560×10^{12} kg is a relatively small reservoir of carbon (Figure 2.2) and 60×10^{12} kg y^{-1} is a relatively large flux. Would you expect large fluxes through a small reservoir to result in long or short residence times?

The terrestrial biospheric carbon cycle is characterised by relatively large global fluxes through small reservoirs of biota (living or dead), with relatively short residence times. Although the average residence time of organic carbon in soils is about 25 years, it varies greatly depending on the composition of the organic matter and the location. Most carbon in fresh litter is decomposed in a year or two, but highly resistant organic matter in the same soil (such as the carbon in plant structural material, **lignin**) may take much longer to decompose. In fact, carbon in anoxic soils may never fully decompose to release CO_2 back to the atmosphere and may be stored as residues in peat bogs, swamps or similar environments, perhaps eventually becoming coal. The amount of carbon preserved in this way is a very small proportion of the total plant debris and very difficult to quantify; it is indicated by x in the 'soil and detritus' reservoir in Figure 2.3.

■ What is the size of the respiration of decomposers flux in Figure 2.3?

■ It must be $(60 - x) \times 10^{12}$ kg y^{-1}.

In other words, Figure 2.3 shows that the carbon in the CO_2 produced by the respiration of organisms decomposing plant debris in the soil over the course of a year must equal the carbon added in plant debris *minus* the amount that has been preserved and effectively removed from the cycle *plus* that carried away in runoff. Both of these values are very small, so the carbon added to the soil in plant debris and the carbon removed from the soil in the CO_2 produced by the respiration of decomposers are therefore more or less equal (both are about 60×10^{12} kg y^{-1}) and together equal the amount of carbon fixed as (gross) primary production (about 120×10^{12} kg y^{-1}).

Although the amount of organic carbon removed from soils in run-off and exported by rivers to the sea each year is small, removal rates are locally higher if there is soil erosion because of exceptionally high rainfall and floods, or as a consequence of human activities such as farming, mining or forestry. This organic carbon may be either **particulate organic carbon** (**POC**) (fragments of soil or organic debris) or **dissolved organic carbon** (**DOC**). This, together with inorganic carbon removed in the weathering of rocks (**dissolved inorganic carbon** (**DIC**); see Box 2.2), gives a runoff flux of about 0.4×10^{12} kgC y^{-1}.

Difficulties in estimating the *sizes* of the different reservoirs (i.e. the amount of carbon stored globally in each of the various categories identified) impose significant limitations on the accuracy of any calculations of residence time. At its simplest, making a global estimate of a carbon reservoir involves multiplying two numbers: the global area (or volume) of the reservoir and the average amount of carbon in a representative area (or volume) of the reservoir. In practice, arriving at these two numbers is difficult and requires making assumptions and approximations. For example, to estimate the amount of carbon stored globally as plant biomass, you could begin by categorising the vegetation of the Earth into a number of distinct ecosystem types (i.e. **biomes**), which characteristically store different amounts of carbon. Figure 2.4 (overleaf) shows the global distribution of terrestrial biomes based on climatological and geographical considerations. In practice, satellite images are often used to determine such distributions (see Figure 1.2). For each biome, you could:

1 Lay out sampling grids (10 m^2, 1 km^2, etc., depending on the type of biome) in numerous different parts of the biome, covering all the known ecological variability.

2 Measure the amount of carbon contained in each of the grid squares.

3 Convert that amount to an average for the entire biome.

4 Multiply that average amount by the global area of the biome.

At best, the final number can only represent an approximate biomass for an extremely varied reservoir. Total global biomass would be obtained by repeating the process for every biome defined and then adding the individual totals together.

In Table 2.1 (overleaf), which shows estimates of terrestrial primary production and net primary productivity, the fourth column shows the results of one such calculation for the total mass of carbon in terrestrial biomass. Here, the land surface is divided into 14 different ecosystem types (column 1), the area of each ecosystem type is calculated (column 2) and then, for each, the mean plant biomass (in terms of carbon) per square metre is estimated (column 3). Columns 2 and 3 are multiplied together to give an estimate of the total mass of carbon in this ecosystem type at any one time. All such estimates may then be summed to provide a total global estimate. The last two columns show the net primary productivity (NPP) for each of the ecosystem types, i.e. the global net fixation of carbon into plant material per year.

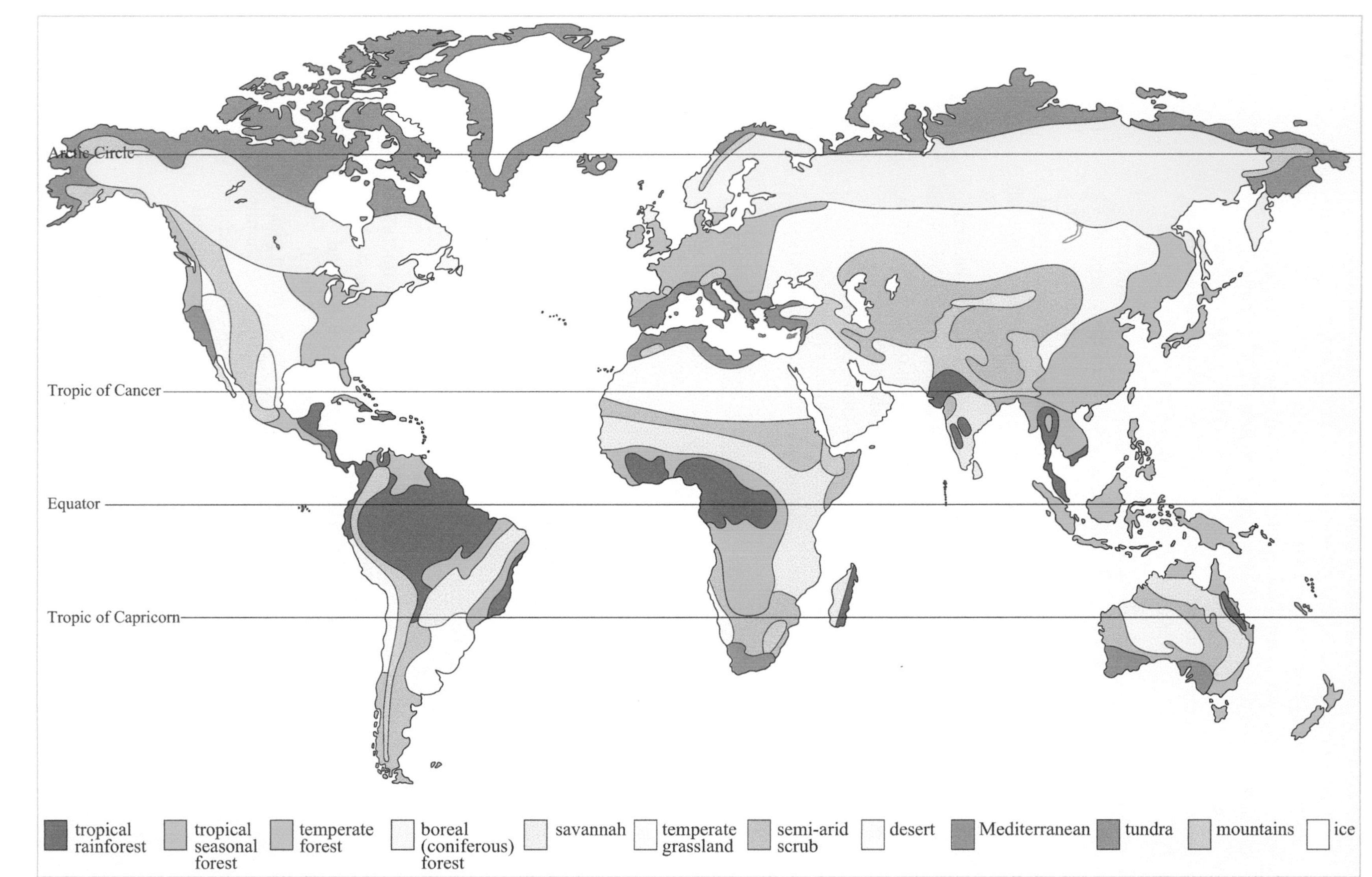

Figure 2.4 Geographical distribution of the major natural regional ecological communities or biomes (compare with Figures 1.1 and 1.2). An annotated version of this figure is in Appendix D. (Cox, 1989)

Table 2.1 Estimates of terrestrial primary production (standing stock) and net primary productivity (i.e. rates of primary production). Mean NPP is to two significant figures. NPP = net primary productivity.

Ecosystem type	Area $/10^6$ km^2	Mean plant biomass per unit area /kgC m^{-2}	Total plant biomass $/10^{12}$ kgC	Mean NPP per unit area /kgC m^{-2} y^{-1}	Total NPP $/10^{12}$ kgC y^{-1}
swamp, marsh	2	6.8	14	1.125	2.2
lake, stream	3	0.01	0	0.225	0.6
cultivated land	14	0.5	7	0.290	4.1
rock, ice, sand	24	0.01	0	0.002	0.04
desert scrub	18	0.3	5	0.032	0.6
tundra, alpine meadow	8	0.3	2	0.065	0.5
temperate grassland	9	0.7	6	0.225	2.0
woodland, shrubland	8	2.7	22	0.270	2.2
savannah	15	1.8	27	0.315	4.7
boreal forest	12	9.0	110	0.360	4.3
temperate deciduous forest	7	14	98	0.540	3.8
temperate broadleaved forest					
evergreen forest	5	16	80	0.585	2.9
tropical seasonal					
forest (deciduous)	8	16	120	0.675	5.1
tropical rainforest (evergreen)	17	20	340	0.900	15.3
total	**150**		**830**		**48.3**
global weighted mean		**5.55**		**0.324**	

According to Table 2.1, the global total for carbon in terrestrial vegetation reservoirs is estimated as 830×10^{12} kgC, which is rather different from the 560×10^{12} kgC used in Figure 2.3. Variations between estimated totals should not be surprising as there are serious difficulties in making the individual estimates in columns 3 and 4 of Table 2.1. The value of 830×10^{12} kgC is in fact at the upper end of the range of estimates for the total global carbon in terrestrial plants; some estimates are as low as $\sim 420 \times 10^{12}$ kgC. All the estimates are, however, of the same order of magnitude and the biggest is only about twice the smallest – quite impressive for a quantity that is so hard to assess. The differences among these estimates are almost entirely due to variation in values for relatively understudied tropical forests.

Even identifying the biomes is not straightforward, as an apparently homogeneous ecosystem may in fact be made up of several subcategories, and errors may be introduced by 'smoothing' the borders between biomes. Other problems include mapping or locating the biomes, measuring the amount of carbon in each grid, choosing representative grid samples, and measuring sufficient samples. In addition, there may be errors in laboratory analysis and inconsistencies arising because different researchers use different techniques.

Other reservoirs are more uniform and easier to estimate. For example, the global reservoir of atmospheric carbon is known to high accuracy because the air in the troposphere is well mixed and therefore fairly homogeneous (in contrast to the very 'patchy' and heterogeneous distribution of vegetation), its volume is known and many direct and precise measurements of atmospheric carbon concentrations (in CO_2 and CH_4) have been taken around the Earth. By contrast, the total amount of carbon in soils and detritus, or in marine sediments, may be significantly larger or smaller than the current best estimate.

Estimating fluxes in and out of a carbon reservoir is even trickier than estimating the reservoir size because it is often difficult to quantify the rates of the processes (such as photosynthesis, respiration and decomposition) that contribute to these fluxes. In practice, many of the processes are measured indirectly. For example, the rate at which various types of organic material decompose may be estimated in the field by the loss in mass over a given time period of a bag containing a known quantity of litter such as leaves; the bag is made of a mesh through which organisms can move freely, decomposing the litter to CO_2. Similarly, gross primary production may be measured by monitoring the CO_2 decrease or O_2 increase in a sealed box containing a plant or leaf (Figure 2.5), and net primary productivity may be estimated by the change in the weight of plant biomass over a given area at the beginning of and at the peak of the growing period.

Before moving on, look more closely at the information in Table 2.1 and see how it relates to the geographical distribution of ecosystems or biomes shown in Figure 2.4.

Figure 2.5 Estimating gross primary productivity by monitoring the decrease in CO_2 concentration in a sealed chamber (supported on a tower) enclosing a branch of Norway spruce (*Picea abies*). (Nancy Dise)

Question 2.3

(a) In terms of net primary productivity per square metre, tropical rainforests (darkest green (TR in Appendix D) shown mainly between the tropics in Figure 2.4) are by far the most productive ecosystems.

(i) What is the annual flux of carbon into and out of tropical rainforests?

(ii) According to the data in the last row and last column of Table 2.1, what percentage of the total global net primary productivity is contributed by these forests?

(b) (i) Area for area, how much more productive are tropical forests than boreal forests (i.e. forests that grow at high northern latitudes, see Figure 2.4)?

(ii) Bearing in mind what you have read earlier, particularly in relation to Figures 1.6 and 1.20 what are the various possible reasons for this difference in productivity?

(iii) How does the residence time of carbon in living plants in swamps and marshes compare with that in boreal forests?

Of course, in reality it cannot be assumed that areas of the globe covered by particular ecosystems remain constant. For example, tropical rainforests are known to be decreasing in area – adding to the widespread concern about global warming. This apart, the assumption that reservoirs and fluxes are constant does not in practice exclude small-scale fluctuations that average out over relatively short periods of time. The amount of carbon stored in living Northern Hemisphere forests for instance, may greatly fluctuate from summer to winter, but over a number of years the average amount stays relatively constant. Likewise, as you will see later in this chapter, the concentration of atmospheric CO_2 fluctuates seasonally.

■ If, on average, the input fluxes currently equal the output fluxes, how do you think these various carbon reservoirs built up in the first place?

□ At various times in the Earth's history the fluxes have not been in balance and inputs have been larger than outputs. For example during the Carboniferous Period (359–299 Ma), the terrestrial reservoir of carbon grew dramatically as plants spread over the land.

2.3.2 Intermediate timescales: the marine carbon cycle

Although the only one-way flux of carbon to the ocean is in rivers and other runoff (Figure 2.2), by far the greatest carbon fluxes into and out of the ocean are those to and from the atmosphere, across the air–sea interface. For this reason, it is convenient to start a discussion of the marine carbon cycle in the atmosphere. It is therefore important to understand how gaseous CO_2 interacts with liquid water.

Carbon dioxide in water

Carbon dioxide is the most soluble of atmospheric gases (Figure 1.41b). This is because some of the CO_2 molecules that diffuse into water then react with it to produce a variety of dissolved ions (sometimes referred to as ionic species). Box 2.2 sets out the details of the reactions that occur between gaseous carbon dioxide and its various aqueous forms, known collectively (for reasons that should become clear) as the **carbonate system**. You do not need to remember all the details of Box 2.2, but you should understand the following two main points.

1 Very little of the CO_2 dissolved in, for example, rainwater or the ocean is in the form of dissolved gas; most is in the form of bicarbonate and carbonate ions (HCO_3^- and CO_3^{2-} respectively) or as molecules of carbonic acid (H_2CO_3).

2 Given time, the carbonate system will always tend to adjust to a state of equilibrium.

Box 2.2 Carbon dioxide and water: the carbonate system

Very little of the CO_2 dissolved in natural waters is in the form of dissolved gas as such. When molecules of CO_2 gas diffuse into water, some react with the water to produce the weak acid, carbonic acid (H_2CO_3), but most occur as hydrated CO_2 (written as $CO_2(aq)$), where each CO_2 molecule is surrounded by water molecules. Since it is difficult to distinguish analytically between $CO_2(aq)$ and $H_2CO_3(aq)$ in practice, dissolved carbon dioxide is normally referred to simply as carbonic acid. Following this convenient shorthand, the chemical equation for the solution of CO_2 gas in water can be written as:

$$CO_2(g) + H_2O \longrightarrow H_2CO_3(aq) \qquad (2.3a)$$

At any particular temperature, the amount of CO_2 that diffuses into the water depends upon the concentration of CO_2 in the atmosphere and the concentration of carbonic acid in the water.

When enough H_2CO_3 builds up in the water, some of the dissolved carbon is released as CO_2 back to the atmosphere in the reverse reaction:

$$H_2CO_3(aq) \longrightarrow CO_2(g) + H_2O \qquad (2.3b)$$

Eventually, the forward reaction (Equation 2.3a) and the reverse reaction (Equation 2.3b) occur at equal rates and a state of dynamic balance or **chemical equilibrium** is established. At equilibrium, *reactants* (on the left-hand side of the equation) are forming *products* (on the right-hand side of the equation) at the same rate as products are decomposing back into their constituent reactants. Although the concentrations of product and reactant do not change, at the molecular level there is a constant and equal exchange between them. To represent this, equilibrium systems are written with two half-headed arrows pointing in opposite directions:

$$CO_2(g) + H_2O \rightleftharpoons H_2CO_3(aq) \qquad (2.3c)$$

The system is, however, more complicated than this. Carbonic acid, like all acids, tends to dissociate, i.e. lose a hydrogen ion. A second equilibrium is established between the carbonic acid and its dissociation products – a hydrogen ion (H^+) and a **bicarbonate ion** (HCO_3^-):

$$H_2CO_3(aq) \rightleftharpoons H^+(aq) + HCO_3^-(aq) \quad (2.4)$$

Furthermore, bicarbonate itself dissociates to form a **carbonate ion** (CO_3^{2-}) and a hydrogen ion. This third equilibrium is written:

$$HCO_3^-(aq) \rightleftharpoons H^+(aq) + CO_3^{2-}(aq) \quad (2.5)$$

So carbon dissolved in water achieves a state of dynamic chemical equilibrium between the CO_2 in the air and that partitioned among three dissolved carbon compounds (shown in bold):

$$\mathbf{CO_2}(g) + H_2O \rightleftharpoons \mathbf{H_2CO_3}(aq) \rightleftharpoons H^+(aq) + \mathbf{HCO_3}^-(aq) \rightleftharpoons 2H^+(aq) + \mathbf{CO_3}^{2-}(aq) \quad (2.6)$$

This dynamic equilibrium among dissolved carbon compounds is called the carbonate equilibrium system, and the dissolved carbon compounds H_2CO_3, HCO_3^- and CO_3^{2-} are collectively termed dissolved inorganic carbon. At equilibrium, the *concentrations* of the different carbon compounds are constant (but not necessarily equal) and, for each pair of compounds, the amounts of carbon *exchanged* are equal. Therefore, the relative proportions of the reactant and product remain constant. You may visualise this by imagining four boxes, each containing a different number of marbles but with a constant exchange of two marbles back and forth between them (Figure 2.6). In the same way (Figure 2.7), at equilibrium the amount of

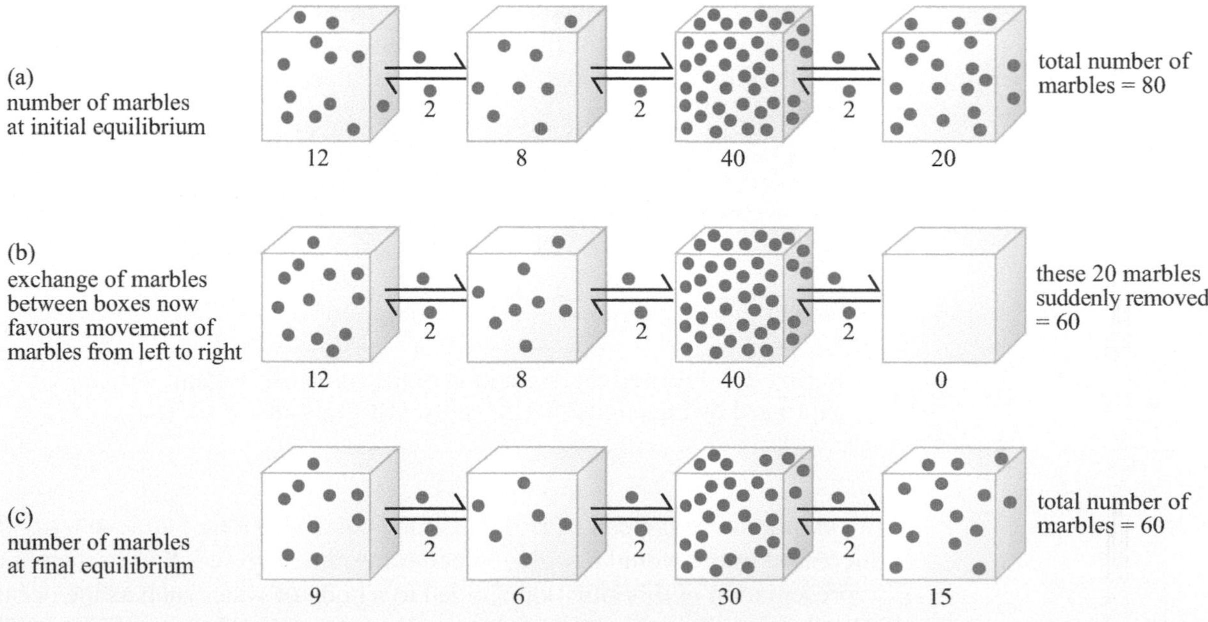

Figure 2.6 (a) Equilibrium among four reservoirs in a closed system; as far as the carbonate system is concerned, the four reservoirs are loosely analogous to atmospheric CO_2, H_2CO_3, HCO_3^- and CO_3^{2-}. (b) If one of the reservoirs is depleted, more reactants will form products in a direction tending to re-establish equilibrium. (c) A new equilibrium is established. The total number of marbles corresponds to the number in the boxes. The marbles on the arrows denote the marbles that will move from one box to another.

CO_2 leaving the surface of the water is the same as the amount going into the water (Equation 2.3c), and the amount of carbonate being formed is the same as that being recombined to form bicarbonate, HCO_3^- (Equation 2.5). Note that the concentration of the components is determined by two equilibrium reactions, Equations 2.4 and 2.5.

When a chemical system in equilibrium is subjected to an external constraint, the system responds in a way that tends to lessen the effect of the constraint; this expression of a fundamental observation is known as **Le Chatelier's principle**, where in a reaction, any factor that removes some or all of the product(s) will cause the reactant(s) to form more product(s) and re-establish the equilibrium.

Returning to the marble analogy, according to Le Chatelier's principle, if the box at the right-hand end is depleted (Figure 2.6b), the exchange of marbles between boxes will begin to favour the movement of marbles towards the right-hand end so that, eventually, equilibrium – a different equilibrium from the first – is established (Figure 2.6c).

■ As far as the carbonate system is concerned, what would be the effect on the rate of diffusion of CO_2 into the water of any process that removes H_2CO_3 from solution?

□ Anything that removes carbonic acid from solution will increase the rate of diffusion of CO_2 *into* the water; hence the rate of dissolution of atmospheric CO_2 will increase.

Conversely, if the concentration of atmospheric CO_2 increased, more carbon dioxide would dissolve in water, forming more H_2CO_3 and ultimately partitioning itself among the various dissolved inorganic carbon compounds (Figure 2.7).

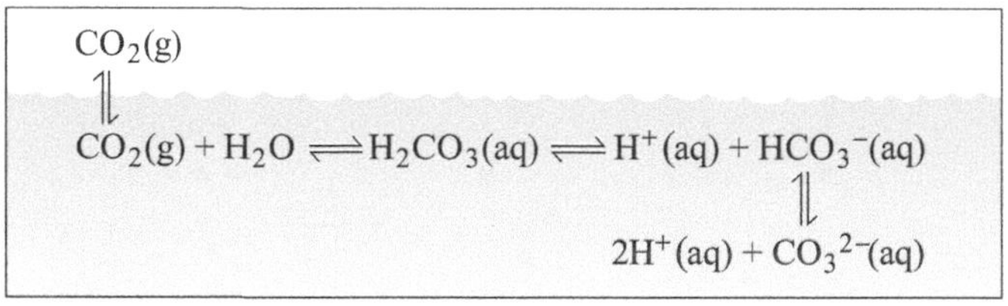

$$CO_2(g)$$
$$\rightleftharpoons$$
$$CO_2(g) + H_2O \rightleftharpoons H_2CO_3(aq) \rightleftharpoons H^+(aq) + HCO_3^-(aq)$$
$$\rightleftharpoons$$
$$2H^+(aq) + CO_3^{2-}(aq)$$

Figure 2.7 Pictorial representation of the carbonate system expressed by Equation 2.6. (In reality, H_2CO_3 is mostly $CO_2(aq)$.)

As mentioned at the end of Box 2.2, if the concentration of atmospheric CO_2 increased, more would dissolve in natural waters. Figure 2.8 is a schematic representation of this situation applied to a body of water such as the ocean. The flux into the ocean would remain greater than the flux out (Figure 2.8b) only until a new equilibrium had been established. Conversely, if the concentration of atmospheric CO_2 decreased, more would come out of solution in the ocean than would dissolve until a new equilibrium had been established (Figure 2.8c).

Question 2.4

The important chemical processes in the global carbon cycle are in approximate balance. What are the equivalent 'reverse reactions' for (a) photosynthesis and (b) weathering of limestones on land?

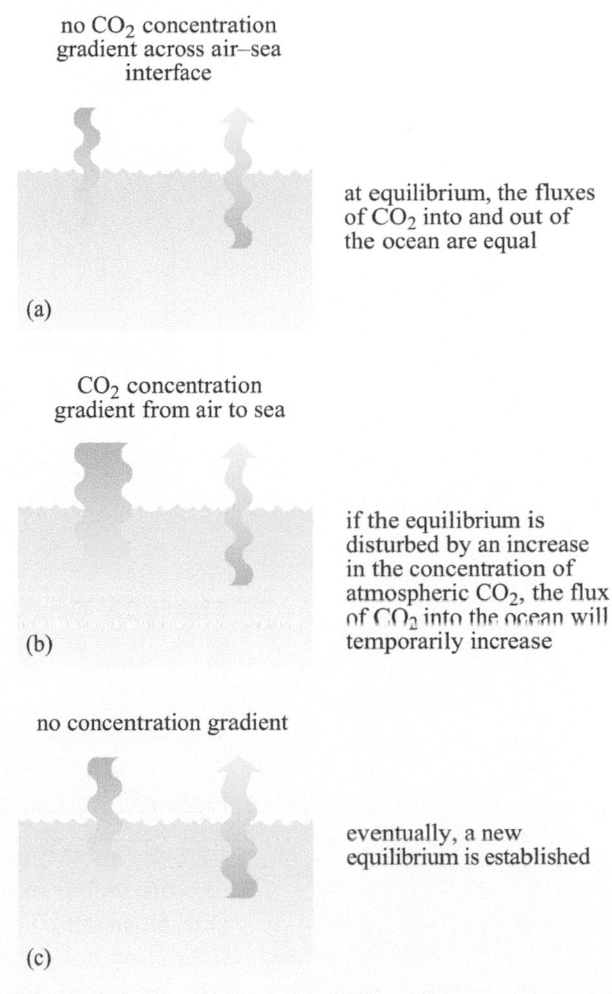

Figure 2.8 Schematic illustration of the change from one state of equilibrium to another. (a) At equilibrium, the fluxes of CO_2 into and out of the ocean are equal. (b) If the equilibrium is disturbed by an increase in the concentration of CO_2 in the atmosphere, the flux of CO_2 into the ocean (i.e. the rate at which CO_2 goes into the ocean) will temporarily increase. (c) Eventually, a new equilibrium is established in which more CO_2 is dissolved in the ocean.

■ Bearing in mind what you read in Section 1.5 about the solubility of gases, suggest another circumstance that would lead to an increased flux of CO_2 from the ocean to the atmosphere.

□ As all gases are more soluble in cold water than in warm water, warming of surface waters would cause more CO_2 to come out of solution, increasing the flux of CO_2 from ocean to atmosphere.

In other words, the state of equilibrium for warmer water has a lower concentration of dissolved CO_2. This new equilibrium would take a while to establish itself and, until this happened, there would be a net flux from sea to air.

A component of the carbonate equilibrium system that has not yet been discussed is the hydrogen ion (H^+). This has an important role as its concentration in solution determines the relative proportions of the different DIC compounds (H_2CO_3, HCO_3^- and CO_3^{2-}). The hydrogen ion concentration in water (often written in square brackets: $[H^+]$) is generally expressed as **pH**, where a low pH corresponds to a high hydrogen ion concentration and vice versa. For example, a pH of 5 represents a hydrogen ion concentration, $[H^+]$, of 10^{-5} mol l^{-1}, while a pH of 10 represents a $[H^+]$ of 10^{-10} mol l^{-1}. Low pH (below about 5) favours the formation of H_2CO_3, while a higher pH (7 to 8) favours the formation of HCO_3^- and still higher values (above 9) favour the formation of CO_3^{2-}. These relationships are shown schematically in Figure 2.9 (overleaf), along with the chemical equilibria that are affected (Equation 2.6).

■ According to Figure 2.9, what are the main contributions to dissolved inorganic carbon in seawater (pH around 7.7) and rainwater (pH 5.6 or less)?

□ In seawater, most (~85%) of the DIC is as HCO_3^- (about 10–15% is as CO_3^{2-}, and less than 1% is as H_2CO_3), while in rainwater about half the DIC is in the form of H_2CO_3 and about half is HCO_3^-. (Remember, however, that the positions of the curved lines can vary – see caption.)

Next, you will see how weakly acidic rainwater is returned to the oceans.

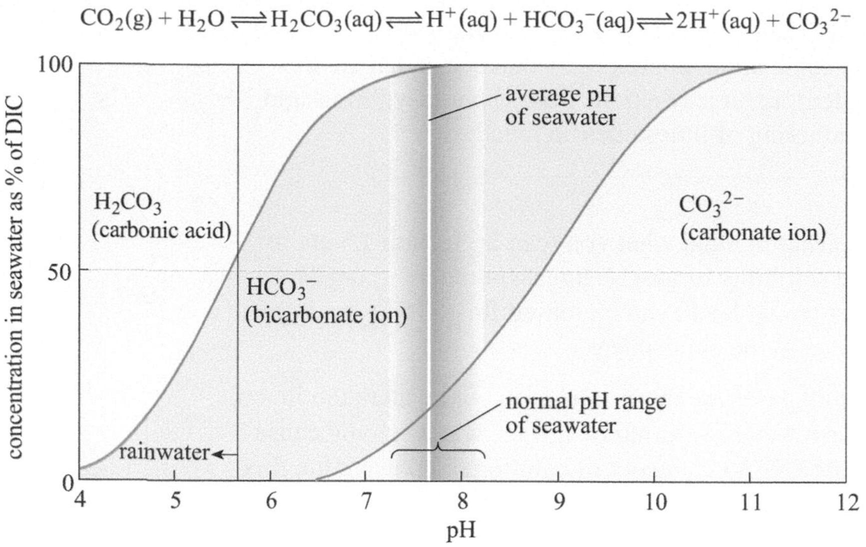

$$CO_2(g) + H_2O \rightleftharpoons H_2CO_3(aq) \rightleftharpoons H^+(aq) + HCO_3^-(aq) \rightleftharpoons 2H^+(aq) + CO_3^{2-}$$

concentration in seawater as % of DIC
100
50
0
H_2CO_3 (carbonic acid)
HCO_3^- (bicarbonate ion)
CO_3^{2-} (carbonate ion)
average pH of seawater
normal pH range of seawater
rainwater
4 5 6 7 8 9 10 11 12
pH

Figure 2.9 Generalised diagram showing approximately how the relative proportions of the three principal components of DIC in the aqueous carbonate system vary with pH in natural waters. The pH of rainwater is 5.6 or less; the pH of seawater averages about 7.7 and can range from about 7.2 to 8.2. The positions of the curved lines can vary with temperature and, in the ocean, also with salinity and pressure.

From rain to soil to river to ocean

Imagine weakly acidic rain falling onto soil where microbes are decomposing organic matter in respiration reactions that use oxygen and release CO_2 (Equation 2.1). Plants are also releasing CO_2 from respiration (most of it through their roots). As a result of both plant respiration and microbial activity pumping CO_2 into pore spaces in the soil, CO_2 concentrations are typically 10 to 100 times higher than in the atmosphere. As a result of this, concentrations of CO_2 in soil water are also raised, moving the reactions in Equation 2.6 to the right, producing additional hydrogen ions. This, plus the addition of various organic acids (e.g. humic acid and fulvic acid) from the decomposition of plant remains, means that soil water is often significantly more acidic than rainwater.

Figure 2.10 Soil forming in a limestone crevice around a newly established thrift (*Armeria maritima*) plant. (Mike Dodd/Open University)

Acidic soil water is the main agent of chemical weathering of minerals in soil and rock. In general, carbonic acid plays a larger role in tropical forests where lower concentrations of organic acids remain after surface organic litter has decomposed, while organic acids dominate the weathering processes in cool temperate forests where weathering processes are slow and incomplete. Thus, especially in cool conditions, weathering rates are increased by the activities of living organisms, including small flowering plants (Figure 2.10), fungi and lichens.

Cations such as Na^+ and Ca^{2+} dissolved from rocks are essential for plant growth. Humic material that collects in a developing soil also plays an important part because the essential elements (i.e. nitrogen (as nitrate), phosphorus (as phosphate) and other nutrients) are retained in the soil and are made available to plants largely through their organic content. As every good gardener knows, a high content of humus also helps retain water in the soil; in addition, the relatively dark colour of soil helps it absorb heat and a warm soil generally increases plant growth rates.

As discussed in Section 1.5, compared with rainwater, river water contains high concentrations of HCO_3^-, Ca^{2+} and SiO_2 that have been released from rocks by chemical weathering. Most of the Earth's crustal rocks are made up of silicate or carbonate minerals, so the two basic chemical weathering reactions can be represented by just two equations. Equation 2.7a represents the reaction of rainwater with calcite, a form of calcium carbonate that is common in sedimentary rocks (especially limestone and chalk), to release calcium and bicarbonate ions in solution. In Equation 2.7b, rainwater reacts with albite, a common mineral in igneous and metamorphic rocks, to produce a clay mineral, along with sodium and bicarbonate ions in solution, plus silica which is only partly in solution (note that the carbon atoms are highlighted in red).

For the weathering of carbonates:

$$\underbrace{CaCO_3}_{\text{calcite (or any carbonate mineral in rock/soil)}} + \underbrace{CO_2 + H_2O}_{\text{from rainwater}} \rightarrow \underbrace{Ca^{2+}(aq) + 2HCO_3^-(aq)}_{\text{in solution in soil/stream water}} \quad (2.7a)$$

For the weathering of silicates:

$$\underbrace{2NaAlSi_3O_8}_{\text{albite (or any silicate mineral in rock/soil)}} + \underbrace{2CO_2 + 3H_2O}_{\text{from rainwater}} \rightarrow \underbrace{Al_2Si_2O_5(OH)_4}_{\text{kaolinite (a clay mineral)}} + \underbrace{2Na^+(aq) + 2HCO_3^-(aq)}_{\text{in solution in soil/stream water}} + \underbrace{4SiO_2(aq)}_{\text{silicate partly in solution}} \quad (2.7b)$$

Albite is a sodium-rich silicate mineral, but similar equations could be written for other silicates, such as calcium-rich anorthite ($CaAl_2Si_2O_8$), potassium-rich orthoclase ($KAlSi_3O_8$), or magnesium–iron-rich olivine ($(Mg,Fe)_2SiO_4$).

Each of these two weathering reactions results in two bicarbonate ions ($2HCO_3^-$) on the right-hand side.

■ How do the two reactions differ in terms of where the two carbon atoms in the $2HCO_3^-$ come from?

□ In the weathering of *carbonates* (Equation 2.7a), one carbon atom comes from atmospheric CO_2 and one comes from the carbonate mineral itself. In *silicate* weathering (Equation 2.7b), *both* carbon atoms come from the atmosphere.

It is important to make a note of this distinction, as it will become crucial later in the chapter.

Soil water, carrying ions released by weathering, collects in streams and eventually reaches the sea via rivers.

■ Apart from being a much more concentrated solution, how, generally speaking, does seawater differ from river water?

□ The relative proportions of dissolved ions in seawater are different from those in river water (and are, in fact, closer to those in rainwater).

As discussed in Section 1.5, river water has relatively high concentrations of HCO_3^-, Ca^{2+} and SiO_2 (Equations 2.7a and 2.7b), whereas the most abundant ions in seawater are Cl^-, Na^+ and SO_4^{2-}. As you will see shortly, the apparent shortfall of HCO_3^-, Ca^{2+} and SiO_2 in seawater is intimately related to the cycling of carbon in the ocean.

Carbon in the oceans

Figure 2.11 shows the marine carbon cycle. Unlike the terrestrial carbon cycle, which is dominated by the formation, decomposition and recycling of *organic* (mainly plant) material (made of molecules built up of carbon, hydrogen and *oxygen*), the marine carbon cycle is dominated by the chemistry of *inorganic* carbon, i.e. minerals (or dissolved ions) containing carbon. The ocean can be envisaged as a soup of living organisms (particularly in the sunlit surface layers) and the chemical transformations involved in the marine carbon cycle take place largely through their activity. Nevertheless, the flux of carbon to and from the atmosphere – in other words, the direction of the net exchange of CO_2 across the air–sea interface (Figures 2.7 and 2.8) – is not primarily regulated by photosynthesis (as it is on land), but by carbonate equilibrium reactions (Equation 2.6) that are affected by both biological and physical factors, which vary from place to place and season to season.

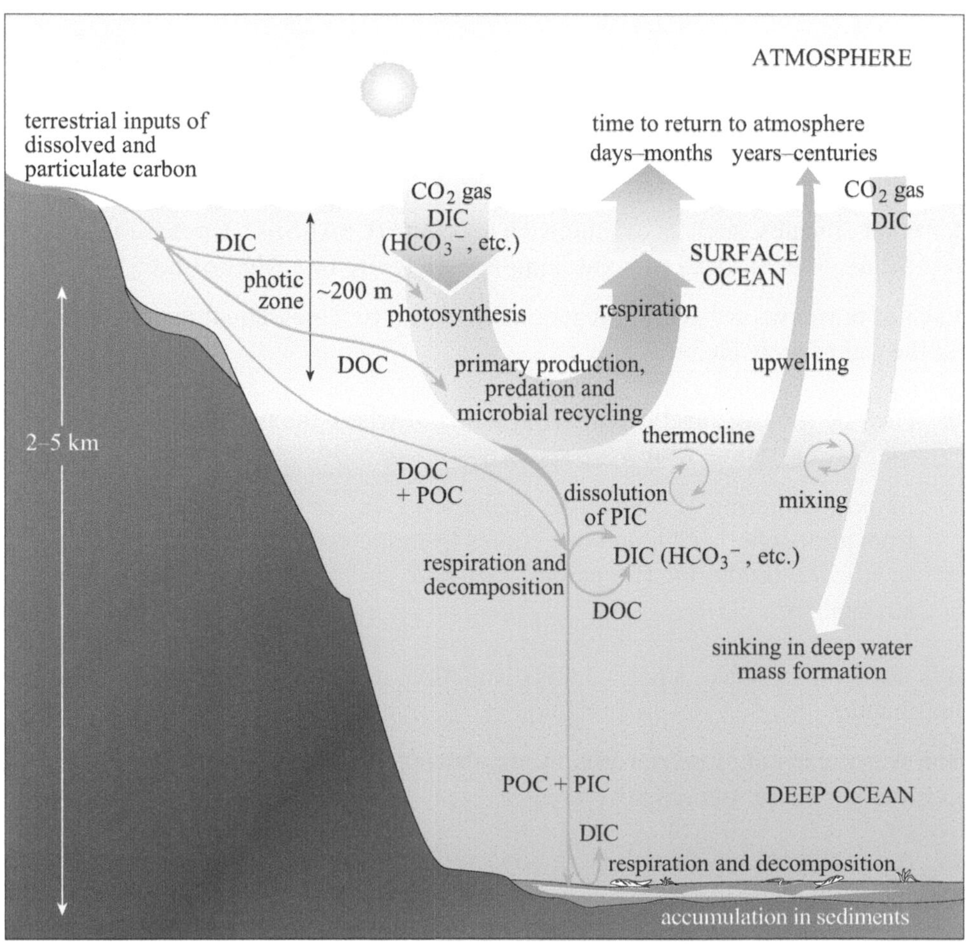

Figure 2.11 Diagram of processes contributing to the marine carbon cycle (not to scale). Remember that, while both physical and biological processes affect the fluxes of CO_2 across the air–sea interface, they occur through the action of *chemical equilibria* (Figures 2.7 and 2.8). Aspects of this diagram not discussed here are addressed later. (Williamson)

The ocean can be divided into two 'layers' (Figure 1.33): warm, well-mixed surface waters that are underlain by the cold, deep ocean. CO_2 exchanges across the air–sea interface (Figure 2.7) at a rate dependent on its concentration in the atmosphere and in the mixed surface layer (Figure 2.11). Windy conditions enhance surface mixing (Section 1.4.2).

Figure 2.12 shows data collected in the North Atlantic in the spring of 1989, when light levels had risen sufficiently for phytoplankton to have begun to multiply (compare with Figure 1.43). The lower (red) plot shows how the concentration of chlorophyll pigment in surface water (an indication of phytoplankton biomass) varied along the ship's track; the highs and lows correspond to fronts, eddies, and so on. The upper (black) plot shows the difference in the partial pressure of carbon dioxide between the atmosphere and the underlying ocean along the same track. Given the complications of estimating the carbon dioxide concentration in water, all you need to know is that the negative values correspond to concentrations in the atmosphere being higher than those in the ocean.

Question 2.5

(a) By reference to the right-hand axis of Figure 2.12, explain whether or not the CO_2 dissolved in the surface ocean was in equilibrium with the CO_2 in the overlying atmosphere at the time in question (Figures 2.7 and 2.8). Was there a net flux of CO_2 across the air–sea interface and, if so, was it into or out of the ocean?

(b) Why do you think the shapes of the two plots are similar?

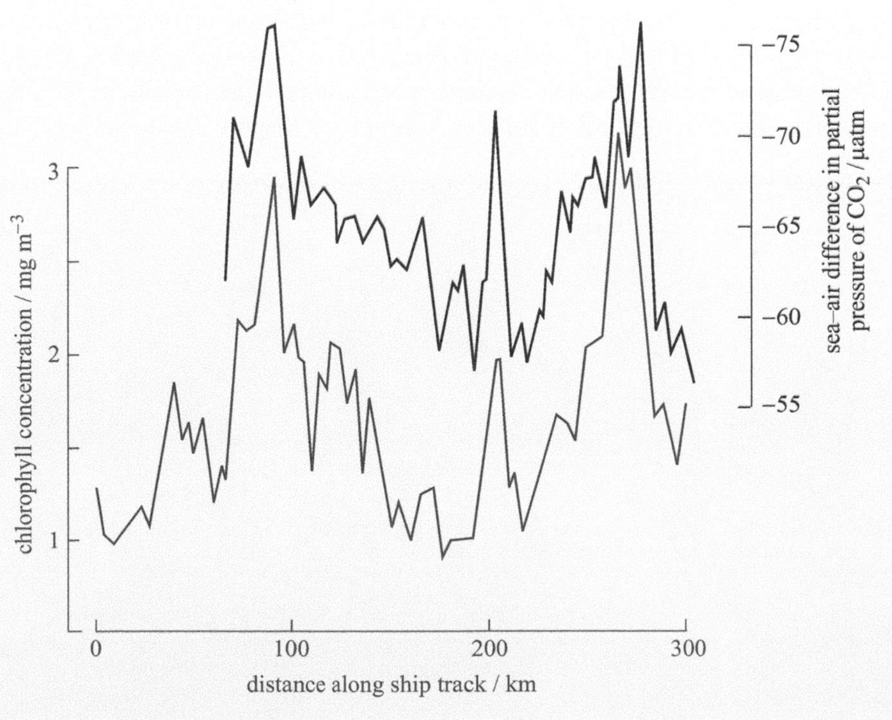

Figure 2.12 Variation in the concentration of chlorophyll in surface water along a ship's track in the North Atlantic (lower red plot) and the difference between the partial pressure of carbon dioxide in surface waters and that in the overlying air (upper black plot), expressed in micro-atmospheres (µatm or 10^{-6} atm), along the same track. Negative values correspond to concentrations in the atmosphere being higher than those in the surface ocean. (Source of data: the North Atlantic Bloom Experiment, 1989, which was part of BOFS, the Biogeochemical Ocean Flux Study)

Phytoplankton populations, therefore, act as *sinks* for atmospheric carbon dioxide, as do microscopic benthic (i.e. bottom-living) algae and seaweeds (sometimes referred to as macro-algae). Table 2.2 includes data on carbon fixation and productivity for both phytoplankton and benthic plants.

Table 2.2 Estimates of marine primary production (standing stock) and net primary productivity (NPP).

Region	**Area /10^6 km^2**	**Mean plant biomass per unit area /kgC m^{-2}**	**Total plant biomass /10^{12} kgC**	**Mean NPP per unit area /kgC m^{-2} y^{-1}**	**Total NPP /10^{12} kgC y^{-1}**
open ocean	332	0.0014	0.46	0.057	19
upwelling areas	0.4	0.01	0.004	0.225	0.1
continental shelf	27	0.005	0.14	0.162	4.3
algal beds and reefs	0.6	0.90	0.54	0.900	0.5
estuaries	1.4		0.63	0.810	1.1
total			**1.76**		**25**
global weighted mean	**361**	**0.0049**		**0.069**	

- Which of the regions listed in Table 2.2 will *not* support any benthic plants?
- Benthic plants can only grow where the bottom is within the photic zone, so the row for 'open ocean' (depth 2000–5000 m) cannot include data for benthic plants.

The photic zone may be as much as 250 m deep in the open ocean, but in turbid coastal waters it is a lot less. Algal beds and reefs only grow within the photic zone and support largely benthic production, while the data for estuaries, upwelling areas and continental shelf will include a variable amount of benthic plants depending on the clarity of the water. Clear nutrient-rich coastal waters can support highly productive stands of vegetation, notably kelp forests (Figure 2.13).

Figure 2.13 A kelp forest, characteristic of nutrient-rich, nearshore waters under the influence of coastal upwelling. (Bob Spicer/Open University)

It is difficult to define satisfactory categories for compilations such as Table 2.2 (e.g. many upwelling areas are over continental shelves) and, like land vegetation, the estimates are subject to considerable error. Nevertheless, Table 2.2 shows that the *net* amount of carbon converted each year into *new* plant (mainly phytoplankton) tissue (i.e. the net primary productivity) is less than that for terrestrial plants: it is only ~25×10^{12} kgC y^{-1}. Furthermore, the actual amount of carbon in the marine plant reservoir at any one time is very small, only ~$1–2 \times 10^{12}$ kgC (see fourth column), indicating a rapid turnover (i.e. a short residence time) of carbon in marine plants.

Phytoplankton are consumed by zooplankton, and both phytoplankton and zooplankton are consumed by larger animals, and all these organisms respire, releasing CO_2 back into the surrounding water, mainly as HCO_3^-. On death, soft tissues are decomposed by bacteria, releasing organic molecules into solution. Bacterial respiration oxidises both tissue and dissolved organic molecules, returning carbon dioxide into solution; thus, decomposition returns both inorganic and organic carbon to solution (both DIC and DOC are produced), which is available for reuse by organisms. In these various ways, most of the carbon in organisms of the upper ocean is recycled many times into dissolved inorganic and organic carbon and back again.

So how is it that the carbon fixed by photosynthesis does not all find its way back into the atmosphere in a very short time via respiration?

The reason is that the recycling system is not 100% efficient and some $4–5 \times 10^{12}$ kgC y^{-1} escapes from surface waters, mostly as particles of organic debris, zooplankton faecal pellets (particulate organic carbon, POC) and skeletal remains. Most of the detritus consists of very small particles (i.e. dead and dying algal cells and bacteria) and would take months to sink to the seabed were it not for the fact it forms clumps or aggregates of fluffy debris, often referred to as **marine snow** (Figure 2.14a and b overleaf). These aggregates sink to the seabed in days or weeks and arrive below where production occurred in surface waters rather than being dispersed by currents. As it sinks, and after arriving on the deep seabed, the particulate organic carbon is decomposed by bacteria, and both bacteria and organic remains provide food for animals in the water column (i.e. *pelagic* animals) and on the seabed (i.e. benthic, or bottom-dwelling, animals) (Figure 2.14c); all of these organisms respire, returning CO_2 to solution. Generally, soft tissue is almost completely decomposed before it reaches the sea floor (Figure 2.15) but, if productivity is very high, as during the spring bloom in the North Atlantic (Figure 1.43a), much phytoplankton debris aggregates into marine snow and sinks out of surface waters without being consumed, providing a food bonanza for bottom-living animals (Figure 2.14d). This transfer of carbon from surface waters to the deep ocean (which means that more has to be supplied from the atmosphere to support plankton growth) is known as the **biological pump**.

Before moving on to look at the *physical* mechanisms that control the flux of carbon into and out of the ocean, another biological process will be considered that uses carbon in seawater. Many, but by no means all, planktonic organisms in the ocean have protective shells or skeletons (as indeed do other larger marine organisms). Many of these hard parts are composed of calcium carbonate

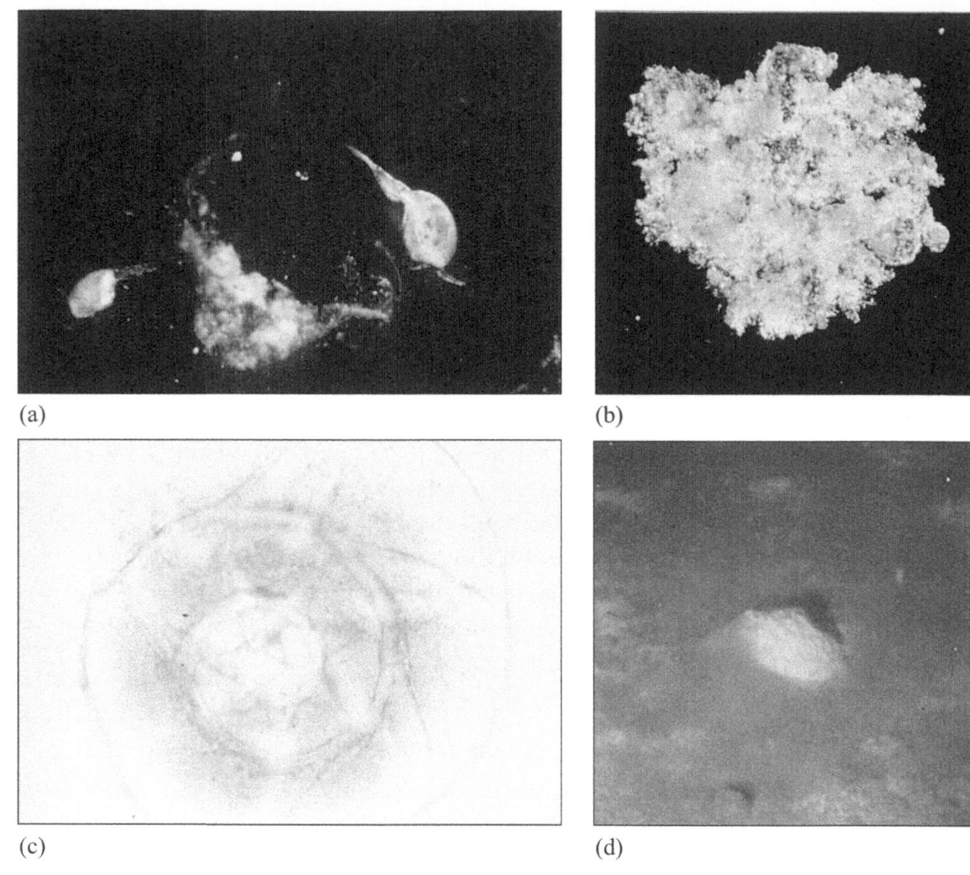

Figure 2.14 (a) Particle of marine snow being consumed by copepods (small planktonic crustaceans). Aggregation of organic material into marine snow speeds up its descent to the seabed, so that there is much less time for it to be decomposed (or eaten) en route. The field of view is ~1–2 cm across. (b) An individual 'snowflake', an isolated aggregate of marine snow ~1 cm across. (c) The ultimate fate of most marine snow is to be consumed by benthic animals. Here you can see phytoplankton remains actually within a benthic foraminiferan (which is about 200 μm across). (d) Marine snow carpeting the seabed and partially burying a mound made by an animal. Some benthic animals time their reproductive cycle so that their young can take advantage of organic debris from the spring bloom in surface waters. The field of view is ~0.5 m across. ((a) Alldredge, A; (b, c and d) NERC)

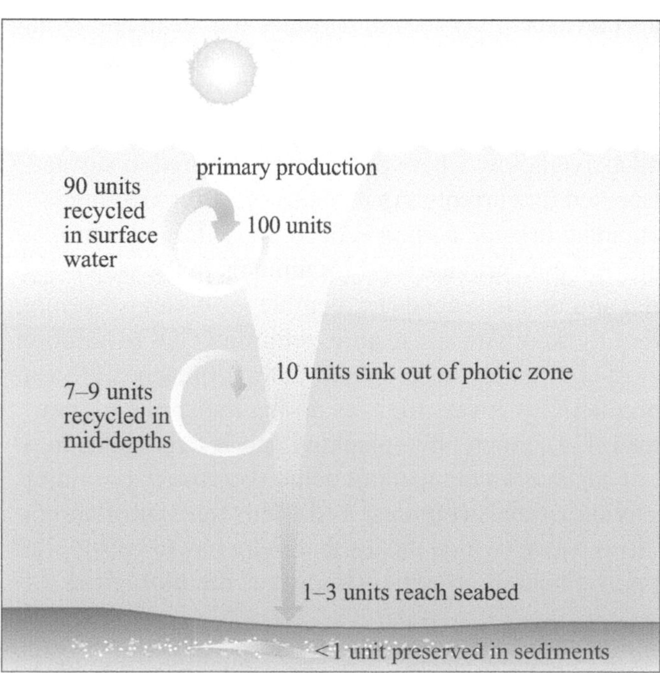

Figure 2.15 Sketch (not to scale) of the progressive decrease with depth of the carbon initially fixed in the photic zone, and its recycling in the upper and mid-ocean. Generally, only 1–3% reaches the seabed and less than one-third of that might be preserved in sediments.

($CaCO_3$) made from bicarbonate ions (HCO_3^-) and calcium ions (Ca^{2+}) extracted from seawater. The chemical reaction for the precipitation of calcium carbonate shells is:

$$\underbrace{Ca^{2+}(aq) + 2HCO_3^-(aq)}_{\text{in solution in seawater}} \rightarrow \underbrace{CaCO_3(s)}_{\substack{\text{precipitated}\\ \text{by organisms}}} + H_2O + CO_2 \qquad (2.8)$$

Note that Equation 2.8 is Equation 2.7a in reverse. The reaction proceeds exothermically in only one direction; the other direction requires energy (i.e. it is endothermic).

- Which direction proceeds exothermically and which proceeds endothermically?

- The weathering reaction (Equation 2.7a) proceeds spontaneously at the surface of the Earth (think of crumbling limestone walls) and so must be exothermic. The biological reaction (Equation 2.8) requires energy – the organism is investing energy for its defence – and so is endothermic.

Two important examples of organisms which secrete $CaCO_3$ are the planktonic algae known as coccolithophores, which secrete plates ('coccoliths') of calcium carbonate (Figure 2.16a and b), and the zooplanktonic foraminiferans (Figure 2.16c). Other planktonic organisms precipitate *silica* from solution to form their shells and skeletons. These include *diatoms*, which are phytoplankton (Figure 1.42a), and radiolarians, which are zooplankton. While organisms that

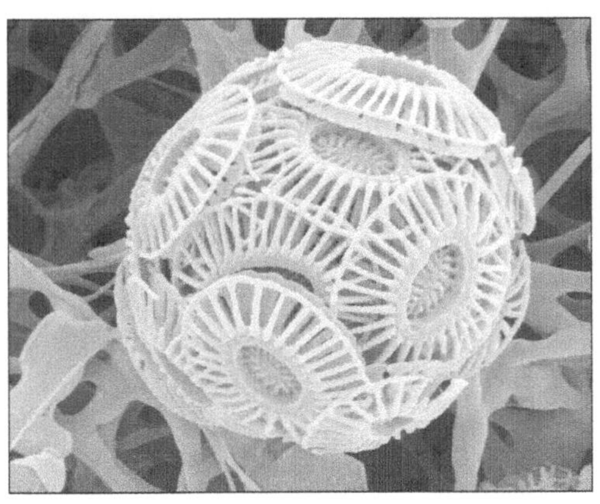
(a)

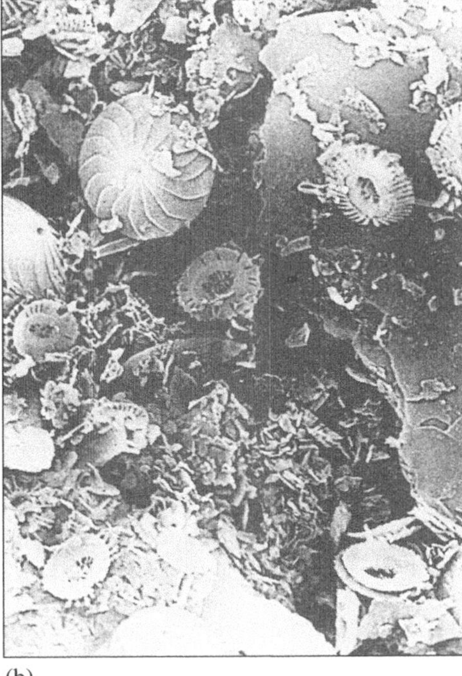
(b)

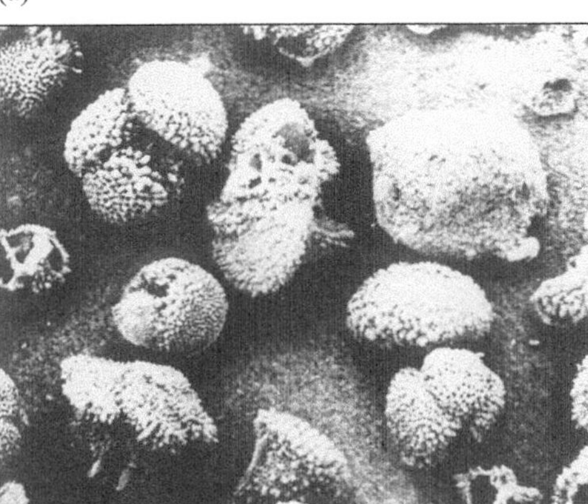
(c)

Figure 2.16 (a) A coccolithophore, which is a single-celled alga ~10 μm across, encased in calcite plates; this species is *Emiliania huxleyi*. Coccolithophores are often the dominant alga in nutrient-poor waters. (b) Debris from within seabed sediments, with coccoliths from several species of coccolithophore (magnification ×2300). (c) Remains of foraminiferans, extracted from seabed sediments. These are ~50 μm across, but some species are much larger. Such shells are often a major carbonate component of sediments below areas of high productivity. ((a) Young; (b) Lefevre; (c) Breger)

build hard parts of calcium carbonate use Ca^{2+} and HCO_3^-, those organisms that have silica shells remove SiO_2 from solution in seawater. Precipitation of shells and skeletons, along with precipitation of similar minerals involving chemical reactions with sediments and rocks on the seabed, is sometimes referred to as *reverse weathering*. (It is processes such as these that help maintain the difference in composition between seawater and river water.)

Despite their long journey to the deep sea floor, the calcium carbonate remains of planktonic organisms are found in sediments over much of the seabed (Figure 2.16b). Whether calcium carbonate is eventually dissolved or incorporated into sea-floor sediments depends on:

- the chemistry of the seawater (its acidity and the extent to which it is saturated with calcium carbonate)
- the speed of sinking of remains and the depth of the sea floor.

Calcium carbonate debris generally begins to dissolve some distance above the seabed, but by that stage of its descent it has already been incorporated into marine snow and is sinking fairly fast, so there is insufficient time for it to dissolve en route. Most dissolution, therefore, occurs on the seabed. The extent to which remains are dissolved before being protected by further layers of sediment (including more calcareous debris) increases with increasing depth.

The depth at which the proportion of calcium carbonate remains falls to less than 20% of the total sediment (biogenic sediments plus terrigenous clays) is known as the **carbonate compensation depth (CCD)**. Under areas of high surface productivity, the 'rain' of calcareous debris is such that carbonate in seabed sediments is preserved at greater depths than it would otherwise be, and so the CCD is also at greater depth and is said to be 'depressed'.

Planktonic organisms are not the only ones to extract constituents from seawater to make hard parts. This process also occurs:

- in shallow nearshore waters, where bivalves such as mussels and oysters often grow together, forming large accumulations of shells of calcium carbonate
- in clear tropical waters, where certain algae form carbonate-rich accumulations, while coral skeletons build up to form substantial reefs.

The physical factors that control fluxes of carbon dioxide across the air–sea interface and its solubility within the ocean are considered next.

■ Referring back to Section 1.5, does the solubility of CO_2 increase or decrease with (i) increasing temperature and (ii) increasing pressure?

■ As discussed in connection with Figure 1.41, the solubility of gases decreases with increasing temperature and increases with increasing pressure.

If you have trouble remembering which way these controls on solubility act, think of the following everyday examples.

1 If you want to check that an electric kettle is working, look at the submerged element: if bubbles of gas are collecting there, air is being driven out of solution as the water warms and the kettle is working.

2 When a can or bottle of gasified drink is opened, it fizzes as gas comes out of solution: the solubility is reduced as the pressure is reduced.

Question 2.6

Figure 2.17 (overleaf) shows fluxes of CO_2 across the air–sea interface in January–March and April–June, in the Northern Hemisphere. Areas acting as a net sink for atmospheric CO_2 (net flux of CO_2 from atmosphere to ocean) are shown in green and blue, and areas acting as a source (net flux of CO_2 from ocean to atmosphere) are shown in brown, red and yellow. Study the map, concentrating particularly on the Atlantic Ocean.

(a) Explain the direction of the CO_2 flux in the northeastern North Atlantic during January–March. (*Hint*: look at Figure 1.33.)

(b) Bearing in mind when the data were collected, discuss possible reasons for changes in the CO_2 flux in the North Atlantic between January–March and April–June. (*Hint*: look at Figures 1.37, 1.43 and 2.11.)

(c) Explain the direction of the flux in tropical and equatorial areas.

The answer to Question 2.6 is illustrated in Figure 2.18, which extends the area of consideration to include the Southern Hemisphere. The main point to appreciate is that net fluxes of carbon dioxide are both into and out of the ocean and that, while regions of deep water-mass formation generally act as sinks, other regions may act as sinks at some times and sources at others, depending on the time of year and the physical and biological processes occurring in surface waters.

While surface waters may be in equilibrium with the overlying atmosphere as far as its gas content is concerned (Figure 2.7), the cold waters filling the deep ocean were last in contact with the atmosphere when they were at the surface hundreds of years earlier.

■ In general, therefore, would you expect deep waters to contain higher or lower concentrations of CO_2 (as dissolved inorganic carbon) than surface waters?

□ Deep water generally contains higher concentrations of dissolved inorganic carbon. When deep water masses were last at the surface, they were at high latitudes where, because of the low temperatures, large amounts of CO_2 could dissolve from the atmosphere.

The decomposition of organic remains as they sink into the deep ocean and the dissolution of skeletons and shells also supply dissolved inorganic carbon to deep ocean waters, making them slightly more acidic than surface waters. This increased acidity, along with increased pressures, partly explains why calcium carbonate debris dissolves more readily at depth.

Deep ocean waters are not an endless sink for carbon dioxide because there is an exchange of water between the surface and the deep ocean. As well as localised regions of downwelling and upwelling (mostly from relatively shallow depths, but also from greater depths at the Antarctic Divergence), there is mixing between

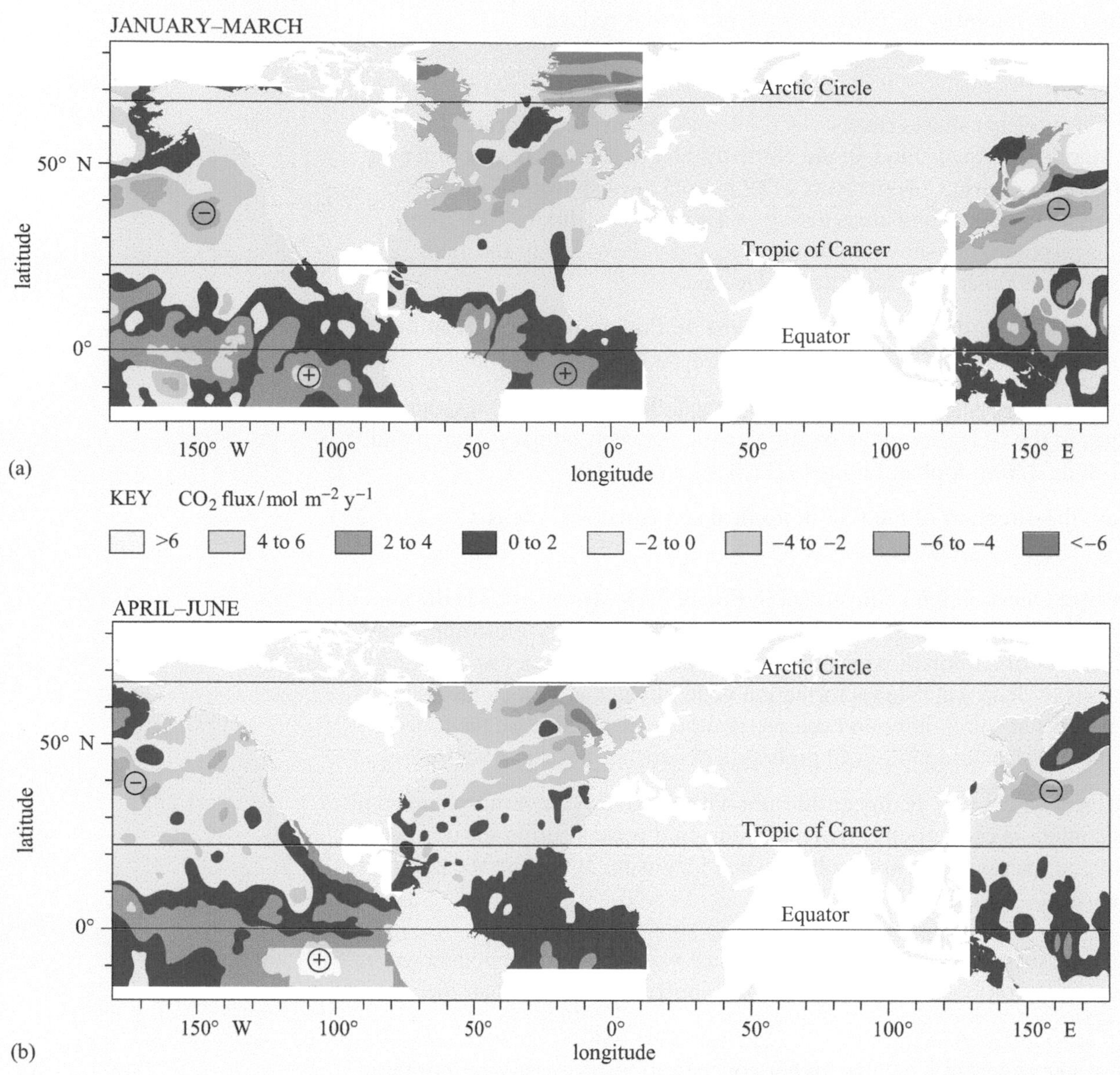

Figure 2.17 Fluxes of CO_2 across the air–sea interface at two seasons of the year: (a) averaged over the months January–March; (b) averaged over the months April–June. Fluxes are given in moles of CO_2 m^{-2} y^{-1}. Positive values (brown, red and yellow, marked ⊕) correspond to a net flux of CO_2 out of the ocean. Negative values (greens and blues, marked ⊖) correspond to a net flux into the ocean. (Lefevre, 1994)

the water masses flowing at various depths in the ocean. To compensate for the large volumes of water sinking at high latitudes, the net direction of mixing in the oceans as a whole is upwards. Due to the general movement of deep water towards the northern Pacific and Indian Oceans in the 'global thermohaline conveyor' (Figure 1.35), a relatively large proportion of carbon-rich water eventually comes to the surface in these regions (as you can see from the yellow and red areas in Figure 2.17). As a result of all the various physical, biological and chemical processes acting on it, a carbon atom will, on average, reside in the oceans for about 1000 years and most of this time will be spent in the deep ocean below the thermocline.

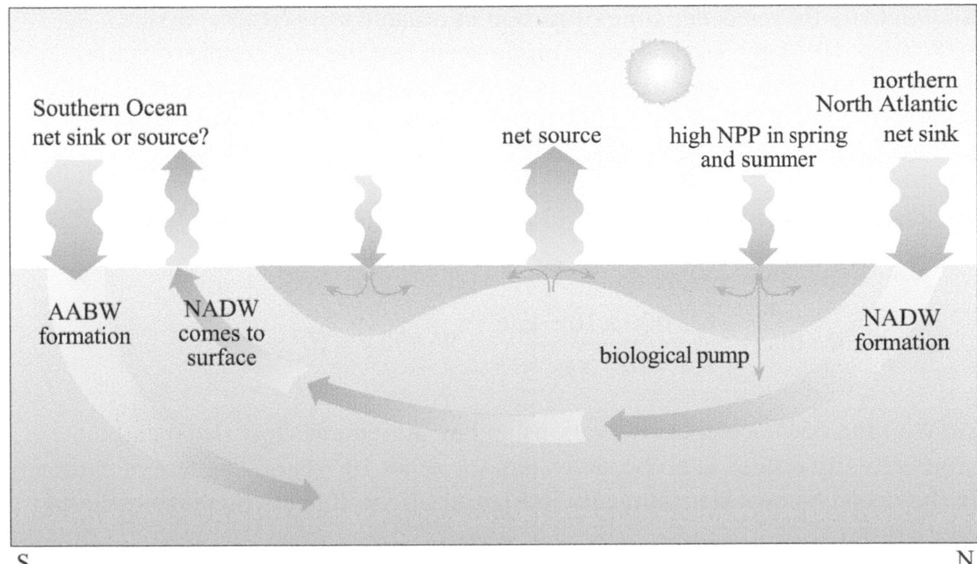

Figure 2.18 Sketch cross-section of the Atlantic Ocean to illustrate the carbon dioxide fluxes across the air–sea interface. Northern and southern high latitudes act as sinks for atmospheric CO_2 because they are regions of deep and bottom water mass formation (for North Atlantic Deep Water (NADW) and Antarctic Bottom Water (AABW) respectively). North Atlantic Deep Water eventually comes to the surface again at the Antarctic Divergence (Figure 1.33), and it is not clear whether the Southern Ocean is a net source or a net sink of atmospheric CO_2. (The diagram is drawn for the northern summer and so high primary productivity, contributing to the drawdown of CO_2, is shown for the North Atlantic.)

Figure 2.19 shows how the marine carbon cycle links with the terrestrial carbon cycle (Figures 2.2 and 2.3). The deep ocean is by far the largest reservoir of carbon, storing some 38 000 × 10^{12} kgC at any one time. The other carbon reservoirs in the ocean are the surface waters, which at any one time contain approximately 1000 × 10^{12} kgC (including 2 × 10^{12} kgC in phytoplankton), and the uppermost seabed sediments, which store some 3000 × 10^{12} kgC. These sediments are sometimes called 'reactive sediments' because the carbon in them may undergo physical, biological or chemical reactions, so releasing it into the overlying water. Also shown in Figure 2.19 are estimates of annual fluxes between the various reservoirs.

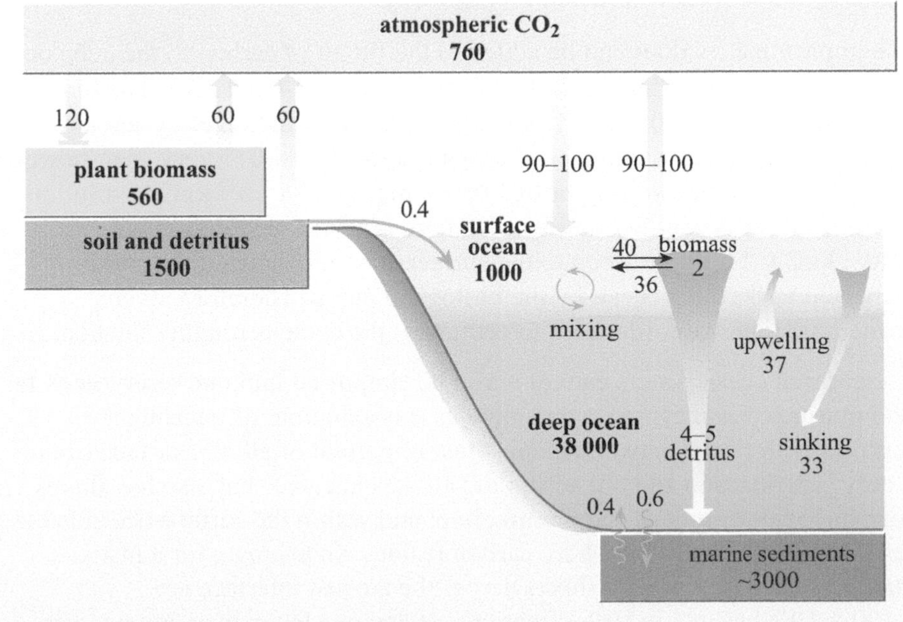

Figure 2.19 The carbon cycle (terrestrial plus marine), including those processes acting over timescales of up to hundreds of thousands of years. Carbon reservoirs are given in 10^{12} kgC and fluxes in 10^{12} kgC y^{-1}.

To calculate the residence time of carbon in ocean waters (reservoir size ~38 000 × 10^{12} kgC; see above), it might seem that the best course of action would be to use the input from rivers as the flux because, although small (only ~0.4 × 10^{12} kgC), the river flux of carbon is one way, while fluxes across the air–sea interface are in both directions.

■ Calculate the residence time using the input from rivers as the flux.

■ Using Equation 2.2:

$$\text{residence time} = \frac{\sim 38\,000 \times 10^{12}\ \text{kgC}}{0.4 \times 10^{12}\ \text{kgC y}^{-1}} \approx 95\,000\ \text{y} \approx 10^5\ \text{y}.$$

As with the deep oceans, observations and measurement show that a carbon atom actually resides in all ocean waters for about 10^3 years. The fate of carbon in the ocean (shown schematically in Figure 2.11) will help you understand this conundrum.

About half the carbon entering the ocean in run-off originates as organic carbon (DOC and POC) and about half is inorganic carbon released in the weathering of carbonate and silicate minerals. Once in the ocean, both DOC and DIC enter into biological cycling and eventually become part of the bicarbonate–carbonate system in equilibrium with atmospheric CO_2 or are precipitated to form shells, skeletons, etc. River-borne particulate material (such as the remains of leaves, dead freshwater plankton and sewage) accumulates in estuaries or nearshore sediments, where the carbon may be released to the overlying water through biological oxidation. Some of the carbon accumulating in these sediments may become deeply buried and eventually isolated from the biospheric carbon cycle. These deposits, together with organic remains buried in deep-sea sediments and buried organic deposits on land (in swamps, etc.), account for the removal of ~0.05 × 10^{12} kgC^{-1} (not shown in Figure 2.19).

Some approximate values can be added to the fluxes of carbon to the deep ocean, discussed a little earlier in connection with Figures 2.14 and 2.15. The biological pump transfers 4–5 × 10^{12} kgC from surface to deep water every year, and downward mixing and sinking adds another 33 × 10^{12} kgC. Every year, however, upward mixing and upwelling bring approximately 37 × 10^{12} kgC in solution up to the surface ocean. In other words, there is a net transfer of about 4 × 10^{12} kgC y^{-1} from the deep to the surface ocean, balancing the carbon transported to the deep ocean by the biological pump. Therefore, over intermediate timescales (decades to centuries) the cycle is roughly in balance.

It is clear that ocean waters cannot really be simplified into one reservoir as far as the marine carbon cycle is concerned as it is a number of interlinked reservoirs with fluxes between them. Most important of all, the surface ocean not only has fluxes in and out across the air–sea interface but also has fluxes across its lower boundary, the thermocline; and within the surface ocean is the reservoir of living biomass where carbon resides, on average, for a matter of days. Moreover, although fluxes across the air–sea interface are 90–100 × 10^{12} kgC y^{-1} in both directions, at any one location at any one time fluxes into and out of the surface ocean are by no means equal. Furthermore, carbon entering the ocean as CO_2 dissolved in cold surface waters at high latitudes is carried straight down into the deep ocean – in fact, at high latitudes the surface and deep oceans are effectively joined (Figure 1.33).

As a consequence of marine-snow formation, the time between the fixation of carbon in phytoplankton and the arrival of organic remains at the seabed can be as little as a few weeks, so this aspect of the marine carbon cycle (i.e. the biological pump) can reflect what is happening in surface waters on very short timescales. Of the $4–5 \times 10^{12}$ kgC y^{-1} transported to the deep ocean in biological debris, about 4×10^{12} kgC y^{-1} is redissolved in deep ocean water, so that only a small percentage of the carbon originally fixed in surface waters is preserved in sediments (Figures 2.11 and 2.15). Of the 0.6×10^{12} kgC y^{-1} deposited or precipitated from deep ocean water within marine sediments, $\sim 0.4 \times 10^{12}$ kgC y^{-1} eventually finds its way back into deep ocean water. This means that about 0.2×10^{12} kgC y^{-1} becomes decoupled from the deep ocean water and, as in the case of carbon buried in sediments on land or in shallow water, this small amount of carbon 'leaks' from the cycle. This 0.2×10^{12} kgC y^{-1} is the amount preserved at the present time but, because of the short timescales involved, short-term changes in biological productivity at the surface are likely to be reflected in the amount of carbon preserved in sediments (although the chemistry of deep ocean waters would also be affected, and this would in turn affect the preservation of material). Over the long term, however, the rate of accumulation of carbon in sediments will reflect its rate of supply in rivers. For this reason, although small, the river flux of carbon is not a trivial player in the global carbon cycle. In the next section, you will consider the fate of the carbon that is preserved.

Question 2.7

(a) According to Table 2.2, what is the approximate total mass of carbon in marine plant material (i.e. the standing stock), and how does this compare with the average standing stock of plant material on land?

(b) According to Figure 2.19, what is the average residence time of carbon in living phytoplankton? Does carbon cycle through the marine biomass reservoir faster or slower than through the terrestrial biomass reservoir?

2.3.3 Long timescales: the geological carbon cycle

The carbon cycle described so far involves cycling from timescales of years or decades (the terrestrial carbon cycle) up to hundreds of thousands of years (the marine carbon cycle). Since both the terrestrial and the marine carbon cycle 'leak' (i.e. they are not closed systems) over long time periods, carbon can accumulate deep within sediments and is removed from these cycles.

As discussed in the previous section, there are two different types of carbon-rich accumulations.

■ What are the two types?

■ Sediments containing carbonate (carbon in *inorganic* form) and those containing soft tissue (*organic*) remains.

The first of these types of sediment is often described as *calcareous*. Such sediments are generally marine deposits containing accumulations of calcium carbonate shells and skeletons, which have been made by organisms using dissolved inorganic carbon, mainly HCO_3^-. They usually also contain variable amounts of land-derived sediments (mainly clay) and, particularly below upwelling regions, the remains of shells of silica. Those with a significant siliceous component are referred to as *calcareous–siliceous* sediments. (There are also sediments that are predominantly clay or predominantly siliceous, which contain negligible amounts of carbonate.)

Question 2.8

Is the following statement true or false? 'Limestone deposits are examples of inorganic carbon, but skeletal remains and shells are examples of organic carbon.' Give a reason for your answer.

Over millions of years, because of high pressures due to increasing thicknesses of overlying sediment and high temperatures due to heat loss from within the Earth, chemical and structural changes occur in the sedimentary accumulations and they become lithified (i.e. converted into rock). The kind of rock that is formed depends on the initial composition of the sediments, for example carbonate-rich sediments become rocks such as chalk and limestone.

The second kind of carbon-rich accumulation containing organic material (i.e. that made up of molecules of carbon, hydrogen and oxygen) is often described as *carbonaceous*. As organic-rich material is covered with a mixture of more organic matter and/or other sediment, the weight of the overlying deposits causes compaction, which squeezes out the water and residual air from the pore spaces. In the resulting oxygen-poor environment, a dense residue enriched in carbon, known as **kerogen**, is formed. Under continued deposition of organic matter and sediments (often related to the subsidence of continental crust), the original material may be buried to depths of several kilometres. As in the case of calcareous deposits, high temperatures and pressures eventually cause the carbonaceous sediments to become lithified. Large accumulations of land plants (e.g. the remains of trees that have accumulated in anoxic swamps) may become coal; marine sediments containing very high concentrations of phytoplankton debris can produce petroleum. If the organic matter is dispersed through sediments, oil shales may result.

At the present time:

- about 0.05×10^{12} kgC y^{-1} of buried organic matter accumulates on land and in the sea
- about 0.2×10^{12} kgC y^{-1} of inorganic carbonates accumulates in shallow waters and the deep ocean.

Over the whole Earth, there is approximately $10\,000\,000 \times 10^{12}$ kgC y^{-1} of carbon in carbonaceous rock (including fossil fuels) and four times that in limestones, chalks, etc. (Figure 2.20). Collectively, rocks are the largest reservoir of carbon on Earth. The residence time for carbon within this reservoir is of the order of 100–200 Ma, but even carbon incorporated into rocks can eventually enter the atmosphere again, as geological processes, particularly mountain

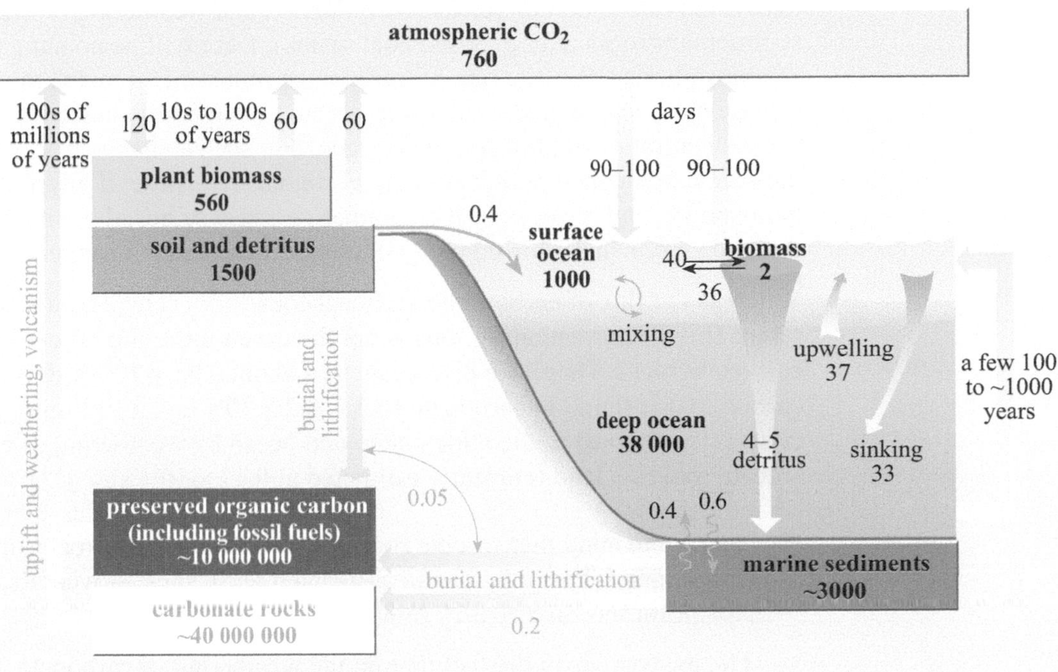

Figure 2.20 Summary diagram of the carbon cycle showing burial and preservation of carbon in sediments (as both calcareous and carbonaceous remains), its eventual return to the atmosphere, and the different timescales involved. Note that the carbonate rocks include lithified remains of calcareous organisms that lived in shallow waters as well as those that lived in the oceans. The approximate mass of carbon in each reservoir is given in units of 10^{12} kgC, and the fluxes are given in units of 10^{12} kgC y^{-1}. (Anthropogenic contributions and current imbalances in the overall system are not included.)

building, bring deeply buried rocks to the surface. The cycle then begins again as the rocks are weathered and eroded by wind, water or biological activity. Section 2.3.2 discussed how weathering of rocks by rain and soil water produces dissolved inorganic carbon, particularly HCO_3^- (Equation 2.7). Carbonaceous sediments exposed to the air may oxidise directly to CO_2, may be oxidised by bacterial respiration, or may be chemically weathered by soil water and rainfall to release carbon (DOC and POC) into streams. Sometimes, the carbon in rocks finds its way back into the atmosphere in volcanic gases (which is discussed in Chapter 3). Either way, the carbon from the rock reservoir is eventually returned to the atmosphere and/or the ocean.

Long-term controls on atmospheric CO_2

What affects the concentration of atmospheric CO_2? As discussed in Section 2.3.1, the main flux of CO_2 into the atmosphere through the terrestrial carbon cycle is through respiration by plants and decomposers, which is almost exactly balanced by that removed by photosynthesis (Figure 2.3), for which the chemical reaction is:

$$\underset{\text{carbon dioxide from atmosphere}}{6CO_2(g)} + 6H_2O \longrightarrow \underset{\text{organic matter}}{C_6H_{12}O_6} + \underset{\text{oxygen}}{6O_2(g)} \qquad (1.2)$$

If the current rate of burial of organic carbon in deep sediments roughly equals the rate of carbon release by oxidation and weathering of organic-rich sedimentary rocks (e.g. exposed coal seams), there will be no net gain or loss of atmospheric CO_2. The rate of burial must more or less match the rate of release by oxidation and weathering because a large imbalance would change the concentration of atmospheric oxygen, and such a change has not occurred in the recent geological past. The early Earth had very little, if any, oxygen in its atmosphere, and it was only the cumulative effect of burial of small amounts of organic carbon that allowed oxygen to build up in the atmosphere.

Whereas the flux of carbon into carbonaceous rocks can be balanced by the return flux from weathering, this is not the case for the flux of carbon into carbonate rocks. There is a discrepancy of about 0.03×10^{12} kgC y^{-1} between the carbon that is stored in carbonate rocks (0.2×10^{12} kgC y^{-1}; Figure 2.20) and that eventually returned to the atmosphere and ocean by weathering of carbonate and silicate rocks on land (currently estimated at 0.17×10^{12} kgC y^{-1}).

■ Bearing in mind that *silicate* rocks as well as carbonate rocks are weathered (Equations 2.7a and 2.7b, and associated text), suggest what the underlying reason for this discrepancy might be.

□ The answer lies in the fact that, in the weathering of carbonate minerals, only *one* of the carbon atoms that forms the bicarbonate $2HCO_3^-$ comes from the atmosphere, the other comes from the carbonate mineral itself; in the weathering of silicate minerals, *both* carbon atoms come from the atmosphere.

So, weathering a carbonate mineral on land removes one atom of carbon from the atmosphere (in a molecule of CO_2) (Equation 2.7a). Precipitating a carbonate mineral in the ocean (e.g. as part of a shell or skeleton) returns that molecule to the upper ocean, which is in equilibrium with the overlying atmosphere:

$$\underbrace{Ca^{2+}(aq) + 2HCO_3^-(aq)}_{\text{in solution in seawater}} \rightarrow \underbrace{CaCO_3(s)}_{\substack{\text{precipitated}\\ \text{by organisms}}} + H_2O + CO_2 \qquad (2.8)$$

Weathering of a carbonate mineral followed by precipitation of a carbonate mineral results in *no net* gain or loss of carbon (in effect CO_2) to the atmosphere.

Now, consider the case of silicate weathering; the reaction for weathering of a silicate mineral (e.g. $NaAlSi_3O_8$, as in Equation 2.7b) requires *two* molecules of atmospheric CO_2. Precipitating calcium carbonate in the sea returns *only one* of those molecules to the upper ocean–atmosphere equilibrium system (Equation 2.8). The burial of this $CaCO_3$ in the sediments removes it from the terrestrial and marine carbon cycles and thus represents a *net depletion* of atmospheric CO_2. These arguments are summarised in Figure 2.21.

Silicate weathering on land followed by carbonate precipitation and burial in the sea removes about 0.03×10^{12} kgC from the atmosphere each year.

The geological record shows that weathering of silicate minerals has occurred on Earth for at least the last 3.8 Ga and that the atmosphere has always contained some CO_2. There must be some return flux of atmospheric CO_2, and

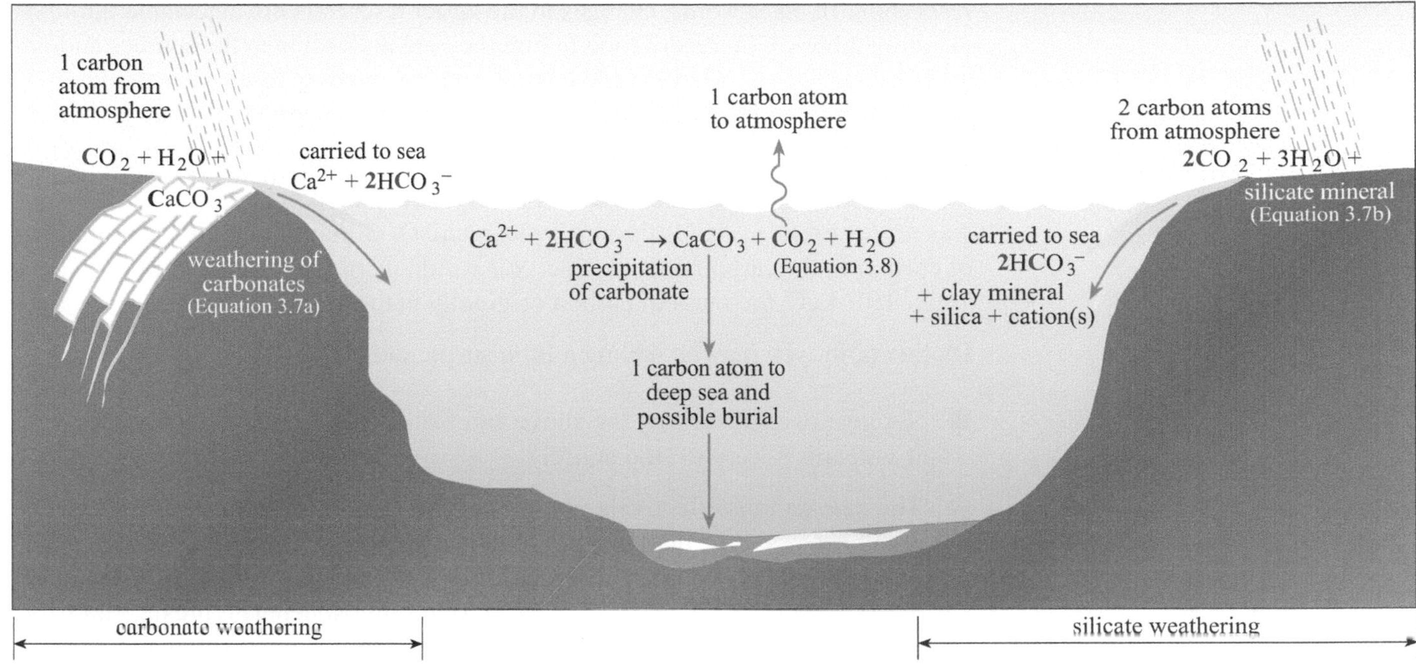

Figure 2.21 Diagram showing why the burial of sediments containing carbon from the weathering of silicates can result in a net removal of CO_2 from the atmosphere, but the burial of carbonates does not. (*Note*: this diagram does not include processes involving *organic* carbon.)

this is thought to be by volcanoes, which are part of the plate tectonic cycle that operates over timescales of hundreds of millions of years (as discussed in Chapter 3). Indeed, about 0.02×10^{12} kgC y^{-1} is estimated to be released to the atmosphere by volcanoes at the present time.

2.4 A system in balance?

You have seen that the global carbon cycle is highly complex: carbon may be in gaseous, dissolved or solid form, and may reside in various reservoirs on timescales ranging from decades (or less) to hundreds of millions of years. With such complexity, is it possible to determine whether the cycle is in balance?

To explore this question, some kind of reference is needed, such as a reservoir that can be used as an 'indicator' of *any changes* in the cycle.

- Would changes in the cycle show up first in a large reservoir or a small one? (Imagine a box containing a thousand marbles and a box containing ten marbles. If five marbles are added to or removed from each of these boxes, in which one would you notice it?)

- The addition or removal of five marbles would be much more noticeable in the box with only ten marbles initially. By extension, changes in the carbon cycle would show up more clearly in a small reservoir.

Atmospheric carbon dioxide is just such a small reservoir. Exchange reactions between the atmosphere and the other carbon reservoirs tend toward equilibrium, with the result that atmospheric CO_2 responds rapidly to changes in the larger reservoirs. In other words, changes in the larger reservoirs are detectable through the atmospheric CO_2 record long before they are noticeable in the sources and/or sinks themselves. Furthermore, the atmosphere is the only reservoir with direct links to the short, intermediate *and* long-term cycles of carbon (Figures 2.19 and 2.20).

The CO_2 in the atmosphere would in theory be depleted in 25 000 years if there was no return flux of CO_2 through volcanism. Over that time period, the amount of carbon in the carbonate rock reservoir would *increase* (in theory) by 760×10^{12} kgC, the mass of carbon originally in the atmosphere.

By this point you may be feeling a little suspicious.

- Suggest one reason why the above calculations would not be accurate, even if volcanism *were* to stop entirely.

- The calculations effectively assume that all the other carbon equilibrium reactions would be unaffected, but this is unrealistic. For example, decreasing atmospheric CO_2 would cause the ocean carbonate equilibria to shift so that more oceanic CO_2 would be released to the atmosphere (Figures 2.7 and 2.8, the latter illustrating the opposite situation), and this would slow the rate of depletion of atmospheric CO_2.

It seems that the concentration of CO_2 in the atmosphere has remained at a few hundreds to several thousands of ppm over the past few hundred million years. This relative stability of atmospheric CO_2 and of the carbon cycle in general is because changes in one reservoir cause repercussions in other reservoirs, and the *negative* feedback loops between them tend to stabilise them.

A property of atmospheric CO_2 that could provide negative feedbacks within the carbon cycle is its role as a greenhouse gas.

Question 2.9

Imagine that the supply of CO_2 to the atmosphere was suddenly increased (e.g. by enhanced volcanic activity); while CO_2 was being added to the atmosphere much faster than it was being removed, its concentration there would be increasing.

(a) Based on evidence in this chapter and by reference to Figure 1.38, suggest at least two ways in which greenhouse warming resulting from such an increase could indirectly affect other processes forming part of the global carbon cycle.

(b) Which, if any, of these processes would result in extra carbon being preserved, and so removed from cycling for hundreds of millions of years?

If CO_2 levels on Earth were, therefore, say ten times higher than they are today, it could be said with confidence that the rate of CO_2 *removal* from the atmosphere would be increased (Equations 1.2 and 2.8). Precipitation and accumulation of carbonate resulting from silicate weathering, and preservation of organic material (mainly in the sea but also on land), both act as long-term sinks of carbon. Although it is not known how rapidly silicate weathering and global net primary productivity may respond to changes in atmospheric CO_2, it can be reasonably ascertained that a 'greenhouse' feedback loop similar to that investigated in Question 2.6 keeps fluctuations of atmospheric CO_2 within a relatively narrow range.

2.4.1 Short-circuiting the geological carbon cycle

If the scale of observation is changed from millions of years to decades, it is clear that the carbon cycle is not in balance today; there is more carbon entering the atmosphere than there is being removed from it. This is one of the few certainties in the global carbon cycle (Box 2.3).

Box 2.3 The record on the mountain

In 1957, the American scientist Charles David Keeling, then still a student, set up two stations for the continuous monitoring of CO_2 in the atmosphere: one in Hawaii on the volcanic peak of Mauna Loa and the other at the South Pole. The results were dramatic. The CO_2 concentration of the atmosphere at Hawaii averaged around 317 ppm but oscillated around that value on an annual basis, increasing in the winter (e.g. due to the decay of vegetation) and decreasing in the summer (e.g. due to photosynthesis); this provided the first clear evidence that life on Earth profoundly influences the atmosphere.

But this was not the end of the story; within about five years, it became obvious that the annual average CO_2 concentration was steadily increasing (Figure 2.22). No natural phenomenon could be found to account for this spectacularly rapid rate of increase (geologically speaking) and the most likely source appeared to be the release of CO_2 from fossil-fuel combustion. Therefore not only was the growth and decay of plants recorded in the atmosphere, but so too was human industrial activity.

The Mauna Loa and South Pole records are continuing to be added to and, over the years, have been augmented by a worldwide network of measurement stations. All the records of CO_2 concentration show the same rising trend. By 2006, the global average atmospheric CO_2 concentration stood at just over 380 ppm, and it is still rising.

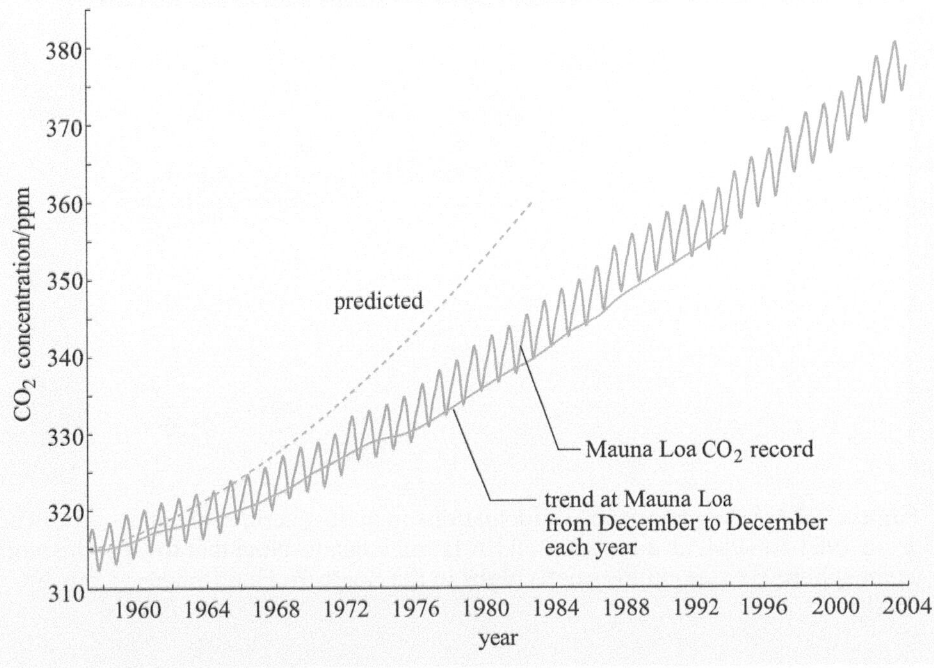

Figure 2.22 Increase in the concentration of CO_2 in the atmosphere, as measured at the Mauna Loa observatory in Hawaii. Also shown is the increase predicted on the basis of fossil-fuel combustion. (CDIAC, 2005)

Whether they are from Mauna Loa, the South Pole or the newer monitoring stations, all records of atmospheric CO_2 concentration show the same striking features (Figure 2.22):

- a strong trend of increasing atmospheric levels of CO_2 over time, with the current rate of increase at about 1.5 ppm per year or about 0.4% per year
- seasonal oscillations, with a CO_2 peak in the winter and a CO_2 trough in the summer.

All monitoring stations do not, however, produce exactly the same results. Figure 2.23 is a composite diagram showing the variation of atmospheric CO_2 concentrations over the course of four years (1981–1984) over all latitudes.

■ Apart from the peaks and troughs, what is the most striking aspect of this diagram?

□ The seasonal oscillations are marked in the Northern Hemisphere, but extremely damped in the Southern Hemisphere with only slight peaks occurring during minima in the Northern Hemisphere.

The peaks and troughs and their global variations both correspond with the issues considered in Question 2.10.

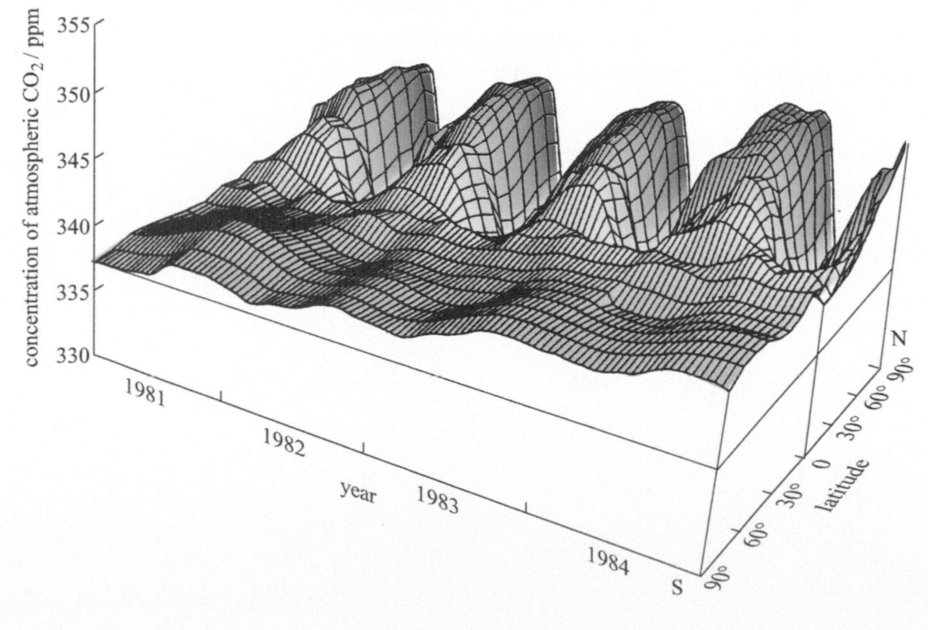

Figure 2.23 Sea-level seasonal fluctuations in atmospheric concentrations of CO_2 from 1981 to 1984, as a function of 10° latitude bands. Note that lows in the Northern Hemisphere correspond to (small) highs in the Southern Hemisphere. (Conway, 1988)

Question 2.10

(a) What is the reason for the seasonal fluctuations in atmospheric CO_2 concentration?

(b) The patterns shown in Figures 2.22 and 2.23 have been dubbed 'the Earth breathing'. State one reason why you think this is, or is not, an appropriate description.

(c) By reference to any global map (e.g. Figure 2.4), suggest why this pattern is dampened in the Southern Hemisphere.

This annual cycle of CO_2 uptake and release emphasises the influence of living organisms on Earth's geochemical cycles. The seasonal variations in CO_2 fluxes into and out of the atmosphere can be quite significant: at Hawaii (latitude 19° N) for example, seasonal variations are sufficient to cause the atmospheric concentration of CO_2 to fall by about 7 ppm over the course of spring and summer, while in Barrow, Alaska (latitude 71° N), the winter to summer decrease in CO_2 is double that.

If seasonal decreases and subsequent increases remained equal in magnitude from year to year, the atmospheric reservoir should remain constant. However, the rising trends apparent in Figures 2.22 and 2.23 indicate that there is a persistent disequilibrium in fluxes into or out of the atmospheric CO_2 reservoir. Analyses of air trapped in ice cores has shown that this rapid increase in atmospheric CO_2 is a modern phenomenon; it began approximately 100 years ago, and has greatly accelerated in recent decades. The current rate of increase is about 1.5 ppm per year, with $\sim 3 \times 10^{12}$ kgC added to the atmosphere each year. This is rapid enough to make textbook values for atmospheric CO_2 obsolete; for example in 1958, atmospheric CO_2 stood at 315 ppm; by 1994 it was about 357 ppm. It is now (late 2007) just over 380 ppm. The overwhelmingly likely cause for the increase in atmospheric CO_2 concentrations is the extraction and burning of fossil fuels.

■ Referring to Figure 2.20, briefly explain *why* the extraction and burning of fossil fuels would cause an increase in atmospheric CO_2 concentrations.

□ Extracting fossil fuels (organic carbon) and burning them rapidly adds CO_2 to the atmosphere, and so the geological carbon cycle is being 'short-circuited' (Figure 2.24 overleaf).

The important point here is not that carbon is being returned to the atmosphere (this would eventually happen anyway), but the *rate* at which carbon is being added. Concentrations in the atmosphere of CO_2 (and other greenhouse gases – notably methane, which also contains carbon; Box 2.1) are increasing faster than fluxes into the other reservoirs (i.e. the major short-term buffering processes) can accommodate them. Simply speaking, human activity has caused an increase in the rate of return of carbon to the atmosphere that was previously stored in the Earth's crust.

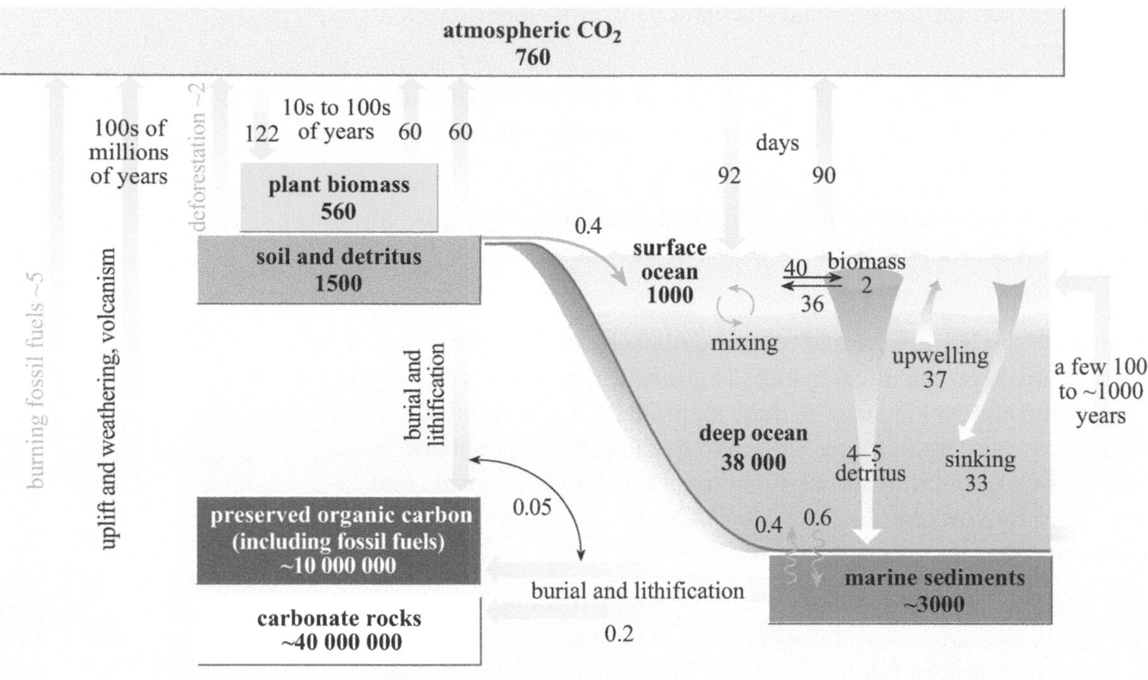

Figure 2.24 Summary diagram of the carbon cycle (Figure 2.20) now also showing the fluxes back to the atmosphere caused by human (anthropogenic) activities. Carbon reservoirs are given in 10^{12} kgC and fluxes in 10^{12} kgC y^{-1}.

As human activity has been directly responsible for this flux, its size is known to a high degree of accuracy and, according to global production figures for coal and oil, it is $\sim 5 \times 10^{12}$ kgC y^{-1}.

■ How does the current yearly rate of increase of CO_2 in the atmosphere (see above) compare with the flux from fossil fuels?

■ The current annual increase is only about 3×10^{12} kgC , which is about two-thirds of that being added from fossil fuels.

So more carbon is being *released* into the atmosphere by human activities than is *accumulating* there – hence the difference between the observed and predicted increases shown in Figure 2.22. This discrepancy is even greater when the amount of carbon added to the atmosphere each year by logging and burning forest vegetation (currently estimated at $\sim 2.0 \times 10^{12}$ kgC y^{-1}) is also included.

Figure 2.24 shows the global carbon cycle as perturbed by human activity; it is similar to Figure 2.20 but now includes not only estimates of the fluxes directly resulting from human activities (fossil-fuel burning and deforestation) but also the changes that might feed through into other fluxes.

■ Referring to Figure 2.24, suggest where the 'missing' carbon might have gone.

■ The only possibilities seem to be into the ocean or into plant biomass (terrestrial and/or marine). (Figure 2.24 shows possible increased fluxes into these reservoirs from the atmosphere.)

As illustrated in Figure 2.8 (and discussed above), an increase in the concentration of atmospheric CO_2 will push the equilibria to the right and cause an increase in the flux in CO_2 into the ocean (although it would take a long time for the oceans to equilibrate with the atmosphere). Computer models of ocean circulation and calculations of carbonate equilibria suggest that an extra 1.6×10^{12} kgC y^{-1} could eventually be removed from the atmosphere to the oceans in this way. When the total fluxes of carbon are calculated, that still leaves an estimated 2.2×10^{12} kgC y^{-1} unaccounted for.

Conclusive evidence as to the identity of this 'missing sink' has so far eluded scientists, partly because, with the exception of the atmosphere, none of the reservoirs in the carbon cycle is small enough to detect a change of the order of 10^{12} kgC. As mentioned above, however, increased rates of fixation of carbon in plant material on land and in the oceans is a possibility. The answer to Question 2.6a shows that an increase in the concentration of atmospheric carbon dioxide can (at least under experimental conditions) increase rates of carbon fixation (i.e. net primary productivity), thus storing some of the missing carbon in vegetation or soil organic matter. Current indications somewhat surprisingly suggest that the forests of Amazonia may be responsible for taking up a large proportion of the 'extra' carbon, both into continuing growth of existing trees and regrowth in cleared areas. Of course, the missing sink may actually turn out to be a combination of sites, including the huge volume of the ocean.

The rapid increase in atmospheric CO_2 observed today simply reflects the fact that the carbon cycle is currently not in balance. Eventually, coal and oil deposits will be exhausted and the rock-to-atmosphere CO_2 'shunt' will be closed. Some time after this, the various reservoirs of the carbon cycle may re-establish equilibrium with one another and atmospheric CO_2 will stop increasing, re-establishing itself at a higher concentration than previously.

Finally, it is worth considering methane (CH_4), whose atmospheric concentration is some two orders of magnitude less than that of CO_2 (Table 1.2). It is in fact a more powerful greenhouse gas than CO_2, but its very low concentration means that it has a negligible effect on global systems. Methane is emitted from swamps, wetlands, rice paddies and cattle. It is also the principal constituent of the natural gas used for domestic heating, and substantial amounts are lost during the extraction of fossil fuels. Since methane is relatively quickly oxidised to CO_2 in the atmosphere, its residence time in the atmosphere is only about 10 years, so at present it is not considered a significant agent of present-day global warming.

However, atmospheric methane concentrations were greater at various times in the geological past, and could be so in the future. Large quantities are presently locked up as **methane hydrates** in frozen wetlands (tundra) and in some deep-sea sediments. There is good evidence that at various times in the geological past (e.g. the Palaeocene/Eocene boundary, 55 Ma ago), this frozen methane released catastrophic amounts of gas into the atmosphere at rates sufficient to cause a positive feedback effect, which exacerbated the warming because of the greenhouse gas characteristics of methane.

Artificial carbon compounds (e.g. chlorofluorocarbons or CFCs), used as spray propellants and refrigerants, are also greenhouse gases but have a much more powerful effect in the stratosphere, where they are the principal cause of the ozone layer depletion that has received so much publicity in recent years. As far as the global carbon cycle is concerned, however, they can be omitted because their atmospheric concentrations are even less than that of methane.

Summary of Chapter 2

1 The chemistry of the carbon atom means that it has a unique role in the living world. Its ability to share electrons with other carbon atoms allows the construction of large, complex molecules of carbon, hydrogen and oxygen of which organic material is built.

2 Carbon fulfils two essential roles in the biosphere: it is the primary component of living tissue (see point 1 above) and, in its gaseous forms (CO_2 and CH_4), it warms the Earth's surface enough to support life.

3 The main biogeochemical connection between Earth and life is the global carbon cycle, i.e. the movement of carbon through the atmosphere, biosphere, lithosphere and hydrosphere. The global carbon cycle involves interlinking cycles over three major timescales:

(i) the terrestrial carbon cycle, driven by biological processes and acting over timescales of months or years to decades

(ii) the marine carbon cycle, involving chemical, biological and physical components acting over an intermediate timescale of up to hundreds of thousands of years

(iii) the geological carbon cycle, involving rocks and sediments, and acting on timescales of up to hundreds of millions of years.

4 The terrestrial carbon cycle is driven by the fixation of atmospheric carbon into organic matter by photosynthesis, and involves large fluxes of carbon between the atmosphere and fairly small organic reservoirs (e.g. living organic matter and organic debris). The average residence time of carbon in plant biomass is about nine years. Estimates of annual net primary production (flux of carbon into the terrestrial biomass reservoir) and of standing stock are problematic, and involve determining the areas occupied by different biomes.

5 The marine carbon cycle is linked to the terrestrial carbon cycle through the carbon carried to the sea in rivers in organic and inorganic form (both dissolved and particulate). The surface ocean is, on average, in equilibrium with the overlying atmosphere and so the fluxes between the two are broadly in balance. The CO_2 flux across the air–sea interface may be driven by physical processes (particularly sinking of cold water masses in high latitudes, and upwelling) and by biological activity (i.e. bacterial and algal photosynthesis which, when combined with rapid sinking of organic matter into the deep sea, constitutes the *biological pump*). In both cases, the mechanism whereby CO_2 is 'pushed out' or 'drawn into' the ocean is chemical. The linked equilibria determine the relative proportions of CO_2 gas and the various forms of *dissolved inorganic carbon*, i.e. H_2CO_3 (carbonic acid), HCO_3^- (bicarbonate ion, the main constituent in seawater) and CO_3^{2-}

(carbonate ion); these equilibria are together known as the carbonate system. Biological activity produces inorganic particulate carbon in the form of shells and skeletons that eventually sink to the seabed; whether they are dissolved or accumulate depends on the chemistry of the deep ocean water and the rate of supply of remains. The depth at which the proportion of calcium carbonate that remains falls to less than 20% of the total sediment is known as the *carbonate compensation depth.*

6 In the geological carbon cycle, carbon in the organic and inorganic products of weathering on land is carried to the sea in rivers, takes part in the marine carbon cycle, and, in the case of a very small proportion, is preserved and buried, eventually to be returned to the atmosphere or ocean through volcanism or weathering and/or oxidation. The geological carbon cycle acts on a timescale of up to hundreds of millions of years, and involves small fluxes between large reservoirs; carbonate and organic sedimentary rocks together store more than 99.9% of the carbon on Earth.

7 Extracting and burning fossil fuels shortcuts the geological-scale return flux of carbon and has currently brought the global carbon cycle into disequilibrium. As a result, the concentration of CO_2 in the atmosphere is rising (as is that of other greenhouse gases), but it is not rising as much as expected. Some of the 'extra' CO_2 has almost certainly been taken up by the ocean; the rest (or at least some of it) is probably being taken up by increased net primary productivity of land plants, notably in forests.

8 The long-term stability of the carbon cycle is controlled by negative feedbacks acting on the atmospheric carbon reservoir through rates of silicate weathering on land followed by deposition and preservation of carbonates in marine sediments, and (possibly) the deposition and preservation of organic carbon, mainly in marine sediments.

Learning outcomes for Chapter 2

You should now be able to demonstrate a knowledge and understanding of:

2.1 The important role carbon plays on Earth, in relation to the systems within the biosphere, atmosphere, hydrosphere and geosphere, and in turn how these spheres are interconnected by various biogeochemical feedback mechanisms.

2.2 The differences between the terrestrial, marine and geological carbon cycles in terms of the timescales involved in cycling carbon between different reservoirs, and the resultant impacts these cycles have on the Earth's climate.

2.3 How changes in the global carbon budget can most easily be identified by investigating variations in a relatively small reservoir (e.g. atmospheric carbon dioxide concentrations), and how this has revealed seasonal fluctuations in concentrations due to natural processes, superimposed on annual increases attributed to anthropogenic activities.

2.4 The negative feedbacks that control the long-term stability of the carbon cycle and how the system increasingly appears to be out of balance due to anthropogenic activity.

Plate tectonics, climate and life

Chapter 3

Before the 1980s, most geoscientists saw life as a mere passenger on a dynamic and evolving Earth. Since then, the view that life is partly responsible for maintaining the conditions necessary for its survival has become widely accepted. In its most extreme form this view, which perceives the planet as a holistic, self-regulating system, has been termed the Gaian hypothesis, after the Greek goddess of the Earth. There is no doubt that life on Earth today plays a key role in the feedbacks that maintain climatic conditions within the limits that life can survive. However, you should not forget that life was created from, and was nurtured by, an inorganic world. This implies there must be inorganic feedbacks that keep the climate sufficiently stable for this to have been possible. Such feedbacks remain very much in play, operating on longer timescales than most biological feedbacks, and underpin the stability of the Earth's climate over the past 4 Ga.

Since its very early history, the underlying geological cycle of the Earth has been driven by plate tectonics. The outermost solid shell, termed the lithosphere, is segmented into several fragments that cover the face of the Earth; each of these is in constant motion, driven by the internal heat of the planet, which, over millions of years, redistributes the positions of the continents. Plate tectonics profoundly affects climate and life in many ways, three of which you will look at in some detail in the following chapters. Firstly, the energy driving the plates is released largely along plate boundaries as seismic and magmatic activity, releasing gases and particles into the atmosphere that impact dramatically on surface temperatures. Secondly, the migrating continents affect the geometry of the world's oceans and therefore their efficiency in moderating temperature differences through ocean circulation patterns; the distribution of ocean basins also affects sea level and the global extent of coastlines, both of which have implications for the diversity of life. And thirdly, the mountain ranges thrown up during the collision between continents affect the rate of rock weathering, which in turn affects the carbon dioxide reservoir in the atmosphere (Chapter 4). In this chapter, you will examine the first two of these issues; the third is the subject of Chapter 4.

3.1 Volcanism and the Earth system

Like the workings of the carbon cycle, most of the processes occurring on and in the Earth are quiet and unspectacular. Volcanism can be very different: Earth's history has been frequently punctuated by massive explosive eruptions, the consequences of which are discernible worldwide. Predictably, these occasionally catastrophic episodes have been linked with everything from the death of the dinosaurs to the triggering of Ice Ages. While some of the more extravagant speculations are hard to sustain, volcanism has played, and continues to play, a crucial role in the evolution of the Earth system.

During volcanic activity, heat and material from the Earth's interior rise to the surface. This movement is triggered by the same internal processes that drive the motions of the Earth's lithospheric plates. Figure 3.1 shows the locations of current or recent volcanic activity in relation to plate boundaries.

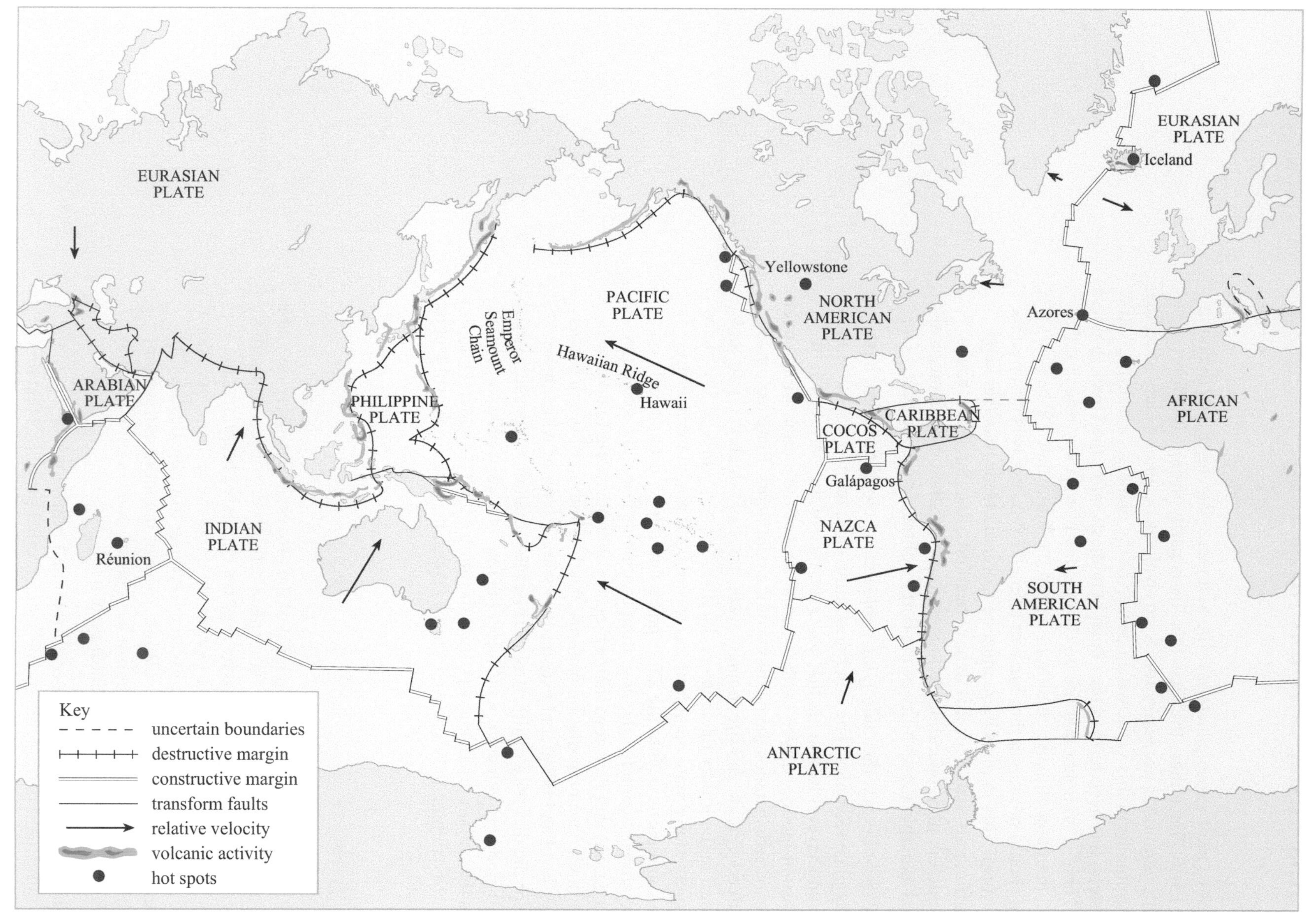

Figure 3.1 The distribution of active (or recently active) volcanic activity in relation to the boundaries of the lithospheric plates, of which there are three types: constructive boundaries (mid-ocean spreading ridges), destructive boundaries (subduction zones), and conservative boundaries (transform faults). The arrows indicate the relative motions of the plates, assuming the African Plate to be stationary; the arrow shown in the key corresponds to a relative velocity of 50 mm y^{-1}.

- Volcanic eruptions related to plate boundaries can be divided into two main groups. What are they?
- Volcanic eruptions at constructive plate boundaries (i.e. mid-ocean spreading ridges) and those in the vicinity of destructive plate boundaries (i.e. subduction zones).

Subduction zone eruptions often take place above sea level, and so they attract more attention than submarine eruptions, particularly as they are intrinsically more explosive. Both types of eruption, however, are important in the Earth system.

At mid-ocean ridges, volcanism is sporadic on short timescales (decades), but over timescales of hundreds of years, ridges are continually active.

A third type of volcanic activity unrelated to plate boundaries occurs at hot spots resulting from mantle plumes. Also shown in Figure 3.1 are the positions of hot spots, which are places where volcanism is triggered by **mantle plumes** (i.e. upcurrents of magma rising from the boundary between the Earth's core and mantle). This source remains fixed as the lithosphere moves over it, so that over time a chain of volcanoes is generated (Figure 3.2). Such hot-spot volcanism is responsible for linear island chains, including those in the northern Pacific, of which the best known is the Hawaiian group (visible in Figure 3.1), with the island of Hawaii itself at the 'young' end of the chain.

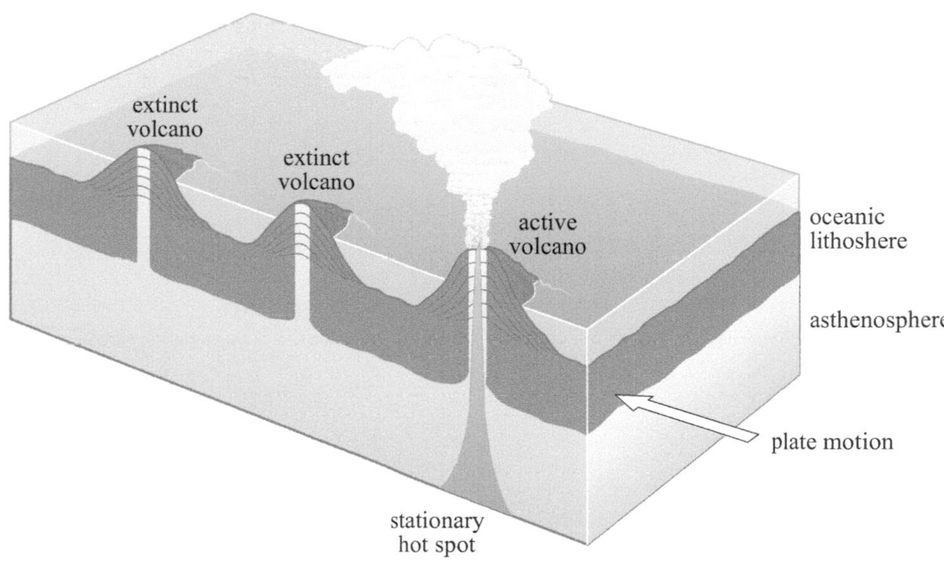

Figure 3.2 Schematic diagram (not to scale) illustrating how a volcanic island chain, such as the Hawaiian Chain, forms over a stationary hot spot. The volcano on the right (the equivalent of Hawaii) is still active, and the age of the volcanoes increases towards the left.

Hawaiian eruptions have been well documented, so they offer a useful starting point for this chapter. Hawaii has two major active volcanoes: Mauna Loa and Kilauea. Kilauea is the world's most persistently active volcano, steadily erupting molten lava into the ocean at about 5 $m^3 s^{-1}$. Such a production rate makes it easy to appreciate the role that volcanism plays in simply forming new land – the island of Hawaii has been built up from the sea floor over the past 1–2 Ma (Figure 3.3). But rocks are only part of the story; Table 3.1 lists the composition of a typical Hawaiian lava at the time of its eruption.

(a) (b)

Figure 3.3 (a) Mauna Kea volcano (altitude 4205 m), seen from the flanks of Mauna Loa, Hawaii. (b) Erupting magma flows into the Pacific Ocean on the southeast coast of Hawaii. ((b) Brad Lewis/SPL)

Table 3.1 Chemical composition of the solid and gaseous components of a typical Hawaiian basalt lava erupted from Mauna Loa in 1859. The data have been presented to show the relative percentages in mass of different elements. Elements are present as silicate or oxide minerals in magma.

Element	Mass/%
silicon, Si	24.15
aluminium, Al	7.31
iron, Fe	8.23
magnesium, Mg	4.35
calcium, Ca	7.55
sodium, Na	1.71
potassium, K	0.35
manganese, Mn	0.13
titanium, Ti	1.24
phosphorus, P	0.10
oxygen, O	44.26
Dissolved gas	
water vapour, H_2O	0.27
carbon dioxide, CO_2	0.32
sulfur dioxide, SO_2	0.18

■ Look at the analysis of the magma in Table 3.1. Which of the constituents or elements are most likely to affect the environment some distance from the volcano?

□ Water, carbon dioxide and sulfur dioxide because they can escape as volatiles (gases) and mingle freely with the atmosphere. (*Note*: the first ten components (Si to P), combined with oxygen in silicates and oxides, all contribute to the rocky lava.)

In Hawaii, the magma originates deep within the mantle so that the carbon dioxide emitted represents a net addition to the atmosphere. Early in the history of the Earth, virtually the whole of the first stable atmosphere is believed to have formed by volcanic outgassing in this manner. First though, the volcanic input of carbon dioxide (which is a potent greenhouse gas) into the atmosphere will be considered.

The total amount of CO_2 liberated to the atmosphere during Mauna Loa's formation so far is about 2×10^{15} kg. If Mauna Loa has been building up over 1.5 Ma, the flux into the atmosphere will have been 1.4×10^9 kg per year (about the same as a 200 MW coal-fired power station) and equivalent to about 0.01% of the annual anthropogenic contribution. Although some volcanoes (e.g. Mount Etna in Sicily) exhale more CO_2 than Kilauea, the total global volcanic contribution to the atmosphere is thought to be about 6.5×10^{10} kg y^{-1}, which is still less than 0.3% of the total anthropogenic (human-generated) flux. This number is a striking example of how human activity is altering the natural environment. It is important to note, however, that although volcanoes presently account for the addition of a relatively small amount of CO_2 to the atmosphere, at certain times in the past the impact of volcanic CO_2 was much greater (this is examined in more detail later in this chapter).

■ Of the three types of volcanism shown in Figure 3.1 (volcanism at mid-ocean ridges, subduction zones or mantle plumes), which will emit the most CO_2 along with erupted magma?

■ It is likely to be volcanism associated with mid-ocean ridges, which is much more widespread, and overall produces much more lava than volcanism associated with either subduction zones or mantle plumes.

Current estimates of magma production by these three modes of volcanism have considerable uncertainties attached to them, but mid-ocean ridges are generally thought to account for around 85% of global magma production. The other two modes, being more episodic in nature, are difficult to estimate, but probably generate roughly equal volumes of magma on average. While both subduction zone and mantle plume volcanism are predominately subaerial and release CO_2 directly into the atmosphere, mid-ocean ridges are located on average at depths of about 2.5 km under water, where pressures of several hundred atmospheres prevent CO_2 escaping from the lavas in gaseous form. Even so, CO_2 is still an important component of the gases dissolved in the fluids that are expelled at hydrothermal vents found wherever there is submarine volcanism. The huge volumes of magma erupted at mid-ocean ridges carry with them correspondingly vast quantities of CO_2, which will eventually be released into the atmosphere after circulating in the ocean in solution for on average about 1000 years. Therefore, on geological timescales, mid-ocean ridges contribute around 75% of the CO_2 from volcanic sources to the atmosphere.

So far, this chapter has concentrated on volcanic emissions of CO_2 *on short timescales*. Sulfur dioxide (SO_2, which is a poisonous gas) has far more dramatic effects on the atmosphere, and this is the subject of the next section.

3.2 Volcanic aerosols and climatic change

> The ashes now began to fall upon us, though in no great quantity … darkness overcame us, not like that of a cloudy night, or when there is no moon, but of a room when it is all shut up and all the lights extinguished.
>
> (Pliny the Younger, AD 79)

This brief extract from Pliny the Younger's classic account of the eruption of Vesuvius in AD 79 is a reminder that volcanic ash clouds can be highly effective at blocking out sunshine. Major explosive eruptions (still called Plinian eruptions to commemorate Pliny's observations) blast huge amounts of volcanic ash high into the atmosphere, often reaching the stratosphere (Figure 3.4a). Awe-inspiring though these eruption columns are, it is not the ash particles that cause the most significant climatic effects. Ash fragments are relatively large (ranging from shards 1–2 μm across to fist-sized lumps of pumice) and they fall back to the ground over timescales ranging from seconds to days. The longest-lasting environmental effects are produced by the volcanic gases.

As shown in Table 3.1, volcanoes produce large quantities of sulfur dioxide as well as carbon dioxide. After eruption, sulfur dioxide reacts with water vapour in the atmosphere to form aerosols (i.e. tiny airborne droplets) of sulfuric acid, H_2SO_4. These acid aerosols form by a complex series of photochemical reactions that may continue for months, replenishing the aerosol cloud so that new particles form while older, larger ones settle out of the atmosphere. Aerosol particles are minuscule (initially only 0.1–1 μm in diameter) so any reaching the stratosphere may remain suspended there for months or years, far longer than solid ash fragments.

Volcanic aerosols often produce gloriously colourful sunsets (and sunrises), which can be enjoyed around the world (Figure 3.4b). During the day, the sky looks blue

(a)

(b)

Figure 3.4 (a) A Plinian-type explosive eruption of Popocatepetl (Mexico) in December 1998; the eruption column is about 5 km high. (b) During the autumn of 1883, after the eruption of Krakatau, spectacular sunsets around the world attracted the attention of artists. This is one of a series of six paintings by William Anscom of the sun setting on a November evening seen from Chelsea, London. It was published in the contemporary Royal Society report on the great eruption. ((a) Wesley Bocxe/SPL)

because gas molecules scatter sunlight in all directions. These molecules are smaller (~ 0.03 μm) than the wavelength of light and scatter blue wavelengths (~ 0.4 μm) more effectively than red (~ 0.7 μm). Thus, blue light from the sun actually reaches your eye even when you are not looking directly at it, whereas the other wavelengths do not. When the size of particles is about the same as the wavelength of light, as in aerosols and smoke, the scattering process is more complex but the dominant effect is a shift to longer visible wavelengths, around orange to red in the visible spectrum. It is less intense than the blue-sky effect, and is most noticeable at sunset and dawn when the sun is low in the sky and its rays follow a much longer atmospheric path (Figure 1.4).

In detail, the physics of light scattering by aerosols is complex, but the overall effect is that when a layer of volcanic aerosols gets between the Earth and the sun, a fraction of the sun's radiation is scattered back to space. In other words, the presence of the aerosols effectively increases the Earth's albedo (Figure 1.12). To estimate the climatic consequences of increased aerosols in the atmosphere, the balance between incoming solar radiation and outgoing longwave (thermal) radiation from the Earth and the atmosphere has to be taken into consideration. As Figure 1.12 shows, a fraction of the incoming radiation is reflected back to space; of the remainder, some passes directly through to the surface of the Earth, and some is **forward-scattered**, also reaching the ground. The ground heated by the sunshine warms up, and reradiates longwave thermal energy, some of which is reradiated back to Earth by the atmosphere and clouds, while the remainder escapes to space. Incoming radiation (shortwave, S_{in}) and outgoing radiation (reflected shortwave, S_{ref}, plus thermal longwave, L) must balance:

$$S_{\text{in}} = S_{\text{ref}} + L \qquad (3.1a)$$

Figure 3.5a applies these symbols to a simplified version of Figure 1.12: part (b) of Figure 3.5 is the same as part (a), but also includes the effect of aerosols, which not only scatter incoming solar radiation back to space but also scatter longwave thermal radiation back towards the Earth's surface. If solar radiation scattered back into space by the aerosol layer (S_{a}) is now included in the equation, along with longwave thermal radiation **back-scattered** to Earth by the aerosol layer (L_{a}), this equation can be modified to read:

$$S_{\text{in}} = S_{\text{ref}} + S_{\text{a}} + (L - L_{\text{a}}) \qquad (3.1b)$$

S_{a} will add a component to the sunlight already normally reflected by clouds, while L_{a} will reduce the amount of longwave thermal radiation escaping to space (Figure 3.5b).

- What would be the overall effect of increasing the term $S_{\text{ref}} + S_{\text{a}}$, bearing in mind that the radiation budget would eventually reach a state of balance?
- In order to maintain balance, the term $L - L_{\text{a}}$ would have to decrease.

$L - L_{\text{a}}$ is the thermal radiation escaping from the Earth; if this term is decreasing, the Earth's global temperature must be decreasing. It follows that increasing the term S_{a} (and hence $S_{\text{ref}} + S_{\text{a}}$) will lead to global cooling. In other words, until proved otherwise, volcanic aerosols will cause global cooling.

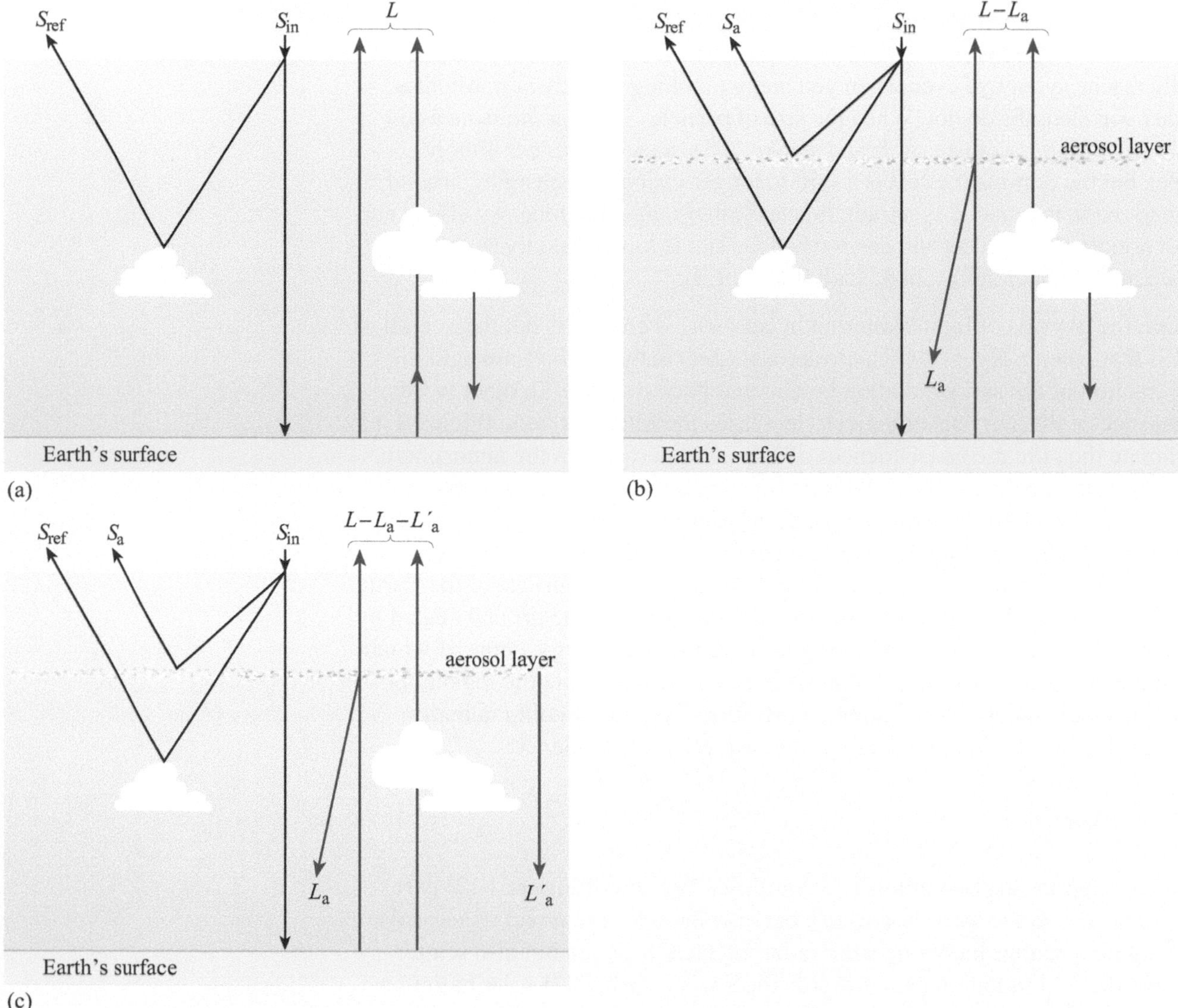

Figure 3.5 (a) Simplified version of Figure 1.12 where S_{in} represents incoming solar radiation, S_{ref} represents the solar radiation reflected by the Earth (i.e. by clouds, the Earth's surface and the atmosphere), and L represents the longwave (i.e. thermal) radiation that escapes to space. (b) If a layer of aerosols is added to the situation shown in (a), the amount of solar radiation reflected back to space becomes $S_{ref} + S_a$, and the amount of longwave radiation that escapes to space becomes $L - L_a$. (c) If the aerosol layer also absorbs outgoing thermal radiation and reradiates an amount L'_a towards the Earth, the amount of longwave radiation escaping to space becomes $L - L_a - L'_a$. Whether the net result is global cooling or global warming depends on the relative sizes of S_a, L_a and L'_a; in general though, S_a is greater than $L_a + L'_a$ and the net result is global cooling.

The details are more complicated and, for a given aerosol layer, the fraction of solar radiation that is actually scattered back into space depends on both the amount of aerosol present and the size of the droplets. In general, an increased albedo caused by aerosol scattering cuts down the sunlight reaching the surface of the Earth, causing cooling. While droplets of all sizes may scatter radiation, they can, however, also absorb it if their diameters are of the same order as the wavelength of the radiation. As reradiated thermal radiation is of relatively long wavelength, it can be absorbed by large aerosols, which will themselves emit longwave radiation, emitting some of this back towards the Earth (Figure 3.5c).

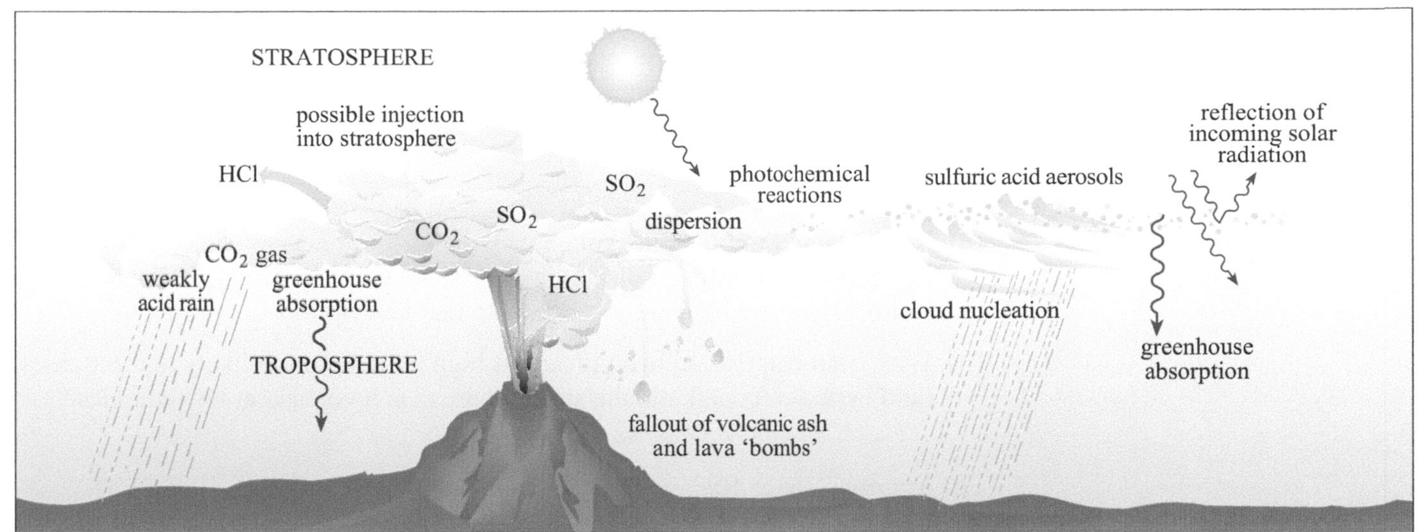

Figure 3.6 Physical and chemical interactions of a volcanic eruption with the atmosphere. The eruption releases sulfur dioxide (SO_2), which is eventually converted to sulfuric acid (through reaction with atmospheric oxygen and water), producing a mist of droplets, or aerosols. These reflect a fraction of the incoming solar radiation, cooling the troposphere; they also absorb some solar radiation, warming the stratosphere. If the aerosols are sufficiently large, the aerosol layer may absorb and reradiate thermal energy, warming the lower atmosphere (Figure 3.5). Note that they are sometimes referred to as *sulfate* aerosols rather than sulfuric acid aerosols. The eruption also produces large amounts of CO_2, which mixes in the atmosphere, forms a much weaker acid with atmospheric water, and does not form aerosols. Smaller amounts of hydrogen chloride (HCl) are also released into the atmosphere, forming hydrochloric acid.

An aerosol layer can therefore act as a 'greenhouse', preventing some thermal radiation escaping and so causing surface warming.

It is difficult to make precise predictions about the effects of an aerosol layer on the Earth's surface temperature. Indeed, the physical and chemical interactions of an eruption are complex (Figure 3.6). For most long-lived volcanoes, however, the volcanic aerosols are relatively small such that they cause a net cooling as you can see from Table 3.2.

Table 3.2 Computed climatic forcings and the average cooling that resulted from some major recent eruptions.

Location of eruption and year	Mass of material erupted /10^{11} kg	Mass of aerosols in stratosphere /10^9 kg	Climatic forcing /W m^{-2}	Observed Northern Hemisphere temperature change/°C
Katamai, Alaska (58° N), 1912	240	25	−3.5	−0.2
El Chichón, Mexico (17° N), 1982	7.5	12	−1.7	−0.2
Pinatubo, Philippines (15° N), 1991	120	30	−4.3	−0.5
Krakatau, Indonesia (6° S), 1883	247	55	−7.8	−0.5
Agung, Indonesia (8° S), 1963	24	20	−2.86	−0.3
Tambora, Indonesia (8° S), 1815	2400	150	−21.4	−1.0

The precise climatic impact of a volcano is related not only to the flux and composition of the volatiles but also to the altitude to which they are ejected and the latitude of the volcano (Figure 3.6). If the character of an eruption is such that the erupted gases are not injected into the stratosphere, then there is little chance of the resulting aerosols having any global effect, as they will be rapidly rained out of the troposphere.

Question 3.1

(a) Referring to Figure 1.20, explain why the effect of aerosols injected into the troposphere from a volcano in (for the sake of argument) the Southern Hemisphere will be largely confined to that hemisphere.

(b) Why is an eruption from a volcano at high latitudes more likely to eject gases (and hence aerosols) into the stratosphere than a volcano at low latitudes?

For example, about 40 days after the eruption of Mount Pinatubo (at ~15° N), stratospheric aerosols had reached high latitudes in both hemispheres. An eruption sending ejecta into the stratosphere at *high* latitudes, however, will have less effect in the other hemisphere than one taking place in the tropics because stratospheric circulation patterns are mainly zonal (east to west, or west to east) and so inhibit the spread of aerosols from one hemisphere into the other (north to south or vice versa).

3.3 Flood basalts and their effects on climate and life

So far you have considered only recent, short-lived volcanic events. Due to the long response time of the Earth system, it is unlikely that such events have had a sustained impact on global climate. A different kind of eruption that might have had more serious environmental effects than the explosive kind is the effusion of huge volumes of lava in the geological past. These are termed **flood basalts**.

Accumulations of flood basalt lava flows cover large parts of the Earth. Some of the statistics are impressive: India's 65 Ma Deccan Traps cover 0.5 million km^2 (Figure 3.7), and may have covered 1.5 million km^2 when first erupted. They have an average thickness of at least one kilometre. Most of their huge volume may have been erupted in less than 0.5 million years (although this is controversial) and they may consist of many hundreds of individual lava flows. On erosion, they acquire a distinctive topography resembling flights of steps. In northwestern USA, the Columbia River Province consists of 240 000 km^3 of flood basalts erupted between 17 Ma and 12 Ma, in individual flows 20–50 m thick, covering an area of more than 200 000 km^2 (Figure 3.8). Approximately 90% of the total volume of the province might have been erupted during a period of less than 1.5 million years between 16.5 Ma and 15 Ma. Work on individual eruptive units of basalt lava in the Columbia River Province has shown that >1000 km^3 could have been erupted in continuous, individual eruptions lasting a decade or longer.

The next step is to consider what impact such massive eruptions would have on global climate. At any given time in Earth history, the amount of carbon dioxide in the atmosphere is the result of the balance of fluxes between sources and sinks in the carbon cycle. If the average global flux of carbon dioxide to the

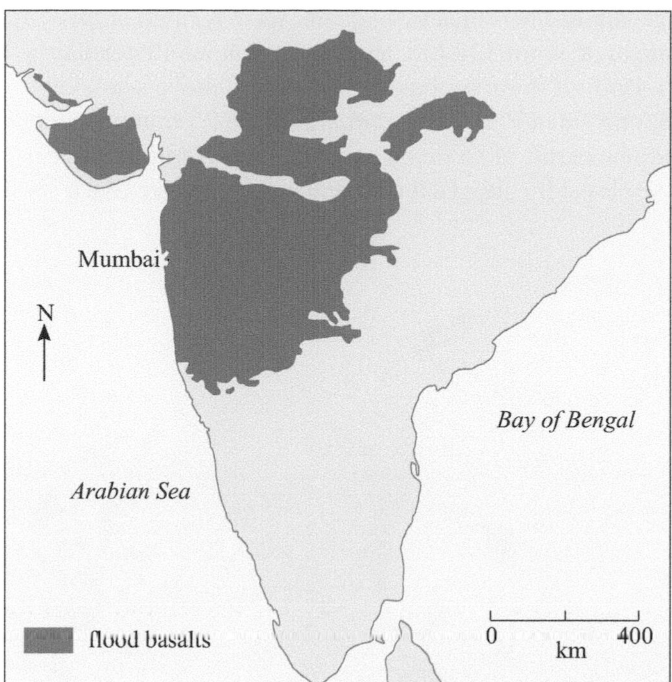

Figure 3.7 The areal extent of the Deccan Traps. A significant fraction of India's land surface is covered by the outcrop of these 65 Ma basalt lavas.

Figure 3.8 Layers of stacked lava flows in the Columbia River Province, northwest USA. The flows shown represent just a small thickness of the voluminous and rapidly erupted Grande Ronde Formation part of the succession. (Steve Self/The Open University)

atmosphere from volcanoes should increase for some reason, and if that increased flux is sustained, then the carbon cycle will settle into a new equilibrium, with important implications for climate (see Section 2.4).

But is there any evidence that the flux of carbon dioxide into the atmosphere has varied over geological time? There are some pointers in this direction. Most attention has been focused on the Cretaceous Period (145–65 Ma), which seems to have been an exceptionally productive period in terms of creation of ocean crust and eruption of massive basaltic plateaux. So much new material was erupted in the Pacific Basin during the Cretaceous that it is difficult to find any older oceanic crust or sediments there – they have been simply swamped by

Cretaceous lavas. Two huge submarine volcanic plateaux were formed at this time: the Ontong–Java Plateau at about 122 Ma, and the Kerguelen Plateau at about 112 Ma (Figure 3.9). Both of these probably grew partly above sea level when they formed. The Ontong–Java Plateau is estimated to be 25 times bigger than the Deccan Traps, and one result of so much new lava being erupted was that it elevated the global sea level by about 10 m, displacing seawater like a body in a bath tub.

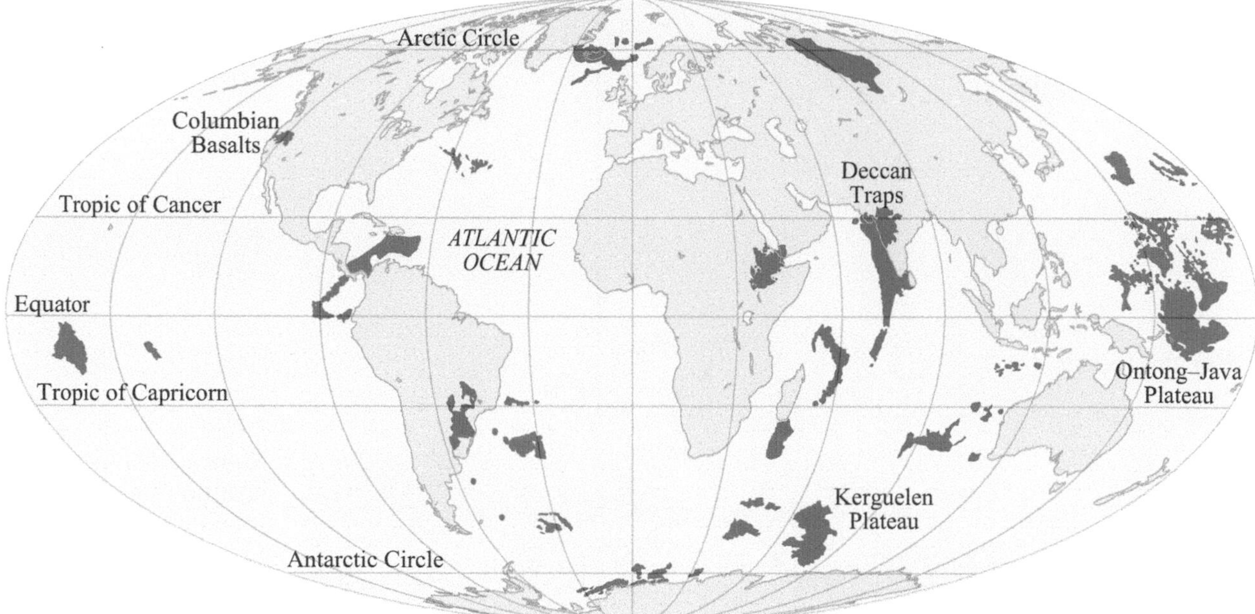

Figure 3.9 Global distribution of large igneous provinces (darker, purple areas), including the large oceanic plateaux. (Coffin and Eldholm, 1993)

Estimates of the average global rate of ocean crust production over the past 150 million years range between 5 and 33 km^3 Ma^{-1}. Estimates of the production rate of the Ontong–Java Plateau on its own are around 12–15 km^3 y^{-1}, i.e. one flood basalt province produced the same order of magnitude of basalt as the entire oceanic ridge network.

From the graph of estimated oceanic crust production rate through time (Figure 3.10), you can see that global rates during the early Cretaceous (145–100 Ma) were relatively high. Such estimates are hotly debated, with some researchers claiming that mid-ocean ridge crust production has been constant for 150 Ma, but the abrupt formation of the enormous Ontong–Java and Kerguelen Plateaux cannot be denied. These two plateaux, and some lesser ones in the southwest Pacific, certainly contributed to the higher rates of crust production at this time.

What were the effects of all this volcanic activity on CO_2 in the atmosphere? There may have been both direct and indirect consequences. As discussed at the start of this chapter, as long as it remained below sea level, the volcanism taking place at mid-ocean ridges, and leading to the creation of new oceanic crust, would have had little immediate effect on the atmosphere. Increased sea-floor spreading, however, has to be balanced by increased subduction – if it were not, the total area of the Earth's lithosphere would increase, so that the Earth would have to expand. When oceanic crust is subducted, carbonate-rich sediments are

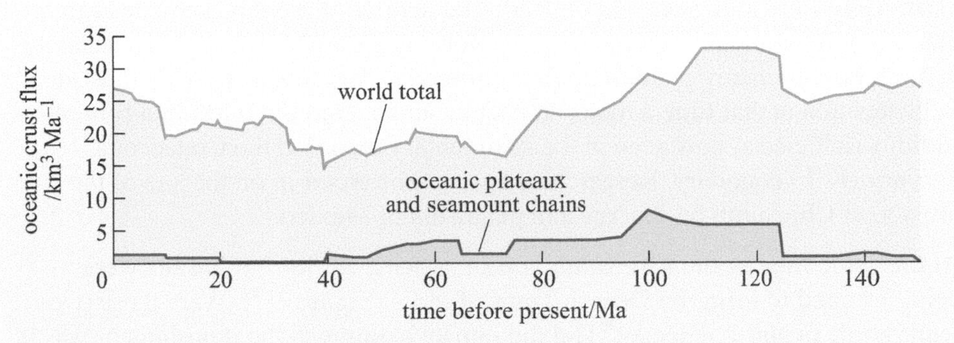

Figure 3.10 Variation with time of the estimated rate of production of oceanic crust. Note the peak between about 120 Ma and 80 Ma. (Cogné and Humler, 2006)

also subducted, and their carbon may ultimately be exhaled as CO_2 at subduction zone volcanoes. More importantly, if mantle plumes lead to the formation of several large volcanoes (e.g. Hawaii) extending above sea level, then a direct flux of CO_2 from the mantle into the atmosphere could be sustained.

Estimates of the abundance of CO_2 during the Cretaceous suggest that atmospheric CO_2 could have reached a level at least four times the modern pre-industrial level of 280 ppm, leading to global greenhouse warming of between 3 °C and 8 °C. This would have been accompanied by other changes, such as an increase in sea level. A runaway greenhouse effect was counteracted by increased deposition and preservation of carbon in the warm oceans and shallow seas.

It is important to consider whether this increase in atmospheric CO_2 concentration could be the direct effect of the eruption of flood basalt or whether other sources need to be considered. Initial estimates based on the (admittedly sparse) knowledge of how much mantle CO_2 is actually released into the atmosphere suggest that these volcanic fluxes are fairly low. For example, the estimated annual flux of CO_2 to the atmosphere from the eruption of 1000 km^3 of basalt over a decade is about 3% of the present-day natural land–atmosphere flux – too small to have a significant effect on the CO_2 concentration in the atmosphere. Such a modest flux would probably allow carbon cycle processes to offset the volcanic addition. This will still result in some global warming, however, and could trigger other mechanisms such as the destabilisation of methane hydrate deposits and the release of methane into the atmosphere. There is no doubt that the short-term impact of aerosol emission (i.e. over a period of a few years) is often one of global cooling (Table 3.1), but there remains considerable uncertainty over the extent of long-term warming (i.e. over a period of millions of years) associated with the eruption of large-scale flood basalts.

Question 3.2

Some volcanologists and climatologists argue that volcanic eruptions usually affect weather, not climate. What aspect of volcanic eruptions underlies this point of view? In what circumstances might it be appropriate to speak of the effects of volcanism on just climate?

It is reasonable to ask how life on Earth was affected during this, and similar, periods of sustained volcanic activity. At the end of the Cretaceous Period (65 Ma) there was worldwide extinction of many animal species, i.e. a **mass**

extinction, both in the sea and on land. The demise of the dinosaurs has captured the public imagination, but many less-notorious groups of species also became extinct, notably many planktonic foraminiferans. An increasing body of evidence indicates that at that time a major impact event covered the Earth in a film of iridium-rich ejecta, now seen in the sedimentary record at the Cretaceous–Tertiary (K/T) boundary. Researchers have even zeroed-in on the site of the impact, at Chicxulub on the Yucatan peninsula of Mexico.

At the same time as the Chicxulub impact, at least 10^6 km^3 of basalts were being erupted to form the Deccan Traps of India (Figure 3.7). Were these two remarkable events connected? Did the impact extinguish the dinosaurs or was it the basaltic flood eruption, or the combined effect of both? Did the impact event trigger the basaltic flood or did the two extraordinary events reinforce each other in some lethal way to cause mass extinctions? These highly contentious issues have sparked vigorous debates for more than 20 years.

As long ago as the early 1970s, some scientists argued that widespread basaltic volcanism might have been implicated in the extinction of the dinosaurs. A round-number estimate of the volume of the Deccan basalts is 10^6 km^3, suggesting that a total of about 10^{16} kg of SO_2 gas was released, which has the potential to generate ~2×10^{16} kg of sulfuric acid aerosols. It seems inevitable that the huge mass of acid aerosols produced by the Deccan eruptions (with each one of the ~1000 eruptions producing ~2×10^{13} kg of aerosols) had far-reaching consequences.

An even greater mass extinction defines the Permian–Triassic boundary, at 251–245 Ma, separating the Palaeozoic ('ancient life') Era from the Mesozoic ('middle life') Era. Huge numbers of species, including trilobites (marine arthropods, probably related to the ancestors of the crustaceans), became extinct by the end of the Permian, while the Mesozoic saw the blossoming of the age of reptiles, including the dinosaurs. Is it purely coincidental that the Palaeozoic–Mesozoic boundary approximately coincides with the vast outpouring of the Siberian flood basalts, currently the largest subaerial flood basalt province known (Table 3.3)? More than 3×10^6 km^3 of basalt may have been erupted within a period of less than one million years, an average rate of more than 3 km^3 y^{-1} (about 100 m^3 s^{-1}).

■ Does the extinction event at the Tortonian–Serravallian boundary (Table 3.3) correspond with a flood basalt eruption, within the limits of error of the ages cited?

■ No. None of the flood basalt eruptions fall in the period 9.5–11.5 Ma (within error of this extinction event). The closest is the Columbia River eruption at 17.0 ±0.2 Ma.

Question 3.3

Which extinction events within the past 300 Ma (listed in Table 3.3) correspond with a flood basalt eruption, within the limits of error of the ages cited?

These correlations indicate that it is plausible that the flood basalts may be implicated in at least some extinction events.

Table 3.3 Table of peak eruption ages for flood basalt provinces and ages of extinction and mass extinction events for the most recent 260 Ma of Earth history. Mean ages are derived mainly from Courtillot and Renne (2003), and stratigraphic boundaries are quoted from Gradstein et al. (2005). M = mass extinction event, E = extinction event.

Flood basalt or large igneous province	**Mean age /Ma**	**± Ma**	**Stage or epoch boundary at which extinction occurred**	**Age /Ma**	**± Ma**
Emeishan Traps	259	3	Guadalupian–Lopingian (M)	260.4	0.7
Siberian Traps	250	1	Permian–Triassic (M)	251	0.4
Central Atlantic Magmatic Province	201	1	Triassic–Jurassic (M)	199.6	0.6
Karoo Ferrar	183	2	Pliensbachian–Toarcian (E)	183	1.5
			Bajocian–Bathonian	168	3.5
			Tithonian–Berriasian (E)	145	4
Parana Etendeka	133	1	Valanginian–Hauterivian	136.4	2
Ontong–Java 1	122	1	Early Aptian (E)	125	1.0
Rajmahal/Kerguelen	118	1			
Ontong–Java 2	90	1	Cenomanian–Turonian (E)	93.5	0.8
Caribbean Plateau	89	1			
Madagascar Traps	88	1			
Deccan Traps	65.5	0.5	Cretaceous–Palaeogene (M)	65.5	0.3
North Atlantic 1	61	2			
North Atlantic 2	56	1	Paleocene–Eocene (M)	55.8	0.2
Ethiopia and Yemen	30	1	Oil event	30	2.5
Columbia River (E)	17	1	Early Miocene–Mid Miocene (E)	16.0	0.1
			Serravallian–Tortonian	11.6	0.3
			Pliocene–Pleistocene (E)	1.81	0.02

3.4 Continental drift and climate

Volcanism is only one means by which plate tectonics affects global climate. The process by which continents are separated is known as sea-floor spreading which occurs at rates of a few centimetres a year, such that major ocean basins open and close on timescales of 100–200 Ma. In other words, continents break up, are carried over the surface of the globe, and eventually are brought together again, at intervals of hundreds of millions of years. Not surprisingly, such movements cause the climate of individual continents to change through time. For example, at 300 Ma the British Isles were located at the Equator, and since that time have migrated northwards to their present position at latitude 50–55° N. During this time, the climate affecting the British Isles has become markedly cooler. Is it possible, though, that *global* climate could itself be affected by the migrating continents?

The periods when the Earth is partially covered by extensive ice sheets are referred to as 'ice ages'. We are presently within an ice age, known as the Quaternary Ice Age: there is a large ice cap over the land mass of Antarctica, another over Greenland, and permanent ice cover over much of the Arctic Ocean. If the present ice age is typical, you can assume that, within ice ages, ice caps grow and retreat many times, as glacial periods alternate with interglacial periods. At the present time, the Earth's ice caps are by no means at their maximum extents and the climate, even at fairly high latitudes, is moderate – in other words, we are currently experiencing an interglacial period.

Over the history of the Earth there have been five, or possibly six, ice ages, three of which were in the last 500 Ma. They occur at intervals of hundreds of millions of years.

Question 3.4

Time and length scales are vital in considering the climatic implications of different phenomena (Figure 3.11).

(a) Where would major periods of extended glaciation (ice ages) plot on Figure 3.11?

(b) Why might this suggest that large-scale changes in the positions of the continents could be at least a contributory cause of the initiation and ending of ice ages?

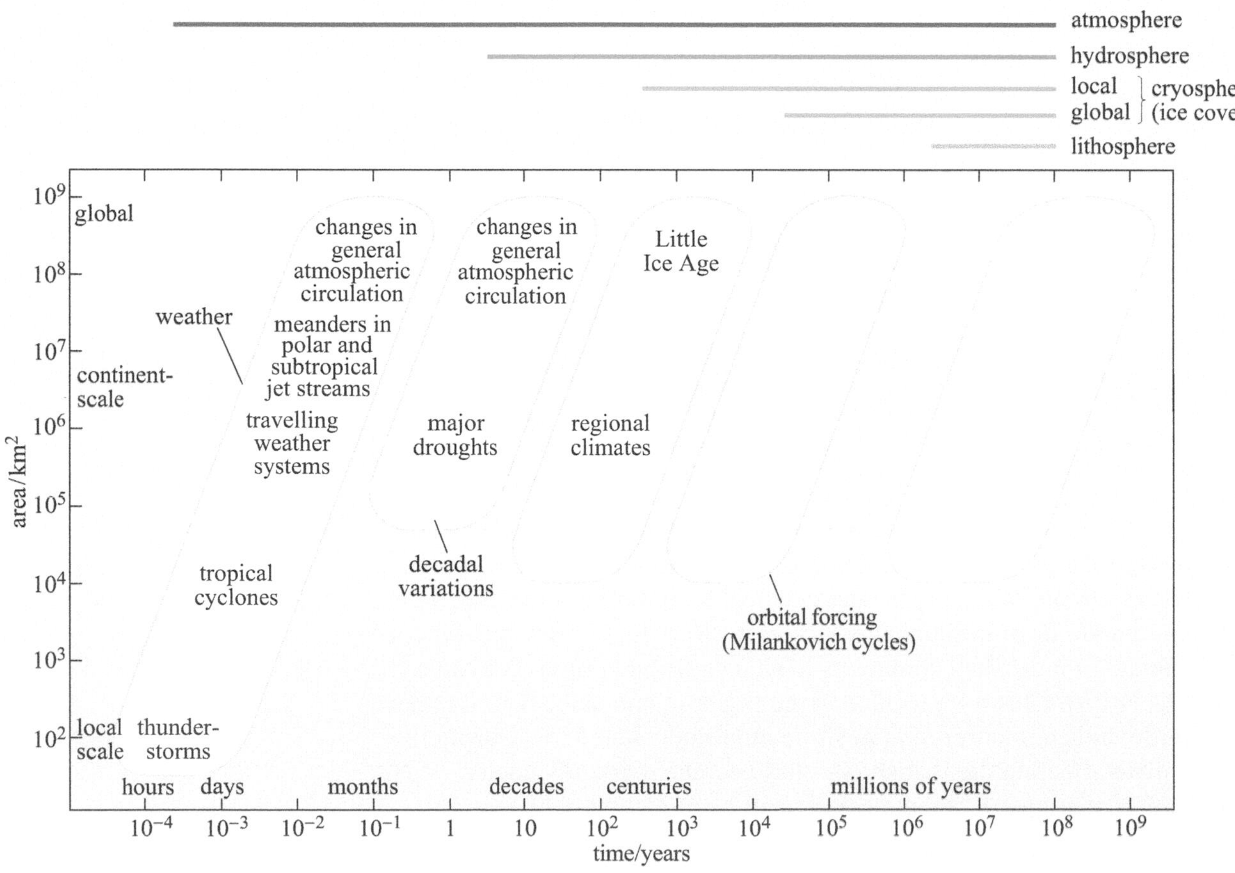

Figure 3.11 Schematic diagram showing the time and length scales of various phenomena affecting conditions on Earth. (The continual increase in solar luminosity since the formation of the Solar System is not included.) The bars along the top show the range of timescales on which each of the major parts of the Earth system vary. These bars illustrate the general principle that while short-term variations are mostly limited to the atmosphere, longer-term variations involve progressively more components of the Earth system. (To be completed in Question 3.4.)

In other words, plate motions would seem to be an obvious candidate for a major forcing factor implicated in the coming and going of ice ages, because they act over a similar timescale. Plate motions are most unlikely to be the only factor, however, as they alone cannot bring about climate change. What they can do, though, is change the distribution and relative positions of continents and oceans in ways that enable other influences to come into play. The ways in which the distribution of land and sea might affect climate can be explored by looking at models that use greatly simplified situations to simulate conditions that might have been obtained in the past, when distributions of continents and oceans were different (Box 3.1).

Box 3.1 Model worlds and their climates

Although contrasting global climates in the geological past appear to be associated with different continental distributions, the only way the relationship between continental configuration and climate can be quantitatively investigated is through the use of computer simulations. Before considering how the two are related, it is worth glancing at the imaginative speculations of Charles Lyell (the 'Father of Geology'). In his book *Principles of Geology*, first published in 1837, Lyell redrew the map of the world to show the present-day continents distributed either in an equatorial/tropical 'ring' (Figure 3.12a) or in two polar 'caps' (Figure 3.12b). This was long before geologists had begun to think about continental drift, but Lyell's vision was sufficiently far-sighted for him to recognise the consequences of shifting continents relative to climatic belts. It was a prodigious intellectual leap, and it is further evidence of Lyell's genius that he also proposed the inverse effect: that shifting continents might themselves be an agent of global climate change.

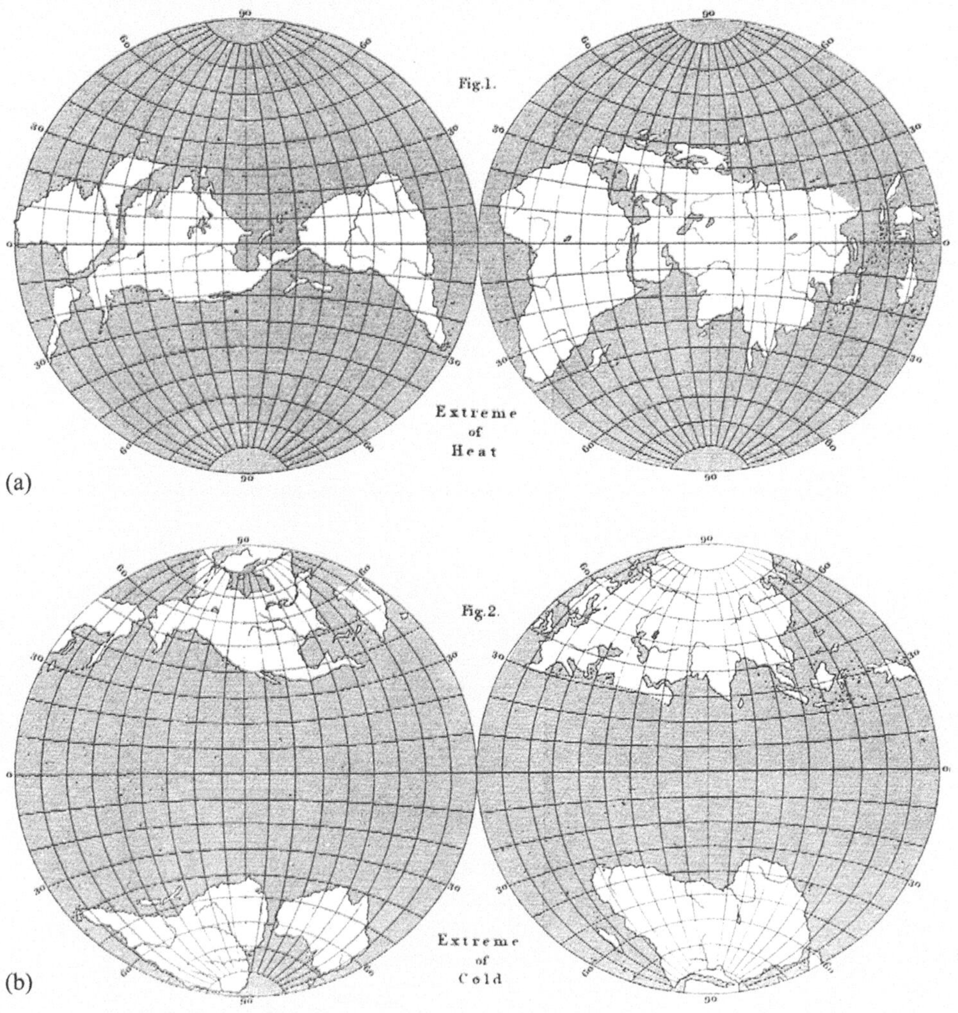

Figure 3.12 Facsimiles of Lyell's maps of present-day continents redistributed to form: (a) a 'ring world' with the continents concentrated in the tropics; (b) a 'cap world' with the continents gathered together around the two poles.

Today, if a problem can be written down mathematically, for instance as a number of simple heat budget equations, it can usually be solved using computers. In Figure 3.13, idealised extremes of continent–ocean distribution have been used to investigate the effects of continental configurations on climate by means of a model consisting of a set of simple, connected balances of heat and moisture in different latitude bands. The model simulates some of the familiar features of the Earth's climate system including seasonal and latitudinal variations of incoming solar radiation, surface temperatures, cloud cover, precipitation and evaporation, and snow and ice cover.

The model was run using two idealised continental geometries based on the present-day total land area:

- two 'caps' of land extending from the poles to 45° of latitude (both with and without ice caps, Figure 3.13a and b)
- a tropical 'ring' of land extending 17° north and south of the Equator (Figure 3.13c).

These idealised continental distributions are nothing like those of today, but the geography of the Earth may have approximated a tropical 'ring world' between 700 Ma and 600 Ma, and could have approximated very roughly a 'cap world' with one polar ice cap in the late Carboniferous at 300 Ma.

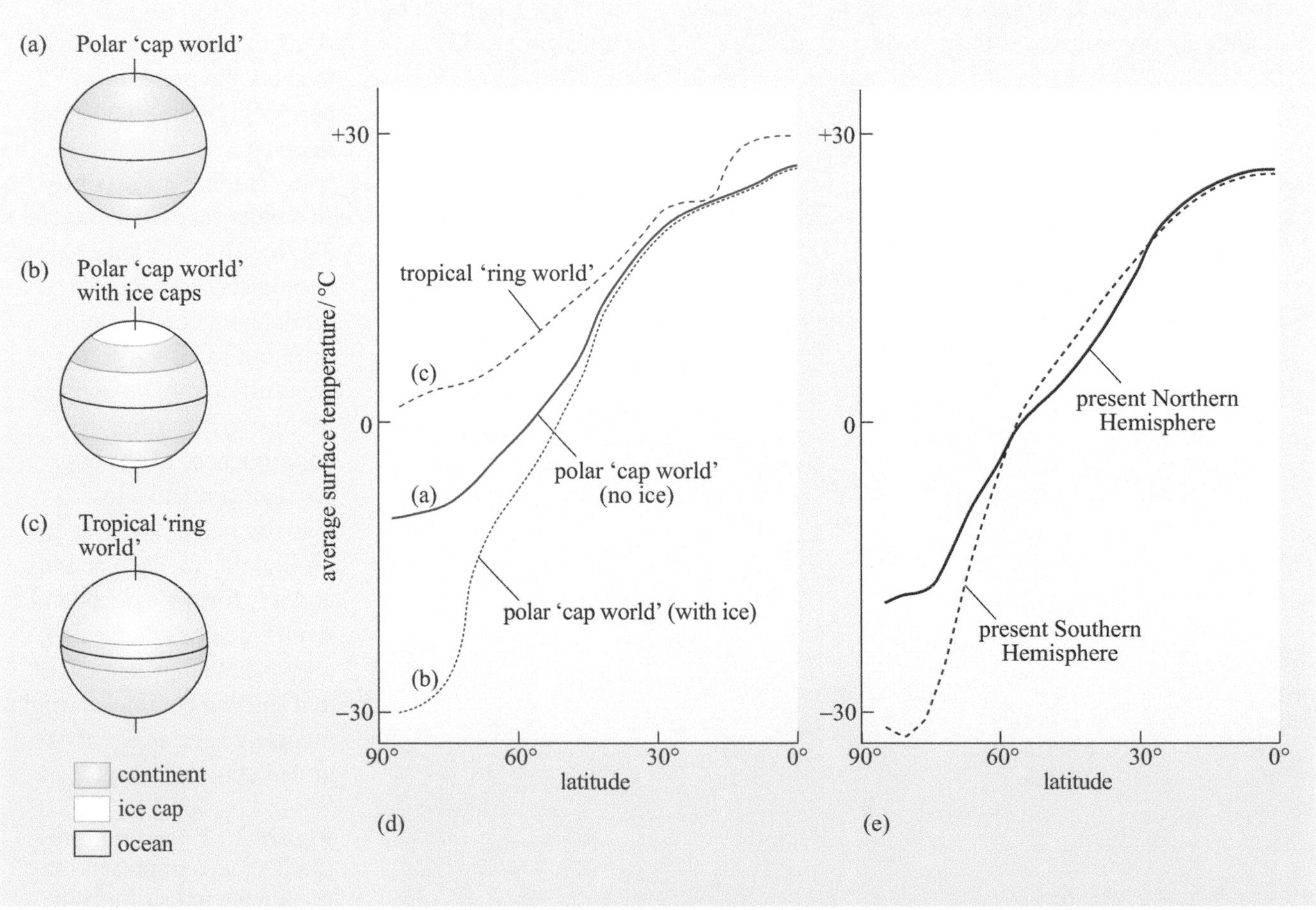

Figure 3.13 Simplified models of the Earth with present-day total land area redistributed to form: (a) a polar 'cap world' without ice caps; (b) a polar 'cap world' with ice caps extending equatorwards to latitude 70°; and (c) a tropical 'ring world'. (d) Mean annual surface temperatures (zonally averaged), simulated for the cap and ring worlds shown in (a). (e) Mean annual surface temperatures (zonally averaged) for each hemisphere, simulated for present-day geography.

As you might expect, for a polar cap world with ice caps (Figure 3.13b) the meridional variation (i.e. variation along a line of longitude) in temperature (Figure 3.13d) looks rather similar to a simulation of the temperature distribution for the present-day Southern Hemisphere (Figure 3.13e) and both of the cap worlds give average equatorial temperatures close to those of the present day. By contrast, the tropical ring world (Figure 3.13c) is significantly warmer than either of the cap worlds.

■ In terms of radiation balance, will the addition of ice caps lead to a cooler world?

■ Ice has a much higher albedo than exposed or vegetated continental crust (Table 1.1), so an ice cap world reflects more solar radiation back into space than an ice-free cap world.

In summary then, a simple model of the kind shown in Figure 3.13a–c can confirm the original supposition that the relative distribution of continents and oceans can influence global climate. This particular model will not be considered further because it ignores important aspects of the Earth's climate system.

As demonstrated in Box 3.1, computer models can be useful in helping to isolate the effects of different factors (in this case, changes in albedo and moisture balance), while keeping other variables (e.g. the total continental area, and total incoming solar radiation) constant. They can also be used to combine the effects of different factors in a quantitative way, and to test hypotheses against physical laws. Climate models are based on a myriad of approximations and, while they can often demonstrate that certain events may possibly have occurred, it is much more difficult to attempt to prove that they must certainly have occurred.

In this case, two important components of the climate system not incorporated into the model are:

- heat transport by surface ocean currents and by the deep thermohaline circulation
- changes in the concentrations of greenhouse gases in the atmosphere.

The second of these factors will be considered later in this chapter, but first it is important to understand how ocean currents might be affected by changes in the global distribution of continents and oceans.

3.4.1 Ocean currents and climate change

Cap, ring and slice worlds (Figures 3.13a–c and 3.14) are very simplistic representations of the Earth, but do help to show how continental configurations might influence ocean current patterns and hence climate.

■ Which configuration – cap world, ring world or slice world – would affect the surface current pattern so as to intensify cold conditions in polar regions?

■ Cap world, where currents carrying heat from lower latitudes would be unable to penetrate to very high latitudes, with the result that there would be a stronger temperature contrast between equatorial and polar regions.

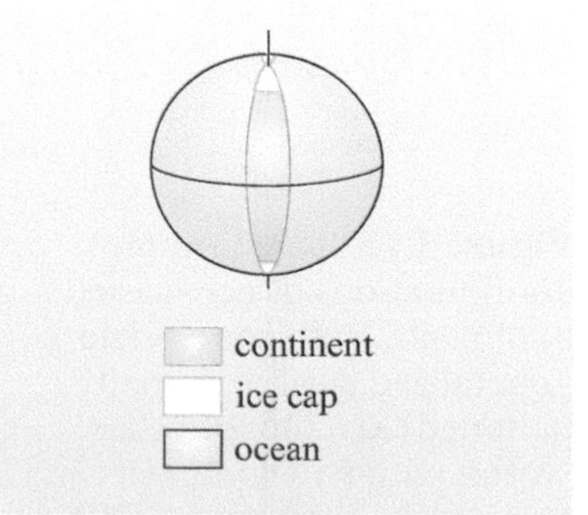

Figure 3.14 A 'slice world', showing the longitudinal configuration of one of the continental areas and parts of two polar ice caps.

In a cap world, the polar regions would be thermally isolated, making the development of polar ice caps more likely: strong eastward currents, comparable with today's Antarctic Circumpolar Current (Figure 1.31), could flow around the polar continents under the influence of westerly winds, further isolating the polar continents from warm currents flowing from low latitudes.

- Look carefully at Figure 1.31. In today's world, are northern polar regions similarly isolated?

- No, but they are nevertheless largely cut off from warm currents flowing from lower latitudes. The Arctic Ocean is almost completely surrounded by land, with the result that the only warm water penetrating the region is the North Atlantic Drift (the downstream extension of the Gulf Stream), which flows northwards through the Norwegian and Greenland Seas.

By contrast, in a tropical ring world (Figure 3.13c), ocean circulation between tropical and polar regions would be possible, so heat could be transferred from mid-latitudes to high latitudes, resulting in lower meridional temperature contrasts.

Figure 3.15 shows three more configurations of oceans and continents, not quite as simple as those shown in Figures 3.13 and 3.14. In the past, the configuration of oceans and continents has been very different, so it is worth using these simple models to aid your thinking about how current patterns may affect the distribution of temperature over the Earth's surface.

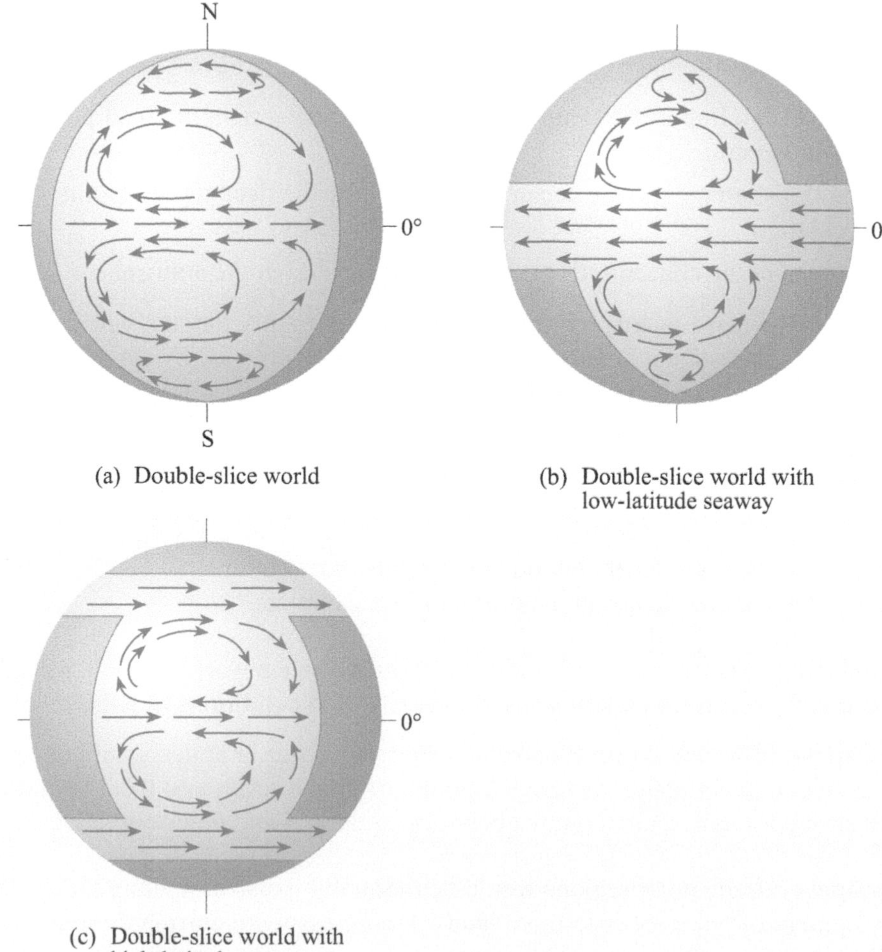

Figure 3.15 Three simple configurations of oceans and continents, with appropriate generalised surface current patterns: (a) a 'double-slice' world, with two continents and two oceans (only one visible here); (b) as (a), but with open oceans around the globe at low latitudes; (c) as (a) but with open ocean around the globe at relatively high latitudes. (Van Andel, 1985)

- On the basis of what you know about winds and currents on the real Earth, you could have sketched surface current patterns similar to those shown in Figure 3.15. Identify features of the circulatory pattern, visible in all three configurations, that are likely consequences of fundamental characteristics of the rotating Earth and its fluid envelopes.

- Features you may recognise as being fundamental to a rotating Earth are: anticyclonic subtropical gyres; cyclonic subpolar gyres; poleward-flowing western boundary currents like the Gulf Stream; westward-flowing North and South Equatorial Currents. Note that only configurations (a) and (c) have eastward-flowing Equatorial Counter-currents, because these are a result of westward flow in the vicinity of the Equator being deflected back *along the Equator* (where the Coriolis force is zero) by the western boundary (which is not present in configuration b).

- Would you expect the temperature distributions over the oceans in Figure 3.15 to be symmetrical, with temperatures along the western boundary the same as those along the eastern boundary at similar latitudes?

- No, the temperature distribution would not be symmetrical in any of the configurations shown. As discussed in connection with Figure 1.31, over much of an ocean, water warmed at low latitudes flows polewards along the western side, while water cooled at high latitudes flows equatorwards in the eastern part of the ocean. This results in the western sides of oceans being generally warmer than the eastern sides (although, at present, the northeasterly flow of the Gulf Stream causes the northeastern North Atlantic to be warmer than the northwestern part).

Of course, discussing the transport of heat around the Earth by means of currents alone is unrealistic as the redistribution of heat by winds, including the effects of evaporation, transport and condensation of water vapour, is being ignored. Nevertheless, the high specific heat of water (Box 1.3) means that heat transport in the ocean is an extremely important influence on climate.

Implicit in the discussion of Figure 3.15 is the effect of gateways on the pattern of surface and deep ocean currents. Gateways are gaps between the continents that permit significant longitudinal or latitudinal connections to be made between oceans. Their opening and closing, and the resulting changes in heat transport, can cause the climate for a particular land mass to change much more rapidly than the slow drift of the continent across climatic belts.

Mention of deep currents should remind you that an extremely important aspect of the oceanic heat transport system has not yet been considered. The effect of wind-driven surface currents in transferring heat from low to high latitudes is reinforced by the transport of cold, deep water away from polar regions in the density-driven thermohaline circulation (Section 1.4.3). It may be facilitated by the opening of gateways in the deep ocean, or impeded by the development of topographic barriers such as mid-ocean ridges and seamounts. Barriers and gateways may be affected by crustal uplift or volcanic activity on the one hand, and subsidence related to lithospheric plate movements on the other.

3.4.2 The break-up of Pangaea

Having used simple models to consider some possible effects of different continental configurations on ocean circulation and global climate, you can now consider how events in the break-up of the supercontinent known as Pangaea (Greek for *all land*) might have influenced climate change. Pangaea formed at ~250 Ma, when what is now North America, most of Europe and Asia came together with the pre-existing southern supercontinent (comprising South America, Africa, India, Antarctica and Australia), which is referred to as Gondwana. Gondwana had been extensively glaciated about 50 million years before (i.e. at 300 Ma) but, by 250 Ma, the Earth was probably already beginning to warm up, and by 100 Ma it was a 'greenhouse' planet some 10 °C warmer (on average) than today. As Pangaea broke up and the continents separated, the Earth subsequently began to cool again, albeit gradually, to arrive at its present 'icehouse' state.

Before Pangaea began to break up, the supercontinent must have been accompanied by a superocean. This superocean (which has been named Panthalassa, Greek for *all ocean*) extended from the North Pole to high southern latitudes (about 50–60° S), and is thought to have extended for some four-fifths of the Earth's circumference around the Equator. There is no reason to suppose that the factors determining the global wind and surface current systems were any different from what they are today, and some likely wind and current patterns have been proposed. If present-day oceanic current patterns are any guide, there could have been a number of linked subtropical gyres in each hemisphere (the present-day Pacific Ocean has a more complex gyral system than the narrower Atlantic; Figure 1.31), and presumably also a number of subpolar gyres.

Figure 3.16 shows a series of maps that illustrate successive stages in the break-up of Pangaea over the past 175 Ma. On the basis of what is known about global winds and the effects of land masses on winds and currents, we can not only infer surface current patterns but also propose regions where deep and bottom water masses might have formed.

The break-up of Pangaea began at about 200 Ma, and an equatorial gateway started to open within the supercontinent. By 175 Ma, in Jurassic times (Figure 3.16a), circum-equatorial flow was blocked by a relatively narrow isthmus at what is now Gibraltar, and the ancestral Mediterranean (known as the Tethys Ocean) had formed. Surface flow is shown as westwards on both sides of this isthmus in the reconstruction. It is possible that, east of Gibraltar, the return flow eastwards was at depth, in the form of a dense water mass.

■ On a relatively warm Earth, how might surface water at low latitudes be made sufficiently dense to sink?

□ High rates of evaporation would remove freshwater and so increase the salinity.

In fact, in today's Mediterranean, a warm saline deep water mass is formed in this way, aided by some cooling in winter. It is thought that by 160 Ma, similar saline but relatively warm water masses could have been forming in shallow

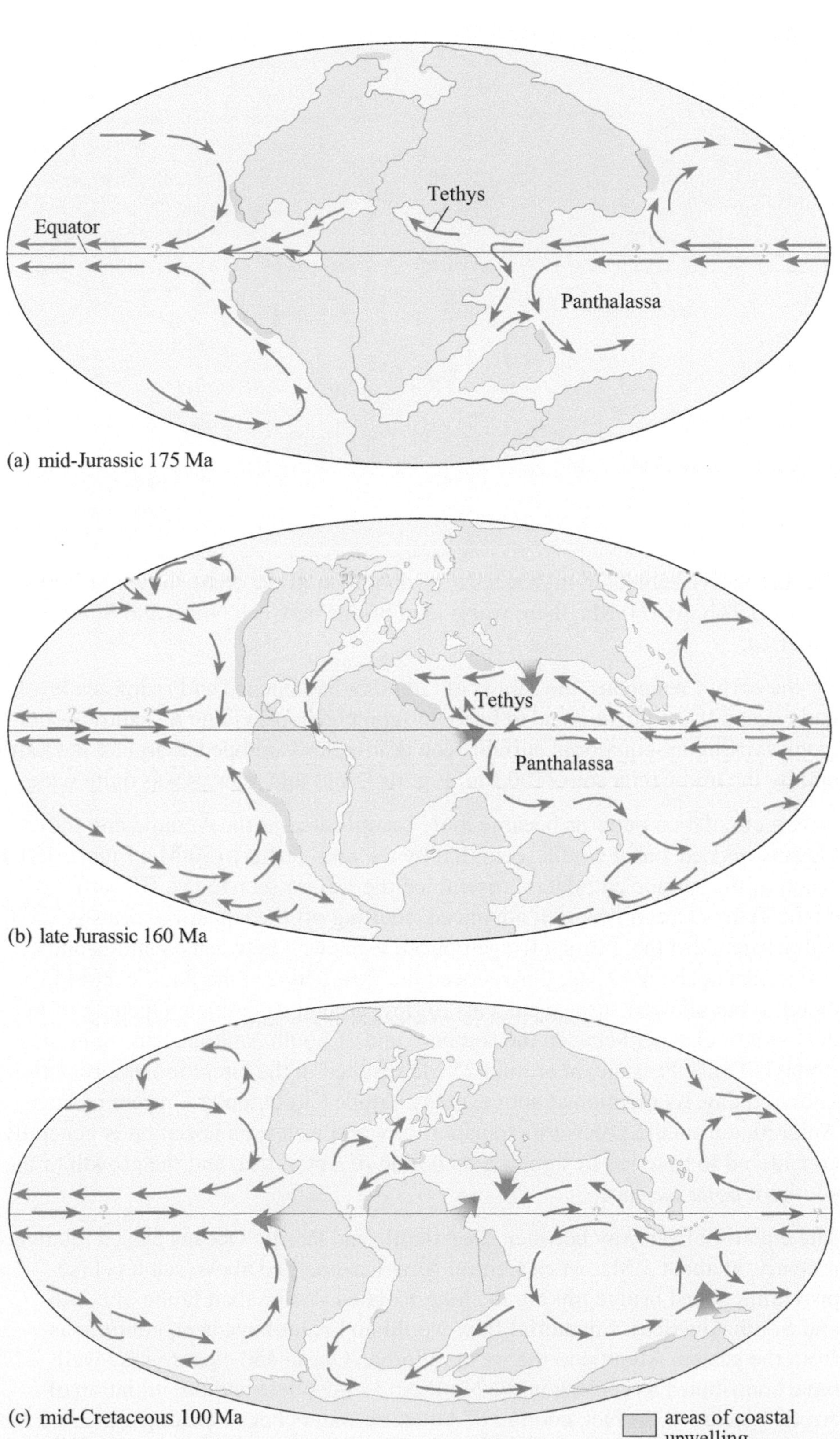

Figure 3.16 Maps showing the changing distribution of the continents and changing surface current patterns during the break-up of Pangaea (no attempt has been made to show actual coastlines or shelf seas). In (b) and (c), broad arrows indicate locations where warm, saline water masses might have flowed down into the ocean from shallow evaporating basins. (a)At 175 Ma, there was no Atlantic Ocean, only the great ocean Panthalassa. (b) By 160 Ma, the North Atlantic was a long narrow ocean but the South Atlantic had barely begun to open. (c) By 100 Ma, the opening of the Straits of Gibraltar and the submergence of 'Central America' had provided a low-latitude seaway. (d) By 30 Ma (overleaf), the oceans were approaching their present configuration. Gyres were now established in the North and the South Atlantic, and currents could flow unimpeded around Antarctica under the influence of westerly winds. The climate was by now cooler, and cold bottom waters formed near Antarctica and flowed north into all the ocean basins – the equivalent of today's Antarctic Bottom Water. (Haq, 1984)

Figure 3.16 (continued)

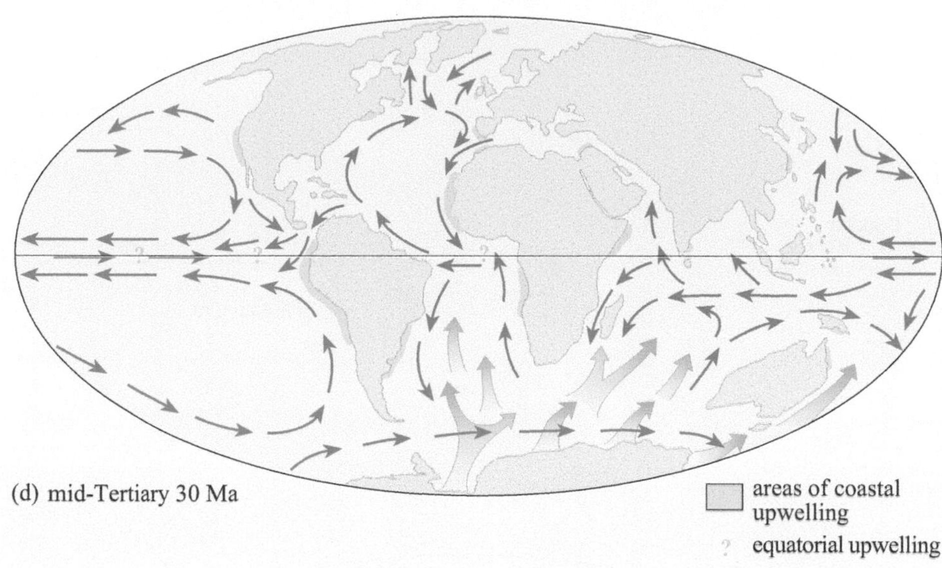

coastal basins, subject to high rates of evaporation in the areas shown in Figure 3.16b. At 160 Ma, there was a land bridge between North and South America.

By the early Cretaceous, the Straits of Gibraltar had opened and rising sea levels had caused 'Central America' to become submerged, providing a shallow-water gateway. Circum-equatorial currents could now flow unimpeded around the Earth, and by the mid-Cretaceous (100 Ma; Figure 3.16c) this seaway was quite wide.

Ocean circulation patterns became more complicated as the Atlantic and Indian Oceans opened, but it would seem that by the mid-Tertiary (30 Ma; Figure 3.16d) much of the surface circulation resembled the pattern seen today. Closure of the Tethys Ocean was well advanced, shutting off the equatorial seaway. Submergence of the Tasman Ridge opened a gateway between Australia and Antarctica at about 40 Ma; this reduced the flow between the Pacific and Indian Oceans, but allowed surface currents to flow around Antarctica. Opening of the deep-water channel between the southern end of South America and Antarctica (today's Drake Passage) at around 25 Ma resulted in the circumpolar circulation known today. As mentioned above, the Antarctic Circumpolar Current isolates Antarctica from the polewards transport of warm water; its initiation is generally considered to have led to significant cooling of Antarctica, and the growth of the southern polar ice cap.

The equatorial gateway between the Atlantic and Pacific Oceans closed relatively recently, at about 3 Ma, when Central America emerged above sea level (so providing a land bridge linking the long-isolated mammalian fauna of North and South America). Equatorial flow would no longer have been continuous from the eastern Atlantic to the western Indian Ocean and closure may well have contributed to cooling in the Northern Hemisphere and the initiation of Arctic glaciation. In fact, cooling of Antarctic waters began as long ago as 60 Ma earlier (Figure 3.17), possibly in response to closure of the 'Tethyan gateway' at Gibraltar. Global temperatures roughly stabilised from then until the early Miocene, when the Drake Passage opened and Antarctica became isolated.

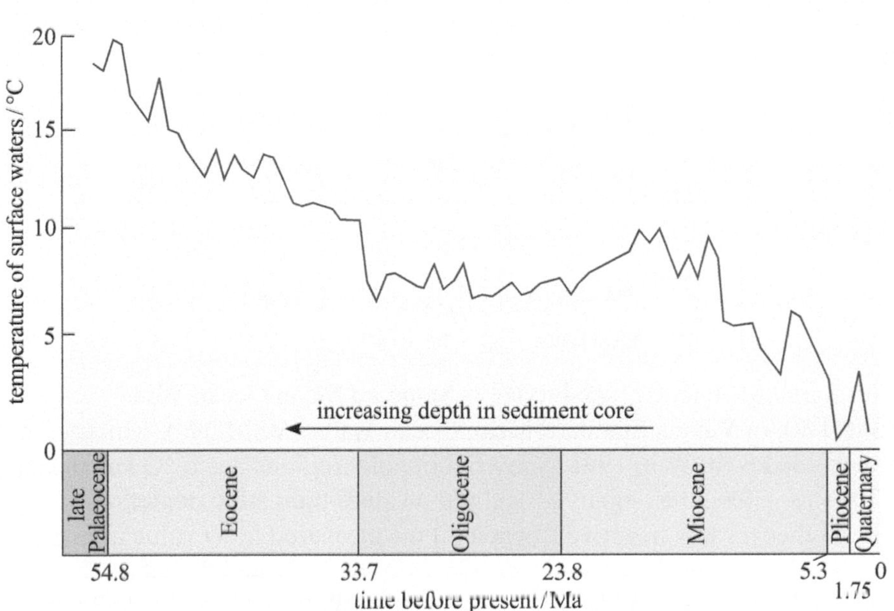

Figure 3.17 Surface water temperatures in the vicinity of Antarctica during the past ~60 Ma, as determined from oxygen isotope ratios of the remains of planktonic foraminiferans. See Box 3.2 for an explanation of the use of oxygen isotopes as a proxy for past sea temperatures. (Kennett et al., 1974)

Box 3.2 Oxygen isotopes and the climate record

Oxygen has three stable isotopes with relative atomic masses 16, 17 and 18. Over 99% of natural oxygen is made up of ^{16}O, with most of the balance being ^{18}O. Water that evaporates from the ocean eventually condenses as cloud and falls as rain or snow. When seawater evaporates from the ocean, water molecules with the lighter oxygen isotope ($H_2{}^{16}O$) evaporate more readily, so atmospheric water vapour is relatively enriched in the lighter isotope. When water vapour condenses and is precipitated back into the ocean, the water containing the heavier isotope ($H_2{}^{18}O$) condenses preferentially. Both processes deplete water vapour in the atmosphere in $H_2{}^{18}O$ relative to $H_2{}^{16}O$. When ^{18}O-depleted water vapour is precipitated as snow in polar regions, the snow will also be depleted in ^{18}O relative to the oceans. The larger the ice caps, the higher the relative proportion of ^{18}O in seawater and the lower the relative proportion of ^{18}O in ice caps.

Marine organisms that form hard parts (shells or skeletons) of calcium carbonate incorporate different proportions of ^{16}O and ^{18}O from dissolved ions in the seawater around them according to the temperature: the lower the temperature, the greater the $^{18}O : {}^{16}O$ ratio in the calcium carbonate secreted. Importantly, all organisms secreting calcium carbonate have higher $^{18}O : {}^{16}O$ ratios in cold than in warm water, although some species shift the isotope ratio by greater amounts than others. This shift in the calcium carbonate $^{18}O : {}^{16}O$ ratio modifies the ratio inherited from seawater, which, as described above, itself varies according to global ice volume.

The fossils in sediment cores used for oxygen-isotope analysis are usually microorganisms with calcium carbonate shells, often foraminiferans. The amount of ^{18}O in their shells is very small, but it can be measured accurately by mass spectrometry. The result is not given as a simple ratio, but as a delta (δ) value, which is determined by comparison of the sample with a standard, and results in a value expressed in parts per thousand (‰ or 'per mil'):

$$\delta^{18}O = \left[\frac{(^{18}O/^{16}O_{sample})}{(^{18}O/^{16}O_{standard})} - 1 \right] \times 1000 \tag{3.2}$$

The standard generally used today is Standard Mean Ocean Water (SMOW), or Vienna Standard Mean Ocean Water (VSMOW), which superseded SMOW in 1995. Snowfall in polar regions has $\delta^{18}O$ values of –30‰ to –50‰, the negative (or 'light') values indicating depletion of ^{18}O. The higher (or less negative, 'heavier') the measured $\delta^{18}O$ value in marine fossils, the greater the enrichment of ^{18}O in seawater and the larger the ice caps on land at the time the organisms were alive.

$\delta^{18}O$ values may be determined for the hard parts of both planktonic and benthic species of foraminiferans.

■ Bearing in mind that oxygen-isotope ratios are affected by the temperature of the water in which an organism lived, which would give the most reliable estimates of global ice volumes over the course of glacials and interglacials: planktonic foraminiferans or benthic foraminiferans?

□ Benthic foraminiferans, because the temperature variation of the cold bottom waters is less than that of surface waters.

Useful information about past climate can also be obtained from planktonic foraminiferans, particularly those living at very high latitudes where seasonal temperature variations are quite small. Their $\delta^{18}O$ values reflect changes in ice volume and global surface temperature and, of course, when global temperatures are lower, ice caps are larger.

As well as changing ocean circulation, the evolution of global climate since the break-up of Pangaea would also have been strongly affected by:

- the concentrations of CO_2 in the atmosphere, influenced by the changing rates of subduction through time, which affects volcanic eruption rates
- the intensity and distribution of rainfall and the hydrological cycle.

These factors in turn influence the distribution of living organisms, which affects chemical weathering rates, a major sink for CO_2 over geological timescales (Section 2.3.3). It has been suggested that during the existence of supercontinents such as Pangaea, there are relatively low levels of volcanic outgassing delivering less CO_2 to the atmosphere. A 'runaway' cooling effect is prevented largely by the arid conditions within the supercontinent, yielding low chemical weathering rates, and hence a lower rate of removal of CO_2 from the atmosphere. During periods of continental dispersal (as today), there are increased levels of volcanic activity but higher levels of continental weathering due to wetter conditions. You will see in Chapter 4 that the topography of the continents also plays a significant role.

Climate models can include all these effects but, because of the vast range of timescales involved, models that accurately represent short-term atmospheric circulations, for example, ignore longer-timescale carbon-cycle processes. Similarly, models that represent changes on very long timescales include only an approximation of faster processes such as atmospheric circulation.

3.4.3 Plate tectonics and life

Plate tectonic processes can have more direct effects than simply those of moving continents relative to climatic belts, transporting species around the globe or causing them to become isolated. Over relatively short periods of time they can also alter the relative proportions of *types* of environment available as living space.

In considering how ocean circulation has been affected by changing continental configurations, one aspect of great importance for life, namely upwelling, was omitted.

■ Why is upwelling of particular importance for life? Suggest how it might be possible to deduce where upwellings occurred in the distant past.

■ Upwelling is important for life as it brings nutrient-rich water up into the photic zone where it can support high levels of primary production, i.e. large phytoplankton populations and hence, directly and indirectly, support other organisms. Areas where it might have occurred in the past could be deduced from the ways in which the positions of the continents affect the wind field (Figure 1.28), and from the distribution of organic remains in marine sediments. High productivity results in large fluxes of organic debris to the seabed, some of which may be preserved in the sedimentary sequence.

Areas where coastal upwelling may well have occurred in the Jurassic and mid-Cretaceous are shown in Figure 3.16a–c. Those along the western sides of the 'American' continents would have resulted from the same equatorward winds (the equivalent of the present-day Trade Winds) that drove the currents along the eastern side of the gyres (Figure 1.36).

Although the open oceans support a large plant biomass, the continental shelves support a much higher biomass per unit area (Table 2.1). Figure 3.18 (overleaf) illustrates how the numbers of families in nine phyla of invertebrate shelf-dwelling animals have varied since the beginning of the Cambrian (~542 Ma). Compare the shape of the curve with the diagrammatic representation of changes in the numbers of continents over the same period. Note that stages D to F correspond to the coming together and break-up of Pangaea, shown in Figure 3.16.

■ Is the information in Figure 3.18 consistent with the hypothesis that the global diversity of shelf-dwelling animals is related at least in part to the availability of shores and shallow seas?

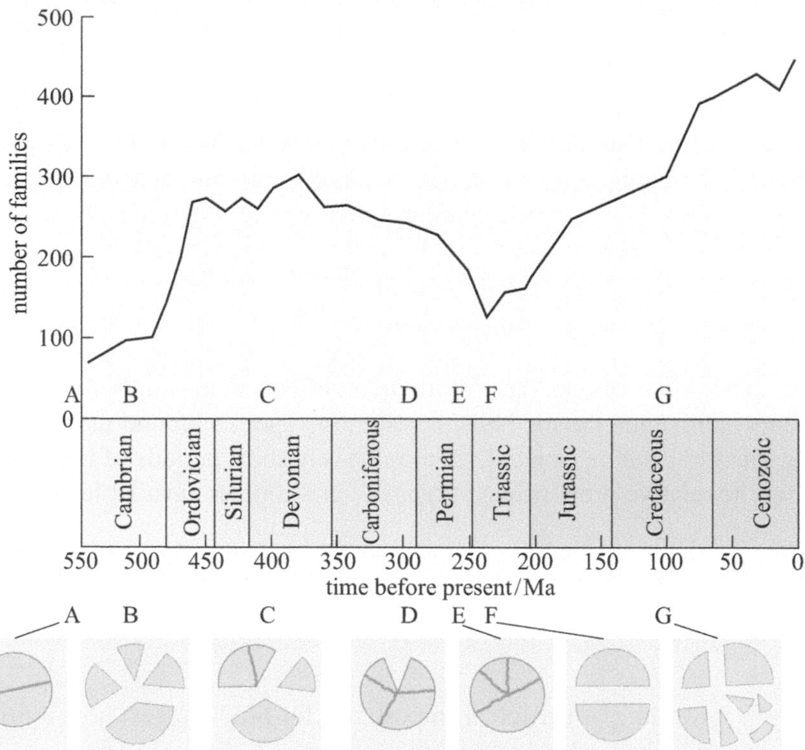

Figure 3.18 Changes in the number of families in nine phyla of benthic shelf-dwelling invertebrate animals with hard parts, and changes in the number of continental fragments (shown diagrammatically), from the Cambrian (~542 Ma) to the present. Positions of letters on the graph indicate approximate times when the continents had the numbers of fragments shown (below). (Valentine and Moore, 1970)

- Yes. In general, the greater the number of continental fragments, the greater the total length of shoreline available for marine animals to colonise, and the greater the number of separate, isolated environments in which endemic coastal faunas could evolve. The number of families increased from Cambrian to Ordovician times as an ancient supercontinent broke up (A, B), remained more or less constant until Pangaea began to be assembled (C, D, E), and decreased dramatically during the Permian, when the total length of shoreline was at a minimum. Thereafter, as Pangaea fragmented (F, G), diversity rose sharply once more.

Clearly, the correlation is a fairly general one and shoreline length is only one possible factor in determining the diversity of shelf-dwelling organisms. It would be unsound, for example, to attribute the great Permian extinction of species (responsible for the dip at F) mainly to this cause, although some scientists think it could have been a contributory factor. Perhaps the most important factor affecting the distribution and nature of shelf-dwelling organisms is sea level, which is examined in the next section.

3.5 Sea-level changes: causes and consequences

Plate tectonics has a profound effect on sea level. It does this in a myriad of ways, such as changing the sizes and distribution of ocean basins, by promoting uplift of land masses and by impacting on climate and so changing the size of the polar ice caps.

- Why is sea level related to ice volume?
- Water evaporated from the Earth's surface and transported polewards may fall as snow at high latitudes, where it can accumulate to form ice caps. Since sea ice displaces its own mass of water, its growth or decline does not alter sea level. By contrast, continental ice sheets such as Antarctica remove water entirely from the oceans as they grow, so sea level falls.

Sea-level changes are made up of two components:

1 **Eustatic changes** of sea level are worldwide changes that affect all oceans and have the potential to cause global climatic changes. Over the past 2 Ma or so, such changes have been caused largely by ocean water becoming frozen into, or melted from, continental ice caps. Over longer timescales, as hinted above, eustatic changes of sea level are caused by other mechanisms (this is discussed in Section 3.5.1).

2 **Isostatic (or epeirogenic) changes** of sea level are caused by vertical movements of the crust. Such movements may be caused by changes in the thickness and/or density of the lithosphere, and by loading or unloading with ice or sediments. Such changes cause the lithosphere to ride higher or lower on the underlying asthenosphere, rather as blocks of wood may float higher or lower in water (Box 3.3).

Box 3.3 Isostasy

Consider a block of wood floating in water (Figure 3.19). Notice the thicker the wooden block, the greater the thickness of wood that emerges above the water. Similarly with an iceberg: the larger it is, the more of it can be seen above the sea-surface. The tendency for the Earth's lithosphere to behave in a similar manner with respect to the underlying asthenosphere is known as **isostasy**.

Continental crust (i.e. the upper part of continental lithosphere) is mostly granitic in composition. Its average thickness is about 40 km, but beneath mountain ranges it can be as much as 90 km thick. Oceanic crust is mostly basaltic in composition with an average thickness of about 7–8 km, and it is denser than continental crust. By analogy with the blocks of wood and icebergs, at isostatic equilibrium, continental lithosphere 'rides' or 'floats' higher on the underlying asthenosphere than oceanic lithosphere, which is why the ocean floors are below sea level.

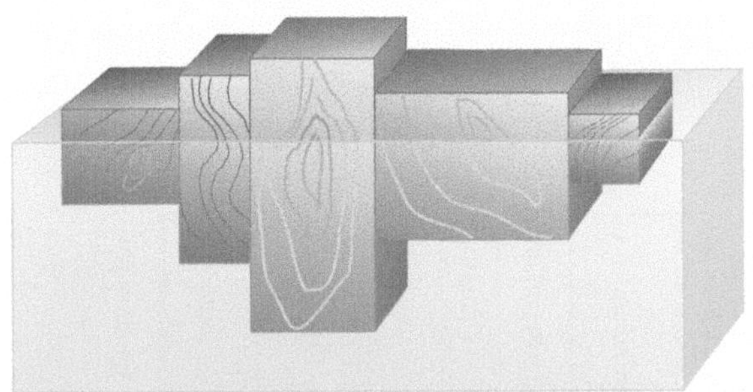

Figure 3.19 Wooden blocks floating in water: an approximate analogy for continental lithosphere of different thicknesses, illustrating why mountains have deep 'roots'.

The analogy with wooden blocks is a simplification because the real lithosphere increases in density with depth but, for the purposes of this discussion, you can imagine that the density of the blocks is equivalent to the average density of the lithosphere.

Crustal thickening occurs during mountain building (through magmatic intrusions and/or convergence of plates at subduction zones). This is why mountains are high. Crustal thinning occurs by stretching during continental break-up, and results in subsidence of the crust which may continue over long periods of time, allowing thick sequences of sediment to accumulate. Loading the lithosphere (by infilling a basin with sediment or by piling ice on top of the crust) will cause its surface to be lowered, and unloading it (by eroding a mountain belt or by melting an ice sheet) will cause it to rise.

Continental shelves and shelf seas

As outlined above, continental crust is thinned by stretching during the break-up of continents, and it therefore subsides. This explains why many continental margins are low lying, with coastal plains bordered by shallow waters (about 200 m deep on average) overlying continental shelves, which in many cases can be extensive enough to be called shelf seas – the North Sea and the Baltic are good examples. Beyond the edge of the continental shelf, where the thinned continental crust ends and oceanic crust begins, water depths increase into the ocean basins up to a few kilometres deep. Substantial thicknesses of sediment can accumulate on continental shelves, deposited from rivers running off the adjacent land. This additional load upon the crust causes it to continue subsiding isostatically, but average water depths remain about 200 m.

Continental shelves are important in the context of sea-level changes because, as they typically have a rather flat topography, relatively small rises or falls can flood or expose substantial areas of shelf.

Figure 3.20 compares changes in sea level and (estimated) global temperature over the past 540 Ma. During past glacial periods, sea level was as much as 150 m below its present level whereas during the Cretaceous it was considerably higher than today. The plots of sea level and global temperature change (warming and cooling) correlate reasonably well during the Jurassic, Cretaceous and Tertiary. Earlier in Earth history, the correlation between the two curves is less good, although cooling during the Carboniferous is initially matched by falling sea level.

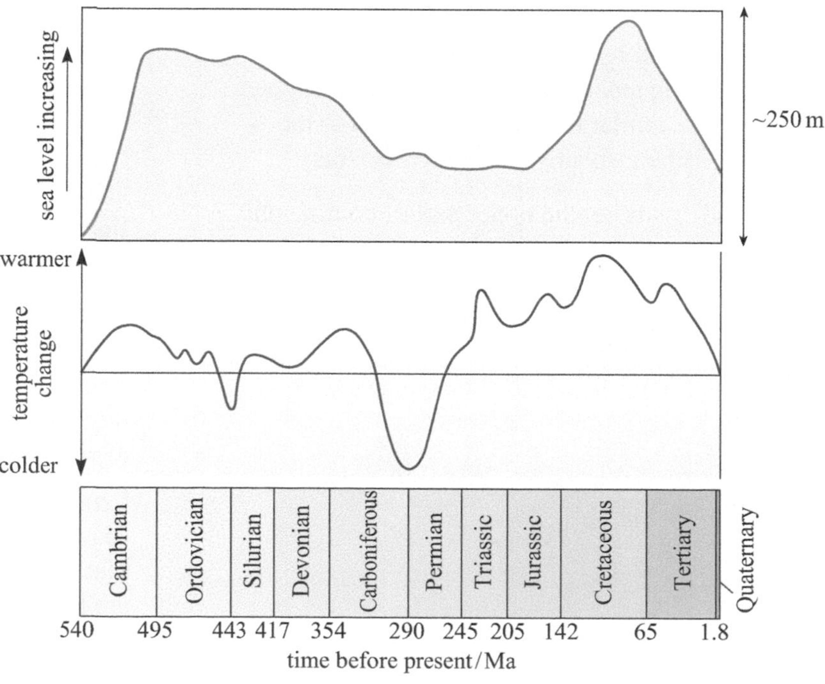

Figure 3.20 Variations in sea level and average global temperature during the past 540 Ma (the Phanerozoic). Sea level has been determined from variations in the extent of shallow-water sediments (limestones, sandstones, mudstones, etc.).

When examining Figure 3.20, you might have wondered whether climate controlled sea level during the Phanerozoic, or whether sea level controlled climate. It is fairly clear that climate has controlled sea level during the last two million years of the Phanerozoic, i.e. during the current Ice Age. Global cooling triggered the growth of polar ice sheets which removed large amounts of water from the oceans to the land, resulting in a drop in sea level. Global warming leads to melting of the ice, and sea level rises again. In addition, changes in the global mean temperature of the oceans will increase or decrease the water volume by thermal expansion or contraction. Expansion and contraction cause much smaller sea-level changes than the formation or melting of ice sheets: for example, an increase of 10 °C throughout the water column in all oceans would cause a eustatic sea-level rise of about 10 m.

An important aspect of sea-level change, however, is that it has a significant effect on the total area of 'emergent' continental crust, i.e. of continental crust forming land above sea level. Today, 30% of the Earth's surface is emergent continental crust, but during glacial periods this increased to as much as about 35%, whereas in the Cretaceous (Figure 3.20) it was probably as little as about 25%. The reason for such large changes in area above or below sea level is the extensive areas of continental shelf and coastal plains, of which large areas can be flooded or exposed by even relatively small rises or falls in sea level.

Question 3.5

How could changes in relative areas of land and sea either reinforce or counteract climate changes caused by other forcing factors? Think back to the discussion in previous sections concerning (a) the opening and closing of oceanic gateways and (b) the Earth's albedo.

Implicit in the answers to these questions is the existence of feedback mechanisms linking sea-level change and climate. The corollary of this is that if sea level were raised or lowered by some process *independent* of climate, then the changing sea level might itself initiate climate change. To explore this aspect, it is important to consider other processes that can affect sea level.

3.5.1 Causes of eustatic sea-level change

There are two basic processes that change global sea level. The first involves changing the volume of water filling the ocean basins, and the second results from changes in shape and size of the ocean basins themselves.

Changes in ocean water volume

The principal cause of changes in ocean water volume is the formation and melting of ice caps. If the entire present-day Antarctic ice sheet melted, global sea level would rise by 60–75 m. Disappearance of the Greenland ice sheet would add about another 7 m. However, over tens of thousands of years, the load of the extra water in the oceans would depress the ocean crust and increase the depth of the ocean basins, so the overall rise would be in the range of 40–50 m. As already mentioned, an increase of 10 °C in the average temperature of water in the oceans would raise sea level by about a further 10 m through expansion of seawater.

Changes in shape and size of ocean basins

It has been proposed that a major cause of rising sea level could be an increase in the rate of formation of oceanic crust at spreading axes, which would increase the volume of ocean ridges, thus displacing water onto the continents. This suggestion needs to be treated with some caution for the following reason. Sea-floor spreading is believed to be the major means of heat loss from the Earth's convecting interior: the rate of heat loss has declined since the Earth formed about 4.6 Ga ago, so any relatively short-term increases in heat loss would have been superimposed on the extremely long-term decrease. None the less, many scientists do consider the expansion of ridges associated with increased rates of spreading to be a major cause of sea-level rise.

Another explanation for rising sea levels – and a much simpler one – involves simply displacing water from a contracting deep basin into a number of shallow ones. Consider the break-up of Pangaea (Figure 3.16a), with new oceans opening between the dispersing continental fragments and the Panthalassa Ocean contracting as the continental fragments disperse. A shrinking deep basin would not become significantly deeper, so water in it would have to be displaced elsewhere. If it could only be displaced into a small number of other basins (newly forming and still shallow) then the net result would be a global rise in sea level. Eventually, deepening of the new ocean basins would become more rapid than shrinkage of the original Panthalassa Ocean, which would approach the dimensions of the present-day Pacific. This probably occurred in the mid- to late-Cretaceous (at about 100 Ma) when sea level was at a maximum (Figure 3.16c).

Distributions of shallow-water marine sediments suggest that at that time global sea level could have been ~200 m higher than at present (Figure 3.20), and some estimates are much higher than that. It seems likely that such high sea levels might have been at least partly caused by production of new ocean floor, despite the reservations discussed above concerning spreading ridges.

The oceanic crust can also grow in response to the eruption of enormous amounts of basaltic lavas to form oceanic plateaux (Figure 3.10) and significant volumes of rock were known to be added to the oceanic crust in this way during the Cretaceous.

- What side effect of the increase in volcanism during the Cretaceous might have had an additional effect on sea level?

- The release into the atmosphere of large amounts of carbon dioxide, leading to 'greenhouse warming'.

The rise in sea level in response to greenhouse warming would have been modest, because it would have been due only to thermal expansion of ocean waters. There would have been no melting of ice caps, as none were present at that time.

Note that, although CO_2 would have been *directly* supplied to the atmosphere by eruptions above sea level (e.g. those forming volcanic islands and subduction zone volcanoes), on timescales of 1000 years or so, much of the CO_2 released into the ocean at mid-ocean ridges (through hydrothermal vents and volcanoes) could also escape to the atmosphere. In other words, on geological timescales it may not matter whether the volcanoes are above or below sea level.

Eustatic sea-level rise can also be caused by the deposition of oceanic sediments transported from continental areas because this will effectively decrease the size of the ocean basins by displacing water (although this effect may be offset by isostatic depression of the crust by the weight of the sediment). Sea-level *falls*, on the other hand, could occur during the aggregation of supercontinents (the reverse of the break-up of Pangaea, discussed above), when continental collisions result in thickened continental crust and the formation of isostatically elevated mountains (Box 3.3).

3.5.2 Sea level, climate and atmospheric CO_2

Earlier in Section 3.5, it was proposed that changes in sea level might lead to secondary effects which would, in turn, affect climate. For instance, changing albedo is a positive feedback, leading to instability in the climate system, while changing currents due to gateways opening or closing could be a positive or negative feedback. Both these effects are essentially physical in nature, so how do chemical and biological consequences of sea-level change compare? In the following discussion and question, three aspects of the Earth system that are biologically and chemically sensitive to changes in sea level will be examined, all of which exert a strong influence on climate.

The first is continental weathering and its effect on atmospheric CO_2 (Figure 2.21). During periods of low sea level, larger areas of continental crust (notably silicate minerals) are exposed to weathering processes, which will tend to increase the rate of removal of atmospheric CO_2, further reinforcing any tendency towards global cooling. The upper parts of mountains are mainly sites of vigorous *physical* weathering. However, the weathering products are carried away (exposing yet more rock surface to be weathered) and are deposited on the warmer lower slopes and coastal plains, where they are subjected to chemical weathering – particularly if there is abundant vegetation.

When sea levels rise, the area of coastal plains available for (chemical) weathering is reduced, so the rate of removal of atmospheric CO_2 will tend to decrease. Consequently, the concentration of CO_2 in the atmosphere would eventually begin to rise again, favouring global warming. (Note, however, that unless carbon added to the ocean through weathering is preserved in sediments such as carbonaceous deposits and limestones, it will be returned to the atmosphere in 1000 years or so.)

■ When sea-level changes affect climate through continental weathering, is this a positive or a negative feedback mechanism, and what general effect might it be expected to have on the climate system?

□ Considered in isolation, the effect of sea-level change on climate through weathering is one of positive feedback. As discussed in Chapter 1, positive feedback mechanisms lead to instability in the climate system.

Question 3.6 asks you to examine two further mechanisms, both of which involve parts of the biosphere, in a similar way to continental weathering.

Question 3.6

(a) Suggest possible ways in which changing the relative areas of land and sea might indirectly influence climate through the resulting changes in:

(i) the areal extent of environments found in low-lying coastal regions

(ii) the areal extent of shallow seas.

(b) Explain whether each of the feedback mechanisms you identified in (a) is positive or negative, and hence what general effect it might be expected to have on the climate system.

Of course, in reality no feedback effect resulting indirectly from sea-level change would be acting on its own, and the real situation would be much more complicated than suggested here. Nevertheless, it is interesting that the two negative feedbacks that are best understood are those that involve the activities (i.e. lives and deaths) of organisms, through their effects on the global carbon cycle. This finding might be consistent with a 'Gaian' view of the world where the activities of organisms contribute to keeping conditions (in this case temperatures) within a range suitable for life.

Summary of Chapter 3

1 Volcanoes are conduits linking deep mantle processes to atmospheric composition and hence climate; they exhale large amounts of water, CO_2 and SO_2. Over geological timescales, volcanoes have an important role in the carbon cycle, but in the recent past their contribution to atmospheric CO_2 has been overwhelmed by the anthropogenic flux.

2 The radiative effects of major explosive eruptions can be large, reducing the amount of solar radiation reaching the Earth's surface by more than 10%. The effects of most explosive eruptions, however, are short-lived (2–3 years), as aerosols fall out of the stratosphere. Due to the long response time of the Earth's climate system as a whole, brief volcanic events do not have a significant effect on climate.

3 Eruptions of flood basalts such as those of the Columbia River Province may involve effusion of more than 1000 km^3 of sulfur-rich basalt lava over periods of 10–100 years. As these eruptions are sustained over longer periods than great explosive events, their radiative and environmental effects may be more profound. The close coincidence in timing between great episodes of flood basalts in Earth history with major mass extinctions, such as those at the Cretaceous–Tertiary and Palaeozoic–Mesozoic boundaries, suggests that mass extinctions may be linked with the environmental and climatic effects of the flood basalts.

4 Climate models using highly simplified continental configurations suggest that a tropical 'ring world' would be significantly warmer than a polar 'cap world'. It is thought that the geography of the Earth may have approximated to the former between 700 Ma and 600 Ma, and to the latter (with one polar ice cap) during the late Carboniferous.

5 It is thought that the changing distribution of continents in response to plate-tectonic processes affects global climate on a million-year timescale, through its effect on the radiation budget, and hence indirectly on the hydrological cycle and weathering (which affects the CO_2 concentration of the atmosphere). Furthermore, the closing and opening of oceanic gateways as continents change their relative positions has a significant effect on shallow and deep oceanic circulation, and can contribute to global warming and cooling on significantly shorter timescales.

6 The break-up of Pangaea began at about 200 Ma. The opening of a low-latitude seaway may have contributed to global warming in the Cretaceous, at about 100 Ma. Around this time, deep water masses were probably warm and very saline, having formed at low latitudes. It is thought that thermal isolation resulting from the initiation of the Antarctic Circumpolar Current at 25 Ma accelerated cooling of Antarctica and the growth of the south polar ice cap.

7 Relative sea-level change results from a combination of eustatic and isostatic sea-level change. Eustatic changes are global in extent, whereas isostatic changes result from local or regional uplift or subsidence of the lithosphere. Eustatic sea-level changes are due either to changes in the volume of ocean waters (resulting from the formation and melting of ice sheets), or to changes in the size and shape of the ocean basins (resulting either from the formation of new oceanic crust, notably as submarine plateaux, or from the replacement of a few large ocean basins by a number of smaller ones). During periods of global warming, some sea-level rise is attributable to expansion of the ocean water; however, an increase of 10 °C throughout the water column in all oceans would lead to a sea-level rise of only ~10 m.

8 Global warming during the Cretaceous may have been related to the addition to the atmosphere of huge amounts of CO_2 as a result of volcanism (as flood basalts and as a consequence of increased rates of sea-floor spreading). However, the warming was counteracted by the removal of atmospheric CO_2, through deposition and preservation of carbon in the ocean.

Learning outcomes for Chapter 3

You should now be able to demonstrate a knowledge and understanding of:

3.1 The influence and importance of different volcanic gases on short- and long-term climate change, and the physical and chemical factors that determine the extent of any such change.

3.2 The potential link between the eruption of flood basalts and mass extinctions at specific points throughout geological time.

3.3 The mechanisms by which the changing positions and configurations of the continents and the oceans have affected global climate over time, as a result of changing circulatory systems, albedo and weathering patterns.

3.4 The positive and negative feedback mechanisms that connect changes in the global temperature and/or eustatic and isostatic sea levels with the preservation of carbon in the atmosphere, oceans, biosphere and geosphere.

Mountains and climate change

Chapter 4

In Chapter 3 you looked at plate tectonics and its influence on the climate and habitability of the Earth. In this chapter you will look at a specific aspect of this relationship, namely mountain building. Mountain ranges attract their own microclimates, which tend to be cooler and often wetter than the lowland areas that surround them.

It may strike you as odd that temperatures drop as altitudes increase; after all, the higher the altitude, the closer the Earth's surface is to the Sun. This is primarily because direct solar radiation causes very little heating of the air; most heating is due to radiation reflected back from the Earth's surface. Furthermore, air forced to rise by the presence of a mountain will expand and cool in response to the decreasing atmospheric pressure. Eventually, any water vapour in the air will condense, forming clouds and precipitation, which at high altitudes may fall as snow. Even at low latitudes, high mountains may be capped with snow or ice (Figure 4.1).

Figure 4.1 Mount Stanley (5100 m) from the Rwenzori Mountains, equatorial Africa. (Keller)

- How can the extent of mountainous terrain at low latitudes affect global climate?
- Through its effect on the global albedo. A large area of highly reflective snow at low latitudes (where a large proportion of incoming solar energy reaches the Earth) could lead to a significant increase in the Earth's albedo as a whole.

It is difficult to estimate what effect low-latitude mountains might have had on the Earth in the distant past, not least because no one knows for sure where the mountains were or how high they were. Nevertheless, it is something that you should bear in mind, particularly when considering what the climate might have been like when a large proportion of the global land area was at low latitudes, as might have been the case during the Precambrian.

4.1 Mountain building and the carbon cycle

As a result of the density of continental crust being lower than that of either oceanic crust or the mantle, where the underlying continental crust is thickened the Earth's surface is elevated due to isostasy (Box 3.3). The world's large mountain ranges, therefore, reflect the thickened continental crust that supports them with wide regions of thickened crust forming as a result of collision between two continental plates (Figure 4.2). Prior to collision, active subduction of oceanic lithosphere generates two sources of CO_2. The dominant source of this CO_2 is volcanism both above the subducting slab (Figure 4.2a) and at spreading ridges into the oceans, while the second source of CO_2 results from metamorphism of the down-going slab.

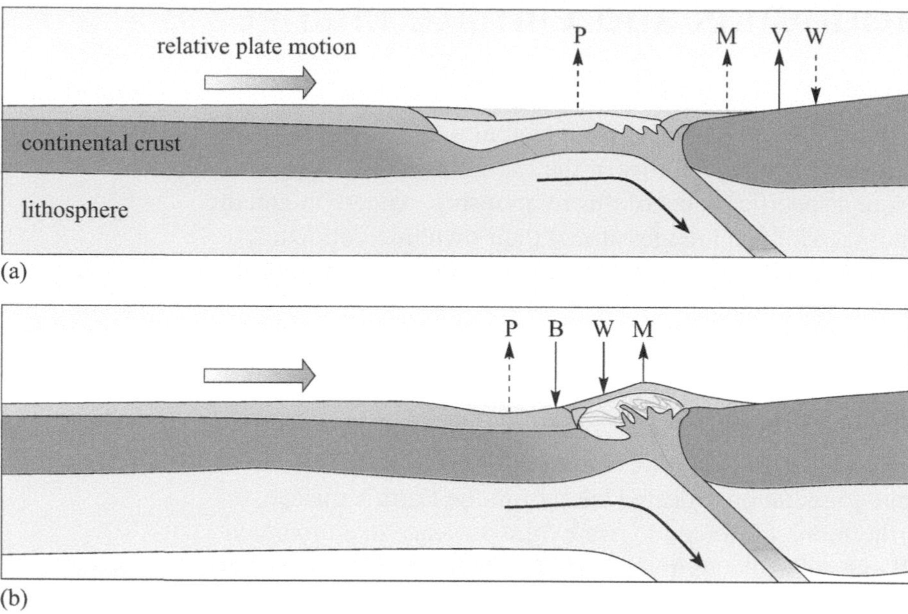

Figure 4.2 Schematic cross-sections showing the formation of a mountain range and the associated carbon fluxes: (a) pre-collision; (b) collision. Increases in atmospheric CO_2 are from volcanic emissions (V), metamorphism (M), and precipitation of carbonates (P). Decreases in atmospheric CO_2 are due to weathering of silicates (W) and burial of organic carbon (B). Dashed arrows indicate relatively minor fluxes.

Eventually, the ocean basin will close due to subduction and the two adjacent continents will collide (Figure 4.2b). Collision will terminate active subduction and the volcanism that results from it. Meanwhile, the CO_2 flux resulting from metamorphism in the thickened crust will probably increase. Throughout the period of subduction in the thickened crust, deep-sea sediments will have been scraped off the surface of the subducting oceanic plate, building up great wedges of sediment at the side of the subduction zone trench, forming accretionary prisms. As the continents collide, these sediments (including calcareous and siliceous remains) will be subjected to increased temperatures and pressures.

If calcium carbonate and silica are heated together at about 400 °C, a **decarbonation reaction** occurs that releases CO_2:

$$\underbrace{SiO_2(s)}_{\text{silica}} + \underbrace{CaCO_3(s)}_{\text{calcium carbonate}} \rightarrow \underbrace{CaSiO_3(s)}_{\text{calcium silicate}} + \underbrace{CO_2(g)}_{\text{carbon dioxide}} \qquad (4.1)$$

■ Where have the silica and calcium carbonate on the descending slab of ocean floor come from?

□ Silica and calcium carbonate are both found in the remains of planktonic organisms (notably diatoms and radiolarians, which are silica organisms, and coccolithophores and foraminiferans, which are carbonate-rich organisms).

Deep-sea sediments are not the only source of calcium carbonate reacting in the decarbonation reactions. At least some products of the shallow-water carbonate factory (e.g. carbonate-secreting algae, corals and bivalve remains) will also be trapped between the colliding continents. Most of the silica in shallow-water sediments is not **biogenic** (i.e. produced by organisms) but in the form of quartz sand, which has been weathered from the land and transported before accumulating on the continental shelf. It is important to realise that decarbonation reactions will only occur when silica and carbonate are mixed together. Although this is the usual situation for planktonic remains in deep-sea sediments, it is not so common for shallow-water sediments where, for example, coral-reef debris is less likely to be intimately mixed with sand. The CO_2 produced by decarbonation reactions due to continental collision is not usually released into the atmosphere via volcanism (as volcanoes are not generally found in such tectonic settings); instead, it seeps out along faults and fractures.

The subduction of sediments of *pure* calcium carbonate (i.e. sediments without silica) will mean that the carbon they contain will not be returned to the atmosphere for millions of years until they are released by weathering processes (after being returned to the surface). Similarly, the 'piling up' of carbonate rocks that occurs during continental collision can mean that large volumes of carbon-containing rocks are buried deep in the crust and are prevented from participating in the global carbon cycle for a very long time.

The most significant aspect of climate change resulting from continental collision results from interactions between the rocks exposed at the surface, the hydrosphere and the atmosphere. Following continental collision, the rising mountain range causes the air masses in the lower atmosphere to rise, cool and precipitate rain or snow. The combination of increased precipitation and steep topography results in high rates of **physical erosion**. At the highest altitudes, rocks are shattered by:

- repeated freezing and thawing of water that seeps into cracks (a consequence of the fact that ice occupies more space than the water from which it forms)
- being crushed and ground into small particles by the action of glaciers.

Vast **alluvial fans** are formed at the outlets of steep-sided mountain glaciers (Figure 4.3). In large mountain ranges, huge volumes of mostly sedimentary rocks are broken up by weathering and erosion, transported by water, ice and gravity and buried. Many of these contain a significant component of organic carbon and the burial of this material removes it from further exposure to the atmosphere, which would result in the oxidation of carbon to CO_2.

Figure 4.3 Fragments of rocks forming a vast alluvial fan at the outlet of a river eroding the Karakoram Mountains, located to the north of the western Himalaya. (Nigel Harris/Open University)

At lower, warmer altitudes, the wet windward slopes of mountains tend to be regions of strong chemical weathering. Rock fragments are carried down in fast mountain streams, accumulate and become chemically altered in reactions with rain and surface waters. The process of dissolving silicate minerals results in the removal of CO_2 from the atmosphere (Section 2.3.2) and, as shown in Figure 2.21, for every two carbon atoms removed from the atmosphere during the weathering of silicate minerals, one carbon atom has the potential to be removed from the atmosphere into seabed sediments, as carbonate remains. By contrast, weathering of carbonate minerals followed by reprecipitation in the ocean does *not* result in any net removal of CO_2 from the atmosphere.

One example of the importance of mountains in enhancing weathering rates is provided by the River Amazon. Even at its mouth, over three-quarters of the dissolved material it carries has been derived from the eastern slopes of the Andes more than 3000 km away – only one-quarter comes from weathering of the vast tracts of bedrock that underlie the Amazon Basin between the source and the mouth of the river.

During subduction, the net effect on atmospheric CO_2 concentrations from processes along the subduction zone is to increase this CO_2 reservoir from volcanic emissions. After continental collision, however, this flux is closed down and the net contribution of this tectonic margin is to deplete atmospheric CO_2 due to enhanced silicate weathering rates and the burial of organic carbon (Figure 4.2). These fluxes are partially offset by increased CO_2 from the metamorphism of carbonate rocks within the thickened crust.

Having looked briefly at the general role that mountains play in climate change, the next section considers a particularly high part of southern Asia, namely the Himalaya and the Tibetan Plateau. You will look at the role the Himalaya and Tibetan Plateau might have played on the climate system during the last few tens of millions of years, and at the effect they have had and continue to have on the global atmosphere and neighbouring seas.

4.2 The uplift of Tibet and the monsoon

The Tibetan Plateau is the highest and largest plateau on the Earth's surface (Figure 4.4). Its southern edge is marked by the Himalaya and several of Asia's other great mountain ranges – the Karakoram, Hindu Kush and Kunlun – decorate its western and northern margins. If continental relief can affect global climate, this is the obvious region to study. This discussion of climatic change over long timescales of millions of years begins with a brief look at climate in this region at the present day.

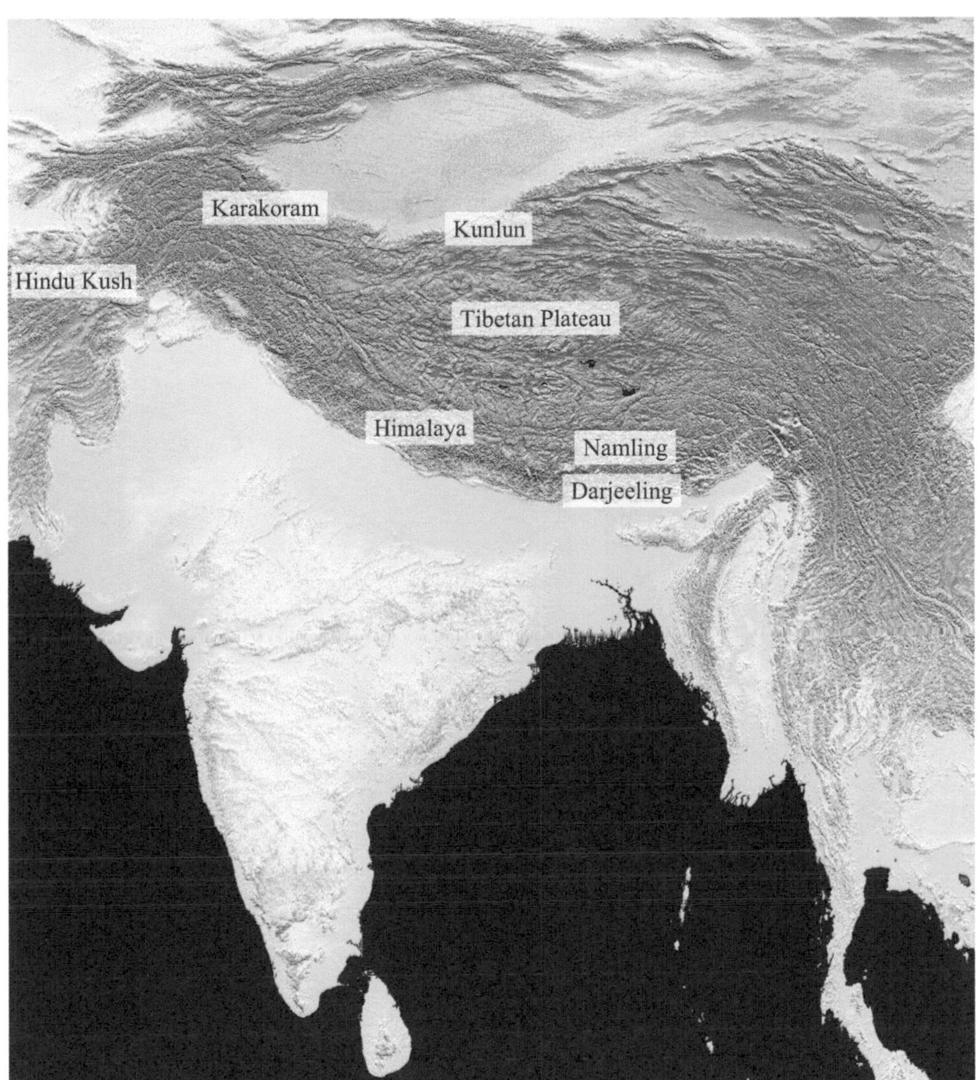

Figure 4.4 Digital elevation map for southern Asia. The Himalaya reach over 5000 m in places; to the north of the Himalaya, the Tibetan Plateau is over 4000 m while to the south, northern India is below 500 m.

4.2.1 Present-day climate across southern Asia

To a large extent, the climate of southern Asia is dominated by the monsoons. In the northern winter, the region is affected by cold, dry northeasterly winds blowing out from the intense high-pressure region over the continent (Figure 1.28). By contrast, in the northern summer when there is a strong low-pressure region over the Eurasian continent, the area receives moisture-laden air in the southwest monsoon.

The whole of southern Asia benefits from the monsoon rains. In particular, the areas of tropical rainforest along southwest India, Myanmar and Sri Lanka (Figure 2.4) are the results of the southwesterlies releasing much of their moisture over high land. On approaching the southern slopes of the Himalaya, the still-moist air mass is driven further upwards, causing summer rainfall over northern India. As a result, annual rainfall in Darjeeling on the southern slopes of the Himalaya is over 3000 mm, of which 88% falls between June and September, i.e. during the southwest monsoon. Less than 100 km away, to the north of the

Himalayan mountains, the Tibetan town of Gyantse receives an annual rainfall of only 270 mm. The contrasting climates result in very different floras and faunas: the southern slopes of the Himalaya are covered by forests (Figure 4.5a) supporting a population density of 200 people per square kilometre; Tibet, which is effectively in the rain shadow of the Himalaya, is characterised by semi-arid **steppe**, and in places is a rocky desert, hardly supporting one person per square kilometre (Figure 4.5b).

(a)

(b)

Figure 4.5 Two locations only 50 km apart but on different sides of the Himalayan watershed: (a) rhododendron forests flourish in abundant rainfall on the southern slopes of the Himalaya; (b) rocky desert conditions on the northern side of the Himalaya. (Nigel Harris/Open University)

As Figure 1.28 shows, the seasonal shift in the Intertropical Convergence Zone (ITCZ) means that seasonally changing winds (monsoons) affect large parts of the globe at low latitudes. The extreme change in pressure over a large part of central southern Eurasia, from intense high pressure in winter (Figure 1.29) to very low pressure in summer, means that seasonal changes in the vicinity of southern Asia and the Arabian Sea are by far the most dramatic. It seems that the reason for this may lie in the existence of the Tibetan Plateau itself.

In 1989, the results of a series of experiments were published using a sophisticated computer model of the global climate (a **general circulation model** or GCM) designed to investigate the effect on climate of such an extensive high-altitude plateau. Starting with a simulation of the present-day climate, the researchers changed just one variable: the topography of present land masses. When the Tibetan Plateau was 'removed', the heavy summer rainfall in northern India all but disappeared. In contrast, an even larger and higher plateau in central Asia greatly *increased* the area of summer monsoon rainfall throughout extensive regions south of the plateau, caused desert conditions over vast areas to the north of the plateau, and *decreased* summer precipitation much further west in the Mediterranean region in Europe.

Of course, such experiments have their critics, many of whom emphasised that no model can take into account all the possible variables. After all, if you think about climate change over the past 60 Ma, uplift of the Tibetan Plateau is not the only change to have taken place within the Earth system. Not only did the Indian land mass move northwards across the globe, traversing climatic belts, but also the global climate was itself changing. Nevertheless, the results of the modelling clearly suggested that the uplift of Tibet could have had a dramatic effect on atmospheric circulation and precipitation throughout much of the Northern Hemisphere, and may well have affected the strength of the monsoon winds, particularly those of the southwest monsoon over southern Asia.

The strength of the southwest monsoon is determined by the pressure difference between the high over the tropical Indian Ocean and the low over the southern part of the continent (Figure 1.28). At the end of winter, the large rocky mass of the Tibetan Plateau heats up fast once its high-albedo covering of snow has melted; the overlying air is warmed and the pressure over the continent falls.

Question 4.1

The winds of the southwest monsoon are initially laden with moisture as they have blown across the Indian Ocean and the Arabian Sea. With reference to Section 1.4 (if necessary), why does the release of this moisture as the Himalayan monsoon rains help to intensify the circulatory pattern shown in Figure 1.29b?

The warming of the air over the Himalaya and the Tibetan Plateau has a particularly large effect because the air is thin at these high altitudes and its temperature, therefore, is more sensitive to changes in heat.

As discussed in the answer to Question 4.1, the monsoon rains indirectly help to warm the air over Tibet. The condensation of moisture to form rain over the southern Himalaya releases latent heat, and so the summer winds driving from the south into Tibet are not only dry but also warm (Figure 1.29). At this latitude and in the absence of a plateau, the air temperature at 5 km above sea level would be around –20 °C (Figure 1.17); as it is, during the summer months, the temperatures rarely drop below freezing. Nevertheless, the dramatic summer storms that produce rain over the Himalaya may produce hail over Tibet (Figure 4.6). Indeed, the Tibetan Plateau is so high that the subtropical jet stream passes either to the north or to the south of it. The climate implications of this are not well understood, but it is clear that the jet stream would not have been diverted in this way before the uplift of Tibet.

(a)

(b)

Figure 4.6 (a) The Royal Crest of the Himalayan Kingdom of Bhutan. The crossed 'thunderbolts' are a common symbol in Tibetan culture, reflecting the power of summer storms in the Himalaya. (b) Dark clouds preceding a violent hailstorm on a Tibetan summer afternoon. High plateaux like Tibet are notorious for their vicious hailstorms, caused by sudden updrafts within clouds under cold conditions. (Nigel Harris/Open University)

The overall effect of the Tibetan Plateau and the Himalaya on atmospheric circulation, therefore, is determined by both their high elevation and their geographical position. Uplift must have caused major changes in atmospheric circulation across the Northern Hemisphere. As far as the southwest monsoon is concerned, because summer heating of the atmosphere over Tibet has increased as the plateau has risen, it is possible that, at some stage during its elevation, a threshold altitude was reached above which the monsoon winds were greatly strengthened. Partly for this reason, scientists have been looking for evidence of climate change in southern Asia that can be linked to the uplift of Tibet. To find such a link, they need to know something about the timing of both the uplift of the plateau and of climate change in southern Asia over the relevant period.

4.2.2 When was the Tibetan Plateau uplifted?

The elevation of the Tibetan Plateau is the result of a head-on collision between the continental margins of two plates: India, which was migrating northwards, and Eurasia, which was stationary. The collision, which has been dated at around 50 Ma, led to crustal thickening and uplift of the Himalayan mountains and the plateau. It is reasonable to infer, therefore, that most of the uplift has occurred during the past 50 Ma. Can modern techniques improve on this very general estimate?

It is difficult to trace the uplift of a plateau through time because, if there is such a thing as a reliable **palaeo-altimeter**, no one has discovered it yet. If an attempt were made to plot the change in altitude of the Tibetan Plateau against time, it would have two (and only two) firm data points:

- the present-day altitude, which on average is 5 km above sea level
- rocks that are now 5 km high were *below* sea level 70 Ma ago because limestones with remains of marine organisms originally deposited in shallow seas at about that time have been found in southern Tibet.

Details of the uplift during the intervening 70 Ma, however, are highly uncertain.

One approach to determine the elevation history of the plateau is to exploit the changes in surface temperature resulting from its elevation. The morphology of successful plant species is quite different in cold and warm climates; because temperature decreases with increasing altitude, the higher the altitude, the more cold-climate species are favoured. Palaeobotanists use this fact to infer altitude changes from fossil flora collected from sedimentary rocks deposited over the past 50 Ma. Work along these lines has led to the publication of several quite different altitude–time paths for the Tibetan Plateau (Figure 4.7).

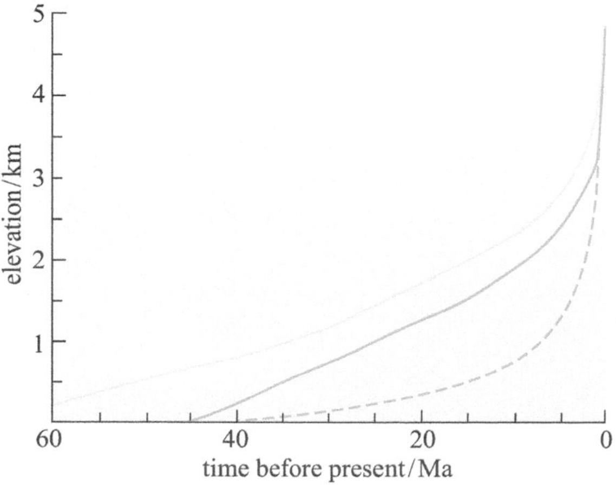

Figure 4.7 The history of uplift of the Tibetan Plateau as determined by studies of three different fossil plant assemblages, using the 'nearest living relative' approach (see Section 7.8.2). (Raymo et al., 1988)

Unfortunately, there are a few flaws in this approach. First, even if the region corresponding to the Tibetan Plateau had not been uplifted at all, its climate would have changed considerably over the course of 50 Ma, not least because of the closure of the Tethys Ocean (Figure 3.16).

Furthermore, within the plateau itself there will have been small-scale climatic variations resulting from local topography, with some areas for example being more sheltered or wetter. To some extent, these latter complications can be allowed for, but there is another problem which is more fundamental. As the fossil species found in sediments older than a few million years are usually extinct, some palaeobotanists have adopted the 'nearest living relative' approach. In practice, this involves identifying the nearest living relative to the fossil species concerned, and then assuming that the climatic conditions under which the ancient plants lived were similar to those of their living relatives.

- Suggest why such an approach might give misleading results.

- Species evolve and adapt to changing environmental conditions, so the extinct species might have lived under somewhat different conditions from its nearest living relative.

Imagine for example that yaks (currently found across the Tibetan Plateau) are extinct but that their fossilised remains are well documented by palaeontologists, who have concluded that cattle are the yak's nearest living relative. The inferred habitat of yaks would then be characterised by a temperate or warm climate, similar to where living cattle are found. The truth, of course, is that yaks are adapted to high-altitude, cold conditions (Figure 4.8). In other words, the 'nearest living relative' approach ignores evolutionary adaptation. Clearly, this approach has inherent problems, and the resulting estimates of uplift rates need to be treated with care.

Figure 4.8 Adult yaks on the Tibetan Plateau. Yaks are an example of adaptation to high-altitude, cold conditions. The location is Namling, which is also the site of abundant leaf fossils. (Nigel Harris/ Open University)

An alternative approach is to examine and classify the shapes of plant leaves that are adapted to different environments (Figure 4.9). By examining the shape, size and margin characteristics of a leaf (**leaf physiognomy**), it is possible to determine the climatic regime under which the plant was growing when it was alive. For example, modern woody, broadleaved flowering plants tend to have smooth leaf margins in warm climates but toothed, jagged margins in cool

Figure 4.9 Fossil leaves from Tertiary sediments on the Tibetan Plateau. (Nigel Harris/ Open University)

climates. Thus, since plants live on the Earth's surface, fossil leaves, along with an independent (e.g. isotopic) dating method, can be used to estimate the height of a land surface at any time in the past. This approach is called **leaf margin analysis** and is discussed in more detail in Chapter 7.

The fossil leaf approach requires very specific field sites that combine abundant well-preserved flora with datable strata, a situation rarely found in nature. So far, only one such locality has been found across the entire Tibetan Plateau (Figure 4.8). From this single site in southern Tibet, 400 specimens, including over 30 species, have been recovered and analysed to yield an estimate of the prevailing temperatures during their growing season. Volcanic strata from above and below the leaf-bearing shales have been dated at 15 Ma using argon isotopes. Finally, a climate model has been run for this period of the Neogene (between 23 Ma and 1.8 Ma) that converted the estimated temperature into an altitude for the latitude of southern Tibet at that time. The result demonstrated that this sedimentary basin in southern Tibet was at an elevation of 4600 ±700 m at 15 Ma, an altitude indistinguishable from its present-day height above sea level (4300 m).

■ Referring to Figure 4.7, how does this conclusion compare with earlier attempts using the 'nearest living relative' approach?

■ The leaf margin analysis indicates that southern Tibet was at an elevation of 4000–5000 m much earlier than was previously thought.

The conclusion from leaf margin analysis that southern Tibet has not changed in elevation for at least 15 Ma has been verified by a second technique that exploits the relationship between the fractionation of oxygen isotopes within H_2O, precipitated as rain or snow (Box 3.2), and the altitude. The details of this new isotopic technique are beyond the scope of this book and, unfortunately, the method also carries large uncertainties. None the less, analyses of oxygen isotopes from carbonates in lake deposits broadly confirm the results from leaf margin analysis.

Our understanding of the uplift history of the Tibetan Plateau remains in its infancy. Present knowledge suggests that it began to rise some time after 50 Ma and reached its maximum present elevation by about 15 Ma in southern Tibet. Although there are no direct estimates of the timing of uplift in northern Tibet, dating structures associated with uplift suggest much more recent uplift, perhaps dating back to the past 10 Ma.

4.2.3 Evidence for climate change across southern Asia

This section looks briefly at some evidence for how climate in southern Asia might have changed over the past 10 Ma, before setting this in the context of global climatic change during the Tertiary Period (i.e. 65–2 Ma).

Evidence for climate change in southern Asia over the past 10 Ma is drawn from a number of different lines of study, including two techniques that use the remains of living organisms:

- zooplankton that lived in the surface waters of the Arabian Sea
- terrestrial plants and the mammals that fed on them.

Look at Figure 4.10, which shows the concentration of phytoplankton in surface waters of the Indian Ocean today (a) during the inter-monsoon period, when winds are generally light, and (b) during the summer monsoon, when winds over the northern Indian Ocean and Arabian Sea are strong and from the southwest.

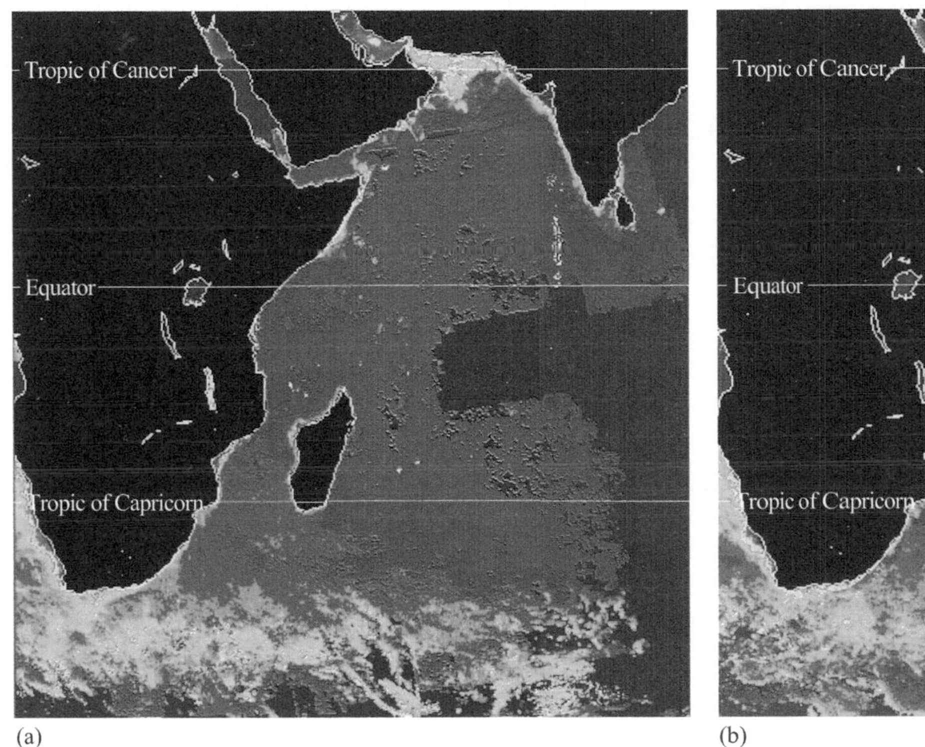

(a)

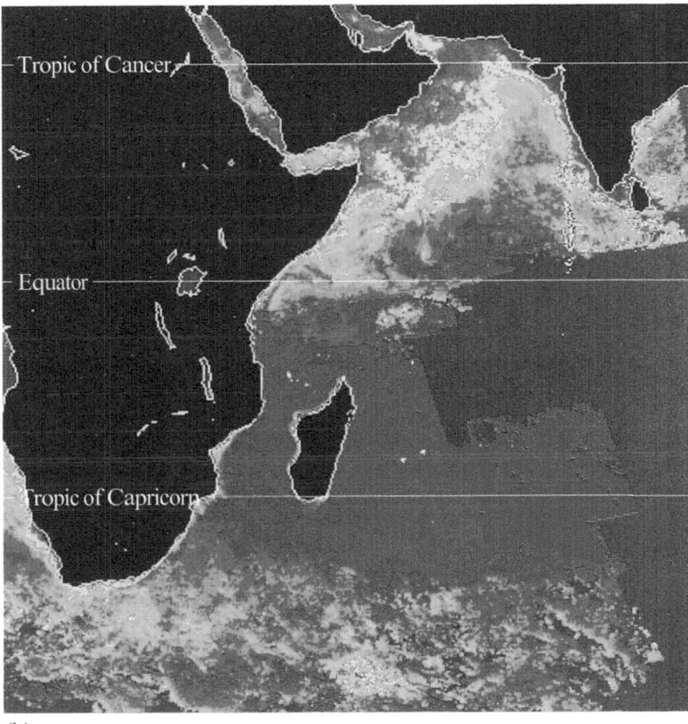

(b)

Figure 4.10 Seasonal variations in phytoplankton concentrations, on the basis of chlorophyll pigment recorded by the satellite-borne Coastal Zone Color Scanner. (a) A composite image for May–June, a period of light winds. The highest concentrations are shown in red (principally in the Persian Gulf and near-coastal areas in the Arabian Sea and around India and Pakistan); the lowest concentrations are shown in pinkish purple (principally in the Arabian Sea and the Indian Ocean area north and south of the Equator from eastern Africa to Sri Lanka). (b) A composite for September–October, during the southwest monsoon. The highest concentrations are shown in red (principally in the southern Red Sea, the Arabian Sea, the Persian Gulf and coastal areas around India, Pakistan and Sri Lanka); the lowest concentrations are shown in pinkish purple (principally in the Indian Ocean south of the Equator to just south of the Tropic of Capricorn). In both (a) and (b) the black areas indicate no data.

■ Without going into details, to what could the different levels of primary productivity in the northern Indian Ocean/Arabian Sea shown in Figure 4.10a and b be attributed?

□ The difference could relate to the fact that upwelling is stronger and more widespread in this region during the southwest monsoon, resulting in higher phytoplankton concentrations.

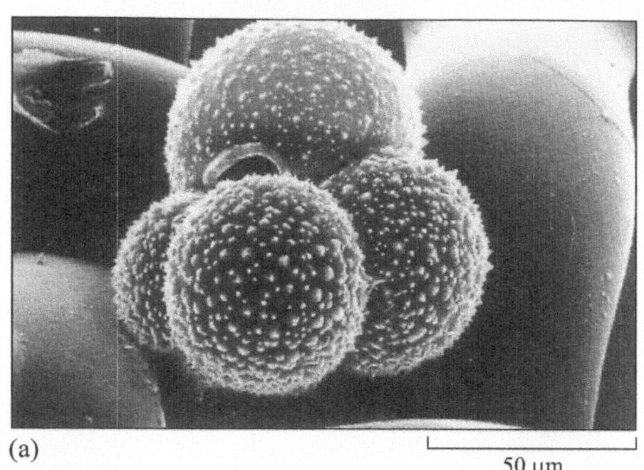

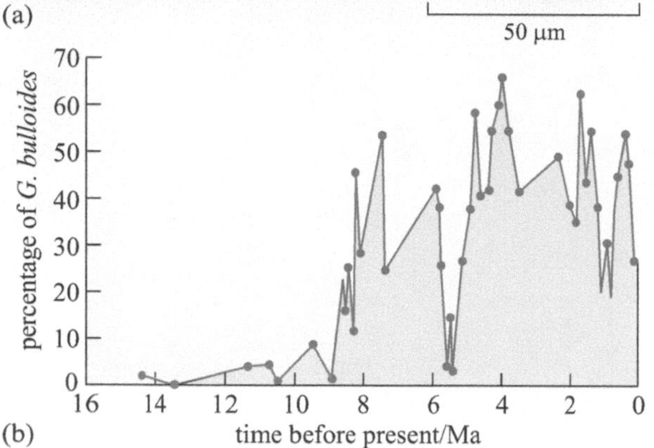

Figure 4.11 (a) The remains of a specimen of *Globigerina bulloides*. The background is the mesh of the sampling net. (b) Variation in the proportion of the microplankton population made up of *G. bulloides* over the past 14 Ma, from a seabed drill core from the Arabian Sea. ((a) Jordan and Smithers; (b) Kroon et al., 1992)

As discussed in earlier chapters, upwelling brings nutrient-rich subsurface water into the photic zone, supporting populations of phytoplankton on which zooplankton and larger organisms feed. If you compare Figure 4.10 with the wind patterns shown in Figure 1.28, you will see that the nearshore upwelling occurring here cannot result from longshore equatorward winds (Figure 1.36a) as it does in the tropical regions of the other oceans. Instead, it mainly occurs where surface waters diverge in cyclonic gyres (Figure 1.36b) and at places where surface currents diverge from the coast.

Now back to the discussion of climatic change. The zooplanktonic organism under consideration is a species of foraminiferan known as *Globigerina bulloides* (Figure 4.11a). Its abundance at various times in the past 14 Ma relative to other species of microplankton (i.e. those between 20 and 200 µm across) has been estimated on the basis of its fossil remains in sediment cores drilled from the floor of the Arabian Sea (Figure 4.11b). *G. bulloides* is presently abundant in nutrient-rich tropical waters, and in the Arabian Sea its abundance (expressed as a proportion of the total microplankton population) increases by three orders of magnitude during periods of upwelling.

Question 4.2

(a) In general terms, what is the variation in the abundance of *G. bulloides* over the past 14 Ma?

(b) Bearing in mind the high levels of primary productivity shown in Figure 4.10b, does the plot in Figure 4.11b suggest anything about changes in the strength of the southwest monsoon over the time period in question (i.e. 14.5 Ma to the present)?

So, the patterns of relative abundance of *G. bulloides* suggest that the southwest monsoon became stronger several million years after the Tibetan Plateau in the south had reached its maximum altitude.

Organic debris in sediments eroded from the Himalaya provides the second clue to past climate: as the Himalaya rose, so the rate of erosion of the steepening slopes increased. Great rivers flowing southwards deposited much of the eroded material into a large subsiding basin, with the result that these sediments from the Himalaya are now exposed in northern Pakistan and India. Their use as climatic indicators lies in the proportion of the different isotopes of carbon they contain, expressed in terms of the ratio **δ^{13}C** (Box 4.1).

Box 4.1 Carbon isotopes and $\delta^{13}C$

Carbon occurs in nature as two stable isotopes: ^{12}C and the much rarer ^{13}C. During photosynthesis, fixation of the lighter $^{12}CO_2$ is favoured over that of the heavier $^{13}CO_2$ because $^{12}CO_2$ diffuses into cells more rapidly and more readily takes part in chemical reactions. As a result of this fractionation of isotopes, organic matter produced by photosynthesis is enriched in ^{12}C and depleted in ^{13}C relative to the inorganic carbon in the atmosphere and hydrosphere (i.e. CO_2 gas plus bicarbonate and carbonate ions in solution). Enrichment or depletion of ^{13}C is expressed using the ratio $\delta^{13}C$, which is calculated in an analogous way to $\delta^{18}O$ (Box 3.2) to give a value in parts per thousand or 'per mil', often written as ‰:

$$\delta^{13}C = \left[\frac{(^{13}C/^{12}C)_{sample}}{(^{13}C/^{12}C)_{standard}} - 1\right] \times 1000 \qquad (4.2)$$

$(^{13}C/^{12}C)_{standard}$ is the ratio calculated for a standard carbonate sample, and is 1/88.99.

- If $^{13}C/^{12}C$ is greater in the sample than in the standard, then the ratio of these ratios will be greater than one, and the expression in square brackets (i.e. $\delta^{13}C$) will be positive.
- If $^{13}C/^{12}C$ is less in the sample than in the standard, then $\delta^{13}C$ will be negative.

Higher values of $\delta^{13}C$ (or less negative values) correspond to a higher proportion of ^{13}C.

The $\delta^{13}C$ value for atmospheric CO_2 is –8‰ and, because of fractionation, plant tissue generally contains a lower proportion of $^{13}CO_2$ by about 20‰. As a result, plant tissue has an average $\delta^{13}C$ value of about $-(8‰ + 20‰) = -28‰$.

■ Does organic material (plant or animal tissue) have a higher or lower value than inorganic carbon?

□ It is always lower (i.e. more negative).

The $\delta^{13}C$ value of a sample reveals more than just whether it is plant derived. As a result of the different mechanisms of photosynthesis that they use, plants and shrubs that flourish under warm conditions tend to incorporate more of the isotope ^{13}C than those that thrive in colder conditions. Furthermore, as a result of plant respiration through the roots and the accumulation of plant debris, soil acquires a $\delta^{13}C$ 'signature' similar to that of the plants that grew in it. In particular, carbonates that precipitate in soil have a $\delta^{13}C$ value somewhere between that of atmospheric CO_2 and that of the living plants and plant debris in the soil. Measurements of $\delta^{13}C$ for carbon (as organic carbon or calcium carbonate) in soils of different ages can, therefore, provide an indicator of the temperatures that prevailed during the growing season.

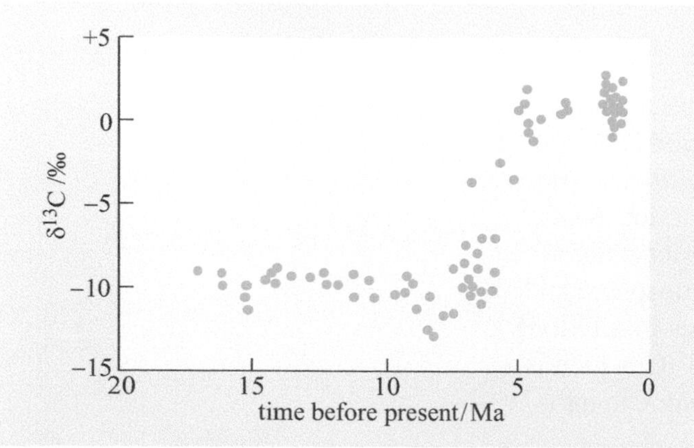

Figure 4.12 Variation of $\delta^{13}C$ with time in carbonates in soils formed from debris eroded from the Himalaya of northern Pakistan. (Quade et al., 1989)

The information that you need to take away from Box 4.1 is simply this: the $\delta^{13}C$ of plant (and hence animal) tissue is always negative, but is higher (less negative) for the kinds of plants that flourish under warm conditions.

The sediments from the Himalaya referred to above eventually formed soil, within which carbonates were precipitated and, as discussed in Box 4.1, the $\delta^{13}C$ of these carbonates reflects the $\delta^{13}C$ of organic debris in the soil. Figure 4.12 is a plot of $\delta^{13}C$ against time for these soil carbonates.

■ According to Figure 4.12, what marked changes in the $\delta^{13}C$ values of organic debris eroded from the Himalaya can be observed during the past 20 Ma?

□ A sharp increase in $\delta^{13}C$ values from −10‰ to around 0‰ occurred between 8 and 5 Ma.

The increase in organic material incorporating more ^{13}C has been interpreted by some palaeoclimatologists as an explosion of plant production during the growing season, resulting from increased summer rainfall because of a strengthening of the southwest monsoon at 8–7 Ma. However, because this increase in $\delta^{13}C$ values has been recognised in plant remains from other continents and is not a peculiarity of southern Asia, other workers in the field have argued that it marks a spread of plants using a slightly different mechanism for photosynthesis, which evolved within the last 14 Ma. Some modern plant groups including savannah grasses rely on this 'newer' mechanism exclusively, and are referred to as **C4 plants**; the majority (including trees and shrubs), using the original mechanism, are referred to as **C3 plants**. Figure 4.12 might, therefore, chart a transition of plant type that could be part of a global response to climate change that may or may not be directly linked to the strengthening of the southwest monsoon.

Interestingly, the remains of fossilised mammal teeth now preserved in the Himalayan sediments also suggest a change in the vegetation in the region at about 8 Ma. The shapes of the teeth suggest that at about this time there was a marked change from browsers (which feed on trees and shrubs, i.e. C3 plants) to grazers (grass-eaters, which feed on C4 plants); furthermore, some identifiable forest-based mammals (e.g. orang-utans) disappeared from the region at that time.

Finally, the *type of sediment* originating from the Himalaya may be used as an indicator of the conditions at which weathering occurred. Before about 7 Ma, the deposits carried down from the Himalaya were predominantly sands and silts; these sediments indicate strong erosional forces (i.e. physical weathering) by freeze–thaw or glaciers acting on exposed rock surfaces that are lacking soil cover. In contrast, sediments younger than 7 Ma include plenty of muds and clays, suggesting chemical weathering, producing thick layers of soil in the source areas of the rivers. Like the $\delta^{13}C$ data, the size distribution of sediments in the sedimentary record points towards a sudden increase in summer rainfall in southern Asia around 7 Ma.

In summary, although it is not possible to be precise about the timing of either the uplift of the Tibetan Plateau or climate change in southern Asia, there is evidence that summer rainfall associated with the southwest monsoon increased between 9 Ma and 6 Ma. By this time, the southern plateau had already stabilised at its present elevation for at least six million years, but northern regions were just reaching their maximum elevation. Geologists have speculated that, by about 9 Ma, the area of the elevated plateau was sufficient to heat the lower atmosphere during the summer months and thus trigger a dramatic increase in the intensity of the monsoon.

4.3 Global climate change during the Tertiary

So far, a possible link between uplift of the Tibetan Plateau and a strengthening of the southwest monsoon has been investigated. Although this may have affected a large part of the globe, it essentially had a regional effect on the climate. If the uplift of this (or any other) high plateau had had an effect on *global* climate, you would have expected the change to have taken place not over the past 10 Ma as the monsoon was strengthening, but over the past 50 Ma, i.e. from the beginning of uplift of the plateau, initiated by continental collision. Before looking more closely at possible global consequences, it is important to understand how global climate has changed since the collision between India and Eurasia.

The variation in global average temperature over the past 120 Ma, deduced from oxygen isotope studies of the remains of deep-sea benthic foraminiferans, is shown in Figure 4.13. Despite the fluctuations, there is a clear downward trend, with a net cooling of nearly 20 °C over the period concerned, with strong independent geological evidence supporting this inference. For example, during the past 50 Ma, the distribution of sediments deposited by glaciers has generally been increasing, with these sediments deposited progressively further away from the present polar regions, reaching their maximum extent during the peak of the present (Quaternary) Ice Age.

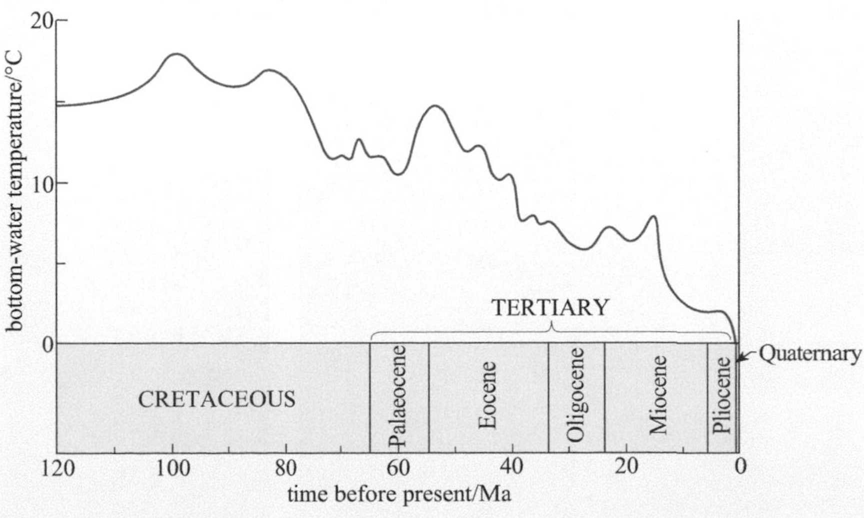

Figure 4.13 Variations in the temperature of ocean bottom waters over the past 120 Ma, estimated using oxygen isotope ratios ($\delta^{18}O$) from deep-sea benthic foraminiferans. (This is an expanded version of the right-hand side of the temperature plot in Figure 3.20.)

What caused this decrease in global temperature? One potential cause may have been the rearrangement of the continents (as discussed in Chapter 3). However, irrespective of how the continents are rearranged, computer-based climate models cannot reproduce the long-term cooling pattern indicated by Figure 4.13. Furthermore, the climate-modelling experiments that investigated the effects of the Tibetan Plateau on global climate failed to show that the uplifting of Tibet could, by itself, cause long-term cooling. It seems that something else is required to explain this trend.

- What other factor might have been involved?
- A change in the composition of the atmosphere; in particular, a long-term *decrease* in the concentration of CO_2.

The reasons behind this are beyond the scope of this book, but you can assume that the temperature variations shown in Figure 4.13 resulted entirely from fluctuations in atmospheric CO_2 concentration. To account for the overall fall in temperature, therefore, the CO_2 concentration would have to have declined from a value of about eight times that of the present day. Many geoscientists believe that the building of large mountain belts in central Asia could well have played a role in bringing about such a dramatic change in atmospheric composition. This particular aspect of mountain building is thought by some to be an important mechanism for changing atmospheric CO_2 concentrations, and hence climate.

4.3.1 The Himalaya, Tibet and atmospheric CO_2

Although bordered by impressive mountain ranges, the interior of Tibet is truly a plateau, with a local relief of generally no more than a kilometre or so (Figure 4.14a). The plateau is the catchment area for many of Asia's great rivers, including the Indus, the Brahmaputra (known in Tibet as the Tsangpo), the Yangtze, and the Mekong (Figure 4.15). These rivers have to descend 5 km before they reach the sea, eroding their way through mountain ranges that are steadily being uplifted (Figure 4.14b).

(a)

(b)

Figure 4.14 (a) The gentle relief typical of much of central Tibet. (b) A deeply incised Himalayan gorge, carved out by a river flowing south from Tibet to join the Ganges on the Indian plains. (Nigel Harris/Open University)

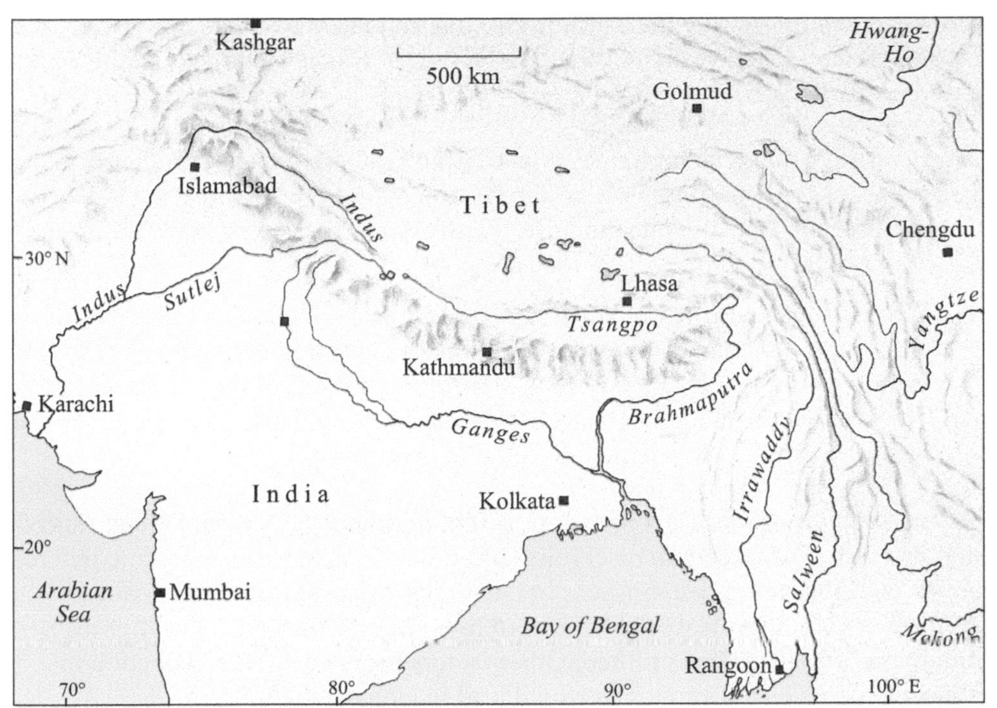

Figure 4.15 The central Asian river system, showing major rivers with their sources on the Tibetan Plateau or in the Himalaya.

Question 4.3

Table 4.1 lists the flux of dissolved material of rivers with a source in the Tibetan/Himalayan region.

(a) The total flux of dissolved material carried by rivers globally is about 2130×10^6 t y^{-1}, while the global flux of suspended sediment in rivers is about $20\,000 \times 10^6$ t y^{-1}. What percentage of these global fluxes of (i) dissolved material and (ii) suspended sediment are contributed by all the rivers in Table 4.1 together?

(b) The catchment area of the rivers in Table 4.1 (Figure 4.15) represents about 5% of the Earth's continental surface. What do your answers to part (a) imply about weathering rates in the Tibetan/Himalayan region in a global context?

Table 4.1 Fluxes of dissolved material in rivers with sources in Tibet or the Himalaya.

River	Flux of dissolved material/10^6 t y^{-1}	Flux of suspended sediment/10^6 t y^{-1}
Ganges	85	538
Yangtze	205	483
Brahmaputra (Tsangpo)	82	668
Irrawaddy	81	260
Indus	28	250
Mekong	124	150
Salween	55	66
Hwang-Ho	19	1 103
Total for Tibetan/Himalayan region	679	3 518
Global flux total	2 130	20 000

It appears that rivers flowing down from Tibet and the Himalaya have great erosive power and carry an unusually large load of fragmented rocks (Figure 4.3) and dissolved material.

- Bearing in mind the high rates of chemical weathering on the southern slopes of the Himalaya, how could the formation of the Himalaya and the uplift of the Tibetan Plateau indirectly affect the concentration of CO_2 in the atmosphere?

- Weathering of silicate minerals, followed by accumulation and preservation of organic carbon and carbonates in the ocean, result in net removal of CO_2 from the atmosphere (Figure 2.20). Therefore, if uplift of the Himalaya and Tibet increased total global weathering rates, it would also have increased the rate at which CO_2 was removed from the atmosphere.

Of course, a change in the concentration of atmospheric CO_2 would affect fluxes into and out of other carbon reservoirs, and it would have been some time before a new equilibrium was established. In fact, it has been estimated that reduced atmospheric CO_2 levels would have been re-equilibrated about 1 Ma after the mountain range had been uplifted. Other factors being equal, this reduction in atmospheric CO_2 concentrations would lead to global cooling.

In climate models proposed in recent years, the role of mountain building in the global climate system has been treated in various ways. Two contrasting approaches will be considered, which differ in their assumptions concerning what primarily determines the CO_2 concentration of the atmosphere, and hence what ultimately drives carbon fluxes between the various carbon reservoirs, namely the atmosphere, the oceans and the Earth's crust.

The **GEOCARB model** (originally conceived by Bob Berner (Berner, 1994) and his colleagues at Yale University) rests on two important assumptions:

1. Global temperatures are determined by the concentration of CO_2 in the atmosphere.
2. The concentration of CO_2 in the atmosphere is determined primarily by the volume of gases emitted from volcanoes.

An *approximate* measure of the global rate of emission of volcanic gases can be obtained from the rate of production of new sea floor: volcanoes above subduction zones release CO_2 directly into the atmosphere, while some of the gases from hydrothermal vents and underwater eruptions also eventually escape from the ocean into the atmosphere. There is evidence to suggest that over the past 110 Ma, the rate of production of oceanic crust has decreased roughly to what it was before the eruption of major flood basalts at about 140 Ma (Figure 3.10). Over the past 110 Ma, less CO_2 has been supplied by volcanism, so, according to the GEOCARB model, there would have been global cooling as a result of lower concentrations of CO_2 in the atmosphere.

The second approach, referred to as the **mountain-forcing model**, is preferred by scientists who suggest that changes in atmospheric CO_2 concentrations are driven by changes in chemical weathering rates. This model was pioneered by two geoscientists from the USA Bill Ruddiman and Maureen Raymo who extended the idea that uplift of the Tibetan Plateau strengthened the southwest monsoon and suggested that high weathering rates over a region of steep topography affected by high summer rainfall were at least partly responsible for the global cooling that followed the collision between India and Eurasia. If this is true, then uplift of Tibet effectively set the scene for the glacial periods that have characterised the climate in recent geological time.

Question 4.4

Figure 4.13 shows an overall downward trend in temperature from about 100 Ma to the present, but there have been fluctuations and the rate of decrease has been variable. On the basis of what you know about the timing of the building of the Himalaya and the uplift of Tibet, along with information in Figure 3.10, to what extent can the shape of the temperature plot in Figure 4.13 be used to support (a) the GEOCARB model and (b) the mountain-forcing model?

The answer to Question 4.4 provides an important insight to the following discussion, which explores in more detail whether weathering of the Himalaya and Tibet *could* have caused global climate change over the past 50 Ma. The first point to remember is that weathering only results in the removal of CO_2 from the atmosphere because some of the carbon carried to the ocean in rivers is *removed* from the oceans and preserved in sediments, i.e. it moves out of the intermediate-scale marine carbon cycle and becomes part of the long-term geological carbon cycle (Figure 2.20). As you may recall, there are two ways in which carbon can be preserved and buried in deep-sea sediments: as carbonaceous sediments (organic carbon) and as carbonates (inorganic carbon).

Beginning with the organic carbon, one factor that would increase the proportion of organic remains preserved in the deep ocean is a large supply of organic debris in rivers. Before the collision of India with Asia, unusually organic-rich sediments were deposited on the continental shelf along the northern margin of the Indian Plate. After collision, the buried sediments were uplifted and exposed at the surface, where they were subject to weathering and erosion. As a result, much of this organic material may have been carried to the sea as particulate and dissolved organic carbon. As discussed in Box 4.1, $\delta^{13}C$ values for organic carbon are lower than those for carbon from other carbon reservoirs. An increase in the flux of organic carbon to the oceans would, therefore, eventually result in a decrease in the average $\delta^{13}C$ value for sediments (both organic and calcareous) being deposited on the seabed. A plot of $\delta^{13}C$ against depth in marine sediments provides a measure of the contribution of organic carbon to the oceans over time. (Note that for reasons that are beyond the scope of this book, marine carbonates always have higher $\delta^{13}C$ values than soil carbonates (Figure 4.12).)

Question 4.5

(a) What would happen to the average $\delta^{13}C$ value of marine carbonates if the global flux of organic carbon from the continents increased significantly?

(b) According to Figure 4.16 (overleaf), when in the past 70 Ma was the largest increase in the flux of organic carbon to the oceans?

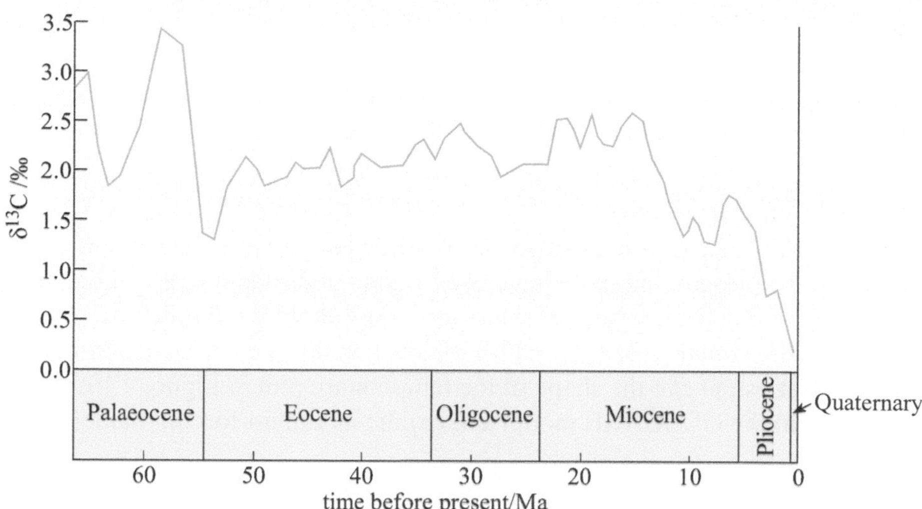

Figure 4.16 Variation of $\delta^{13}C$ in marine carbonates laid down over the past 65 Ma or so, i.e. during the Tertiary and Quaternary Periods.

■ Why might an increase in the rate of continental weathering result in an increase in the accumulation and preservation of both inorganic and organic carbon?

□ Increased chemical weathering will increase the concentration of a number of constituents in seawater, including the following nutrients: dissolved nitrate, phosphate and silica, and (perhaps more importantly) micronutrients such as iron. An increase in the supply of nutrients to the oceans could result in increased primary productivity of phytoplankton, both with and without calcium carbonate hard parts. This would result in an increased accumulation of both organic carbon and inorganic carbon.

An increase in primary productivity in response to increased nutrient supply could also result from increased oceanic upwelling. Vigorous upwelling occurs during the southwest monsoon, so if the uplift of the Himalaya and Tibet caused the monsoon to strengthen then, at the same time, it could have enhanced the rate of preservation of carbon in the deep sea.

In fact, some geochemists regard the preservation of organic carbon as more important than the preservation of carbonates. However, it is hard to determine the extent to which the accumulation of organic carbon is driven by weathering rates, especially as marine primary productivity is greatly affected by other influences, notably wind-driven upwelling, and the supply of nutrients generally.

Ironically, the very fact that the uplift of Tibet and the Himalaya appears to have such a strong effect on the levels of atmospheric CO_2 could potentially be a problem for supporters of the mountain-forcing model, because it seems to leave open the possibility of runaway cooling occurring, for which there is no evidence in the relevant geological records despite the overall decline in global temperature during the past 100 Ma. According to the GEOCARB model, levels of atmospheric CO_2 would be maintained at more or less the same level over time by a negative feedback process as follows: if more CO_2 is released into the atmosphere as a result of increased rates of production of sea floor, global temperatures will rise. This temperature rise would lead to increasing weathering rates, which in turn would *remove* CO_2 from the atmosphere, thus *decreasing* and returning the temperature towards its original value; and so on (Figure 4.17).

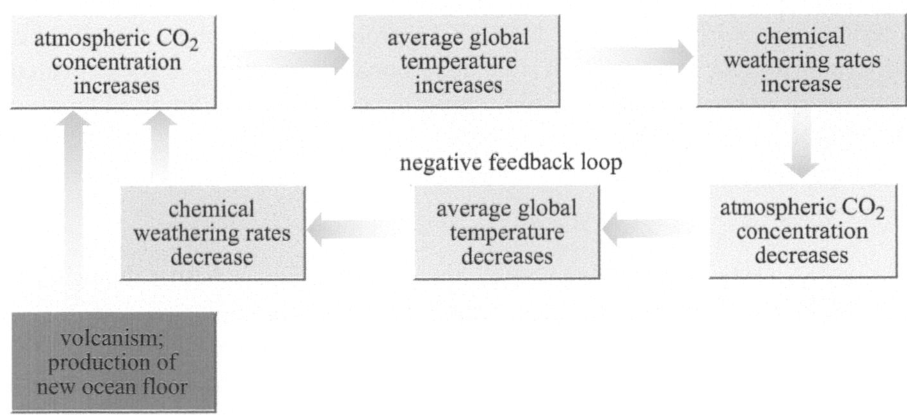

Figure 4.17 The negative feedback loop in the climate cycle derived from the GEOCARB model. If atmospheric CO_2 concentrations *rise* because of *increased* rates of ocean-floor production, then global temperatures will *rise*, leading to *increased* weathering rates. This, in turn, would *remove* CO_2 from the atmosphere, thus *decreasing* the temperature towards its initial value. Similarly, if atmospheric CO_2 concentrations should *fall*, then global temperatures will fall, leading to *decreased* weathering rates. This, in turn, would *remove less* CO_2 from the atmosphere (there would still be various sources of atmospheric CO_2), thus *increasing* the temperature towards its initial value.

On the other hand, if (as assumed by the mountain-forcing model) the concentration of CO_2 in the atmosphere is primarily controlled by mountain uplift, then there is no *direct* link between the rate of operation of the CO_2 sink (i.e. preservation and burial of carbon in the oceans) and the rate of operation of the CO_2 source (i.e. volcanic emission). As a result, there can be no direct stabilising feedback loops of the type described above to prevent runaway cooling.

Estimates of the rate at which the weathering of Tibet and the Himalaya has depleted CO_2 in the atmosphere according to the mountain-forcing model show that weathering alone would exhaust all the CO_2 in the atmosphere in only a few million years – something that obviously has not happened. It is important to remember, however, that the processes that eventually lead to mountain building can, at the same time, provide a source of atmospheric CO_2.

- What is the nature of this source of atmospheric CO_2?
- *Decarbonation* of carbonates mixed with silica from shallow-water carbonates deposited at the continental margins as well as deep-water carbonates on top of the subducting slab, descending beneath the collision zone.

Overall, during the lifetime of a particular evolving mountain belt, the CO_2 flux from lithosphere to atmosphere as a result of subduction, collision and volcanism will replenish a significant proportion of the CO_2 lost from the atmosphere through weathering. It is important to note, however, that although the operation of source and sink are generally related through the rates of plate convergence at the destructive plate margin and collision zone, decarbonation and weathering are *not* closely coupled within individual zones of mountain building: CO_2 gain to the atmosphere will occur early in the history of the mountain belt when carbonates and silicates are first heated; CO_2 loss will predominate later, with the formation of high mountains and rapid weathering.

A further source of CO_2 associated with the collision of continents is the oxidation of organic carbon in the crust. As mentioned above, after the collision of India and Asia, buried carbon-rich sediments were uplifted and eroded. Once exposed at the surface, this organic carbon was oxidised to CO_2 through bacterial activity and by simple chemical reactions with atmospheric oxygen.

In summary, there are at least two mechanisms that would have counteracted any tendency towards a runaway loss of CO_2 from the atmosphere through the weathering of Tibet and the Himalaya: decarbonation reactions in rocks heated up as a result of continental collision within the crust, and oxidation of organic carbon at the surface of the crust.

4.3.2 Testing the models

So far, it has not been possible to comment definitively on whether the original GEOCARB model or the mountain-forcing model is more appropriate for the estimation of changes in atmospheric CO_2 concentrations over the past 50 Ma. When scientists need to decide between alternative models, they usually examine specific outcomes or predictions to see how they differ. As far as these two competing models are concerned, a significant difference arises in the contrasting way in which they link climate change and weathering rates.

- According to the GEOCARB model, how will a high rate of production of new ocean floor affect the rate at which the continents are weathered?

- According to this model, global temperatures are determined by atmospheric CO_2 concentrations, which in turn are determined primarily by volcanic emissions. *Increased rates of production of new ocean floor will, therefore, lead to higher CO_2 levels and higher temperatures*. They will lead to an increase in rates of continental weathering because chemical reactions occur more rapidly at higher temperatures (Figure 4.17).

Implicit in the GEOCARB model, therefore, is the assumption that high rates of weathering of continental rocks would result from global warming.

- What would be the relationship between global temperatures and high continental weathering rates if the mountain-forcing model were correct?

- According to this model, atmospheric CO_2 concentrations are driven primarily by continental weathering rates. As weathering of the continents removes CO_2 from the atmosphere, *high weathering rates should lead to global cooling*.

The GEOCARB model and the mountain-forcing models (at first sight at least) have contrasting implications for the relationship between global temperature, atmospheric CO_2 concentrations and weathering rates. The GEOCARB model involves an association between periods of high rates of chemical weathering and *high* global temperatures (high atmospheric CO_2), whereas mountain forcing predicts an association between high rates of chemical weathering and *low* global temperatures (low atmospheric CO_2). This is a significant point because it means that if the rate of continental weathering through time could be measured and compared with the variation in global temperature (Figure 4.13), it might help scientists to decide whether the principal control on climate change (at least over the past 50 Ma) has been the rate of CO_2 released as a result of either the production of new ocean floor or the formation of high mountain ranges.

Fortunately, geochemists have discovered that certain isotopic ratios in marine sediments go some way towards providing an approximation of rates of weathering of continental rocks. Before illustrating this point, it is necessary to understand some aspects about strontium (Sr) isotopes (Box 4.2).

Box 4.2 The isotopic composition of strontium

The element Sr occurs in nature as several different isotopes. Geochemists are mainly concerned with two of these: ^{86}Sr and ^{87}Sr. The first, ^{86}Sr, is stable and is not the decay product of any other isotope. The other isotope, ^{87}Sr, is known as **radiogenic strontium** because it is the product of radioactive decay of one of the isotopes of rubidium, ^{87}Rb.

Most rocks contain varying amounts of both elemental Sr and Rb. Sr includes both ^{87}Sr and ^{86}Sr; the amount of ^{87}Sr is continually increasing due to the decay of radioactive ^{87}Rb, but the amount of ^{86}Sr remains unchanged. In other words, the $^{87}Sr/^{86}Sr$ ratio will increase with time in any rock that contains Rb. As a result, geochemists generally refer to the $^{87}Sr/^{86}Sr$ ratio rather than to the absolute concentration of ^{87}Rb or ^{87}Sr because the isotopic ratios of different rocks (and fluids) provide more insight about the history of the rocks (and fluids) than the total concentration of elemental Sr.

Rocks that make up the continents have variable $^{87}Sr/^{86}Sr$ ratios, but these values are all greater than the $^{87}Sr/^{86}Sr$ ratio of rocks from the upper mantle. This is because the rocks that form the continental crust have much higher elemental Rb/Sr ratios than those that make up the mantle.

The $^{87}Sr/^{86}Sr$ ratio of present-day seawater represents a combination of two components. The first, with a high average value of about 0.7118, comes from rivers that have entered the oceans after flowing over the continents. The second, with a low average value of 0.7035, comes from hydrothermal fluids that have circulated within the upper mantle and escaped at vents along ocean ridges; this lower value is similar to that of rocks of the upper mantle. These two fluxes combine to yield the present-day $^{87}Sr/^{86}Sr$ value for seawater, i.e. 0.7092, which is an approximation for all the world's oceans because circulation of the ocean water masses is more rapid than the residence time of Sr in seawater and so has homogenised the Sr isotopic ratio by mixing processes.

By measuring the isotopic composition of marine carbonates of known ages, geochemists have calculated the $^{87}Sr/^{86}Sr$ ratio of seawater over the past 100 Ma. Their results are plotted in Figure 4.18.

- ■ What has been the overall trend in the $^{87}Sr/^{86}Sr$ ratio of the ocean since the Himalayan collision ~50 Ma ago?

- □ There has been a marked increase.

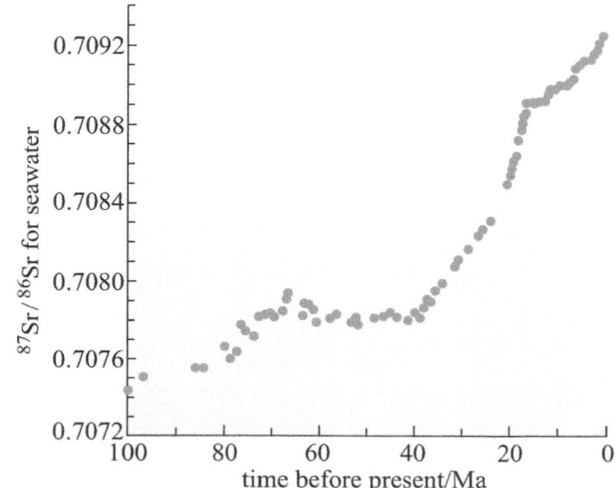

Figure 4.18 Variation in $^{87}Sr/^{86}Sr$ ratio for seawater over the past 100 Ma. (Richter, 1992)

This means that either the rate of supply of hydrothermal fluids (with low $^{87}Sr/^{86}Sr$ ratios) has decreased or that the flux of Sr from rivers (with high $^{87}Sr/^{86}Sr$ ratios) has increased, or perhaps a combination of the two. Hydrothermal circulation is driven by the heating associated with sea-floor spreading and other volcanic activity at the seabed. As ocean-floor production has been slightly increasing over the past 40 Ma, the trend in Figure 4.18 is unlikely to reflect a marked decrease in hydrothermal circulation. On the basis of the data plotted in Figure 4.18, therefore, it appears that the change in $^{87}Sr/^{86}Sr$ in seawater must relate to an increased flux of high $^{87}Sr/^{86}Sr$ from rivers into the ocean. This, in turn, could be interpreted as evidence for the rate of weathering of continental crust increasing since the Himalayan collision.

- If this interpretation is correct, does it support the mountain-forcing model?
- Yes, because the model predicts that weathering of silicates results in global cooling, and there has indeed been global cooling over the period in question (Figure 4.13).

However, the evidence presented so far linking the uplift of the Himalaya and the Tibetan Plateau with weathering rates via seawater chemistry is entirely circumstantial. More direct evidence can be gleaned from examining the Sr-isotope geochemistry of present-day rivers.

If the $^{87}Sr/^{86}Sr$ ratios of the world's largest rivers are plotted against the reciprocal of the concentration of Sr (i.e. 1/[Sr]), most data points are positively correlated, implying they are essentially mixtures of two contrasting endmember components (a low $^{87}Sr/^{86}Sr$ and a high $^{87}Sr/^{86}Sr$ component) (Figure 4.19). In this case, rocks dissolving into the rivers are either carbonates, which have low $^{87}Sr/^{86}Sr$ ratios due to their low Rb/Sr compositions and dissolve rapidly so contributing abundant Sr, or silicates, which have high $^{87}Sr/^{86}Sr$ ratios due to their high Rb/Sr compositions and dissolve slowly so contributing low concentrations of Sr. Of all the world's major rivers, there are two exceptions: the Ganges and the Brahmaputra.

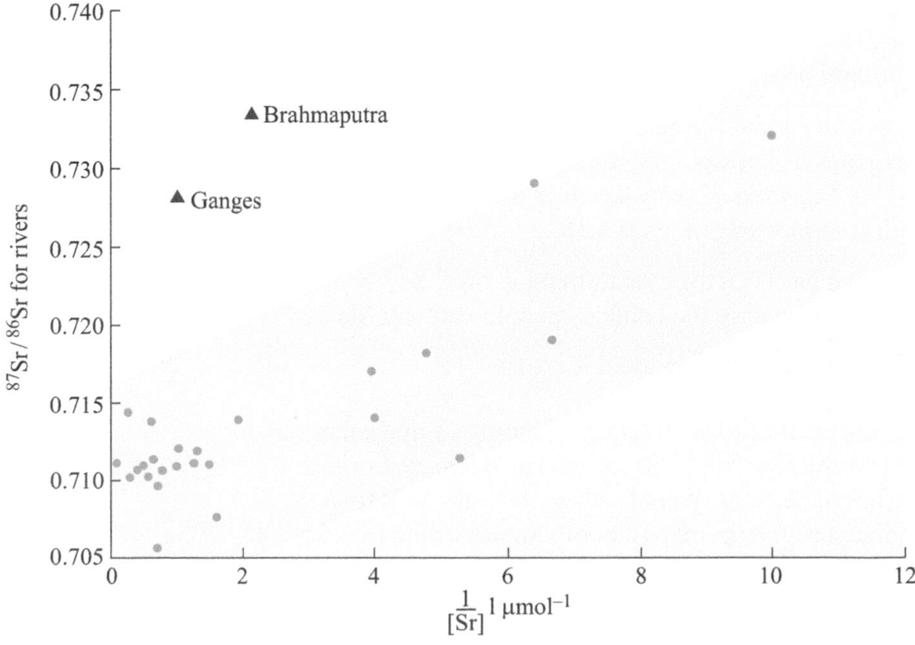

Figure 4.19 The $^{87}Sr/^{86}Sr$ ratios of the world's major rivers plotted against 1/[Sr].

The Ganges and Brahmaputra rivers both flow from South Tibet, across the Himalaya, into the Bay of Bengal (Figure 4.15). Their waters are fed by the southwest monsoon, and their high Sr-isotope ratios, combined with high Sr concentrations, indicate that their addition to the oceans has a marked impact on the Sr-isotope seawater curve (Figure 4.18).

Although the interpretation of the Sr-isotope proxy is complicated by the presence of some unusual rock compositions in the Himalaya, detailed studies of Himalayan rivers suggest weathering rates that are at least triple those of the global average. This conclusion has been taken as evidence to support the hypothesis that the uplift of the Himalaya and Tibet intensified the monsoon system and so caused a major increase in the chemical weathering flux of silicates, leading to increased rates of the removal of CO_2 from the atmosphere, thus providing key evidence for the mountain-forcing model.

It is important to emphasise that the Tibetan Plateau and the Himalaya appear to play quite different roles in the climate system. The strengthening of the monsoon seems to be the result of a *large high plateau* being uplifted, as it was this that affected the atmospheric circulation. In contrast, the link between changes in atmospheric CO_2 concentrations and weathering requires increased weathering rates, and it is over the *Himalaya* that these rates become particularly high and not on the Tibetan Plateau, which experiences little rainfall (much of which evaporates) and has relatively modest relief (Figures 4.6b and 4.14a).

So, which is right: the GEOCARB model or the mountain-forcing model? Well the answer is that they both are! The GEOCARB model for the carbon cycle is a **steady-state model,** which means that it assumes the system has attained equilibrium, e.g. it assumes the flux of CO_2 into the atmospheric CO_2 reservoir equals the flux out of it. If such a steady-state system is perturbed, then steady-state conditions will no longer apply (this perturbation is in effect a forcing function) and the system will adjust until the balance of fluxes is re-established. Uplift of the Himalaya and the Tibetan Plateau can be considered a transient event that disturbed the steady-state carbon cycle. It is believed that, during this transient event, the chemical weathering rates rather than the rates of production of new ocean floor determined atmospheric CO_2 levels.

So the GEOCARB model provides a satisfactory mechanism for long-term climate change over hundreds of millions of years, while the mountain-forcing model is appropriate for the relatively short-term disturbance to global climate caused by the uplift of Tibet and the Himalaya over a period of several million years. Although the two models have been presented as opposing each other, more recent revisions of the original GEOCARB model now include a feedback link between Tertiary mountain uplift and global cooling. In other words, the two models are converging despite operating on different timescales, thus providing a good example of how science advances through the testing of competing hypotheses.

Taking all the potential sources and sinks of CO_2 into account, current estimates suggest that the overall effect of the mountains and plateau of southern Asia is to cool global climate by about 3 °C. Although the role in the climate system of the uplift of the Himalaya and the Tibetan Plateau have been emphasised so as to provide a focus for this discussion, it is highly improbable that a single mechanism could be generally responsible for global climate change. This is not least because the atmosphere is a relatively small reservoir of CO_2 and its size depends on differences between large fluxes. Subtle changes in these fluxes may cause significant climate change. The best that can be said for any supposed cause of climate change is that any specific mechanism *could* result in an observed trend in the climate record; this is not to say that it was unaided, or even that it was the most significant player in such a complex game.

Summary of Chapter 4

1 Over geological timescales, the uplift of mountains as a result of continental collision can affect the operation of the global carbon cycle, and hence the amount of CO_2 in the atmosphere, in various ways. Subduction of ocean-floor sediments and collision of continental masses result in decarbonation, in which silica and calcium carbonate react together to produce CO_2, which eventually escapes to the atmosphere. Mountains are also sites of vigorous erosion and weathering: physical erosion at high altitudes and chemical weathering lower down. Chemical weathering (followed by the accumulation and burial of carbon in the ocean) results in a *loss* of CO_2 from the atmosphere. In addition, continental collision may result in large volumes of carbon in rocks being removed from contact with the atmosphere for many millions of years.

2 Mountain ranges and plateaux affect the climate physically by redirecting air masses around and/or over them. Moisture-laden winds release their precipitation on the windward side, whereas the leeward side (in the case of the Himalaya, the Tibetan Plateau) is dry. The rise of the Himalaya and Tibet is believed to have intensified the strength of the southwest monsoon by providing a source of heat (including *latent* heat) at a critical position in the atmospheric circulation.

3 The rate of uplift of the Tibetan Plateau is not well established; attempts to measure it have made use of fossil plants. There is evidence that the southern plateau reached its present elevation by 15 Ma ago.

4 Attempts to determine climate change in southern Asia include studies of foraminiferans in the Indian Ocean, studies of $\delta^{13}C$ (to investigate plant growth) and analysis of the types of sediment eroded from the Himalaya. These provide evidence to suggest strengthening of the monsoon between 9 and 6 Ma ago.

5 The Earth's climate has cooled markedly over the past 50 Ma. The mountain-forcing model postulates that the uplift of the Himalaya and Tibet imposed cooling on the global climate, by strengthening the southwest monsoon and increasing rates of chemical weathering: increased weathering of silicates followed by the accumulation and burial of carbonates and/or organic carbon in the ocean is assumed to result in the long-term removal of CO_2 from the atmosphere. To prevent 'runaway cooling', the rapid removal of CO_2 by weathering would need to be partially compensated. Possible mechanisms for replenishing atmospheric CO_2 include decarbonation at subduction zones, and the oxidation of organic carbon from exhumed sediments.

6 The steady-state carbon cycle model (i.e. the original GEOCARB model) interpreted the role of mountains as providing a negative feedback loop that stabilised fluctuations in temperature resulting from variations in volcanism associated with the production of new sea floor.

7 The isotope ratio $^{87}Sr/^{86}Sr$ of the world's oceans has increased over the past 50 Ma. This has been interpreted as indicating increased weathering rates due to the uplift of Tibet and more especially the Himalaya. This interpretation is supported by the Sr fluxes and isotope ratios of rivers currently eroding the Himalaya.

Learning outcomes for Chapter 4

You should now be able to demonstrate a knowledge and understanding of:

4.1 How the global carbon cycle is affected by plate tectonic processes and in particular continental collision and the uplift of mountains.

4.2 The influence mountain ranges have on the physical characteristics of the climate, altering past and present atmospheric circulation patterns and, in turn, how this has influenced the location and intensity of precipitation.

4.3 The various lines of evidence that can be used to determine climate change, including studies of foraminifera relating to nutrient levels and periods of upwelling in the oceans, changes in $\delta^{13}C$ values associated with rates of plant growth, and the types of sediments being eroded, indicative of physical and/or chemical weathering and changing temperature.

4.4 How the mountain-forcing and GEOCARB models can be used to model changes in the Earth's climate over different timescales, and how their apparently conflicting predictions can be deployed to test different feedback mechanisms.

4.5 The isotopic, palaeontological and geological evidence used to demonstrate how and why the Earth's climate has cooled over the past 50 Ma, and the different positive and negative feedback roles that subduction, ocean-floor formation and mountain building have played.

The emergence and persistence of life

Chapter 5

> I look at the natural geological record as a history of the world imperfectly kept, and written in a changing dialect; of this history we possess the last volume alone.
>
> (Darwin, 1859, p. 311)

One aspect of Darwin's disappointment with the then known fossil record of evolutionary change was the apparent absence of Precambrian fossils. The only hint of such ancient life that he knew of, when he made the statement quoted above, was 'the presence of phosphatic nodules and bituminous matter' in them. Nevertheless, he firmly believed that there had to have been a long history of evolution prior to the seemingly abrupt first appearance of marine shelly fossils just over 500 Ma ago, as the latter already represented many different major groups of animals. His conviction has been vindicated, for there is a fairly extensive, if partial, Precambrian fossil record stretching back to Archaean times – although the story it tells might have come as something of a surprise to Darwin himself. It turns out that animals (multicellular organisms that ingest their food) do seem to have made a relatively late appearance in the **Neoproterozoic**, followed by the evolutionary explosion of shelly forms that announced the dawn of the Phanerozoic. Yet it was during this long earlier history of the Archaean that the fundamental pattern of feedbacks between life and the Earth, particularly the atmosphere, became established, and the precursors to the later forms of Phanerozoic life, the **eukaryotes** (whose cells contain nuclei), evolved. In Chapter 1 you learned about some of the factors of the Earth system that regulate the habitability of the surface; in this chapter you will look at some of the history of life on Earth and its evolution.

The Precambrian or 'Cryptozoic' is the period of the geological timescale from the formation of Earth (around 4.6 Ga) to the evolution of abundant macroscopic fossils, which marked the beginning of the Cambrian (~542 Ma).

5.1 Former worlds

As you try to understand the past conditions on Earth you must remember that the environment has apparently stayed within modest limits for most of the Earth's history, although some have questioned this, postulating mean temperatures of over 60 °C and even higher oceanic temperatures for the Archaean. Other aspects – most notably atmospheric composition – have undergone radical changes (see Box 5.1 overleaf). Life has apparently been present throughout virtually all the time considered from near the beginning of the Archaean, and it has been implicated both in the regulation of global conditions and in their changes. Individual complex multicellular organisms are nevertheless of relatively recent vintage, arriving approximately 600 Ma ago during the beginning of the Phanerozoic. Today, they present a kaleidoscopic web of interactions, both with each other and with their environments.

Box 5.1 Oxygen through time

The concentration of oxygen in the Earth's atmosphere has not remained constant through time. Prior to about 2.4 Ga, evidence suggests that the atmosphere was essentially oxygen-free (although some researchers doubt this). The presence of reduced minerals such as pyrite and uraninite in riverbeds older than 2.4 Ga attest to **anoxic** conditions. The early geological record also contains banded iron formations (BIFs), which are thought to have been formed by reduced iron that welled up from the anoxic deep ocean and oxidised in surface waters. BIFs begin to disappear from the geological record at about 1.8 Ga. Their continuation for about 600 Ma after the first rise in oxygen may reflect the time it took for the deep oceans to become oxygenated.

One conundrum with this view of the early Earth is the fossil evidence for oxygen-producing life before 2.4 Ga. Fossil stromatolites, carbon isotopic ratios consistent with oxygenic photosynthetic organisms, and **biomarkers** (specific biological molecules) suggest that oxygenic photosynthetic life had evolved by 3.5 Ga. Why was the rise in oxygen delayed long after oxygen-producing life appeared? One solution to this apparent problem is that in the early stages much of the oxygen being produced reacted with reductants in the atmosphere such as hydrogen and hydrogen sulfide (H_2S). Only when the concentrations of these reductants declined could the net atmospheric concentration of oxygen increase.

The lack of oxygen in the early atmosphere would have meant that there was no stratospheric ozone to screen ultraviolet radiation between 200 nm and 290 nm, resulting in potentially higher radiation damage to surface-dwelling organisms. In the worst case, UV radiation would have been about one thousand times more damaging to DNA than it is today, although some researchers have suggested that the early atmosphere might have contained a hydrocarbon smog, which could have provided UV shielding.

A second rise in oxygen can be recognised from the geological record at about 800–600 Ma. Like the first rise in oxygen about 2.4–2.3 Ga ago, this transition was associated with a global freezing or 'snowball' event. One explanation for these dramatic environmental changes associated with the rise in oxygen is the consumption and conversion of the greenhouse gas methane (CH_4) by oxygen into CO_2 and H_2O, causing a global cooling. The period following the second rise in oxygen also coincides with the Cambrian explosion and the proliferation of animal diversity.

Investigating what controls the concentration of oxygen in the atmosphere, past and present, and how this might be linked to biological evolution is an important challenge in understanding the habitability of the Earth.

Question 5.1

What conditions would you expect to find if you were to visit a mid-latitude, lowland area on Earth at (a) 4 Ga, (b) 2.5 Ga, (c) 500 Ma, (d) 100 Ma, and (e) 100 000 years ago? To help you think about your answer, decide which of the following items you would need to take with you: (i) breathing apparatus (with a supply of oxygen), (ii) suncream (factor 100), (iii) a thermally insulated self-contained capsule, (iv) a crabbing net, (v) weapons for fighting large predators, (vi) sandwiches, and (vii) a good book.

For the first three-quarters of life's 3.8 billion-year history, there is frustratingly little fossil evidence (Figure 5.1a), despite its profound influence on the composition of the atmosphere, but what there is shows surprisingly little change in outward form and diversity. Yet this was a time of fundamental innovations in cell structure, giving rise to the first eukaryotes around ~2.7 Ga. Much later, at 1.2–1.0 Ga, a prolific burst of eukaryote diversification ensued. Such meagre fossil evidence as there is, together with other kinds of chemical biomarker, judiciously spiced with speculation, is the diet on which our understanding of life on the early Earth is based.

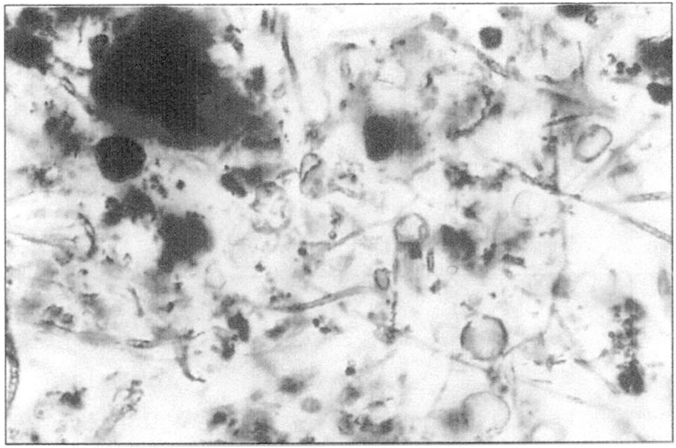

(a)

(b)

Figure 5.1 (a) Early Proterozoic fossil assemblage. Photomicrograph of prokaryotic cells in chert from the Gunflint Iron Formation, Ontario, Canada. Spheres are about 10 μm across. (b) An assemblage of shells of the trilobite *Redlichia* from the Lower Cambrian of Kangaroo Island, Australia. The fossil at bottom centre is 5 cm long. ((a) Andrew Knoll; (b) Peter Crimes)

From around 540 Ma, a proliferation of shelly animals began to litter the sea floor with their remains (Figure 5.1b), commencing the rich fossil record of the Phanerozoic. The Phanerozoic has yielded a sharply focused picture of patterns in the history of life relative to the Precambrian record. Very uneven rates of turnover (e.g. appearance and disappearance of fossil groups) have been revealed, with mass extinctions episodically cutting across long phases of slower evolutionary change. One eminent palaeontologist aptly commented, 'the history of any one part of the Earth, like the life of a soldier, consists of long periods of boredom and short periods of terror'. Indeed, this episodic pattern was widely acknowledged by geologists from the beginning of the 19th century and was utilised as a basis for the relative geological timescale that is still in use today:

- the most devastating mass extinction was used to mark the close of the Palaeozoic Era (i.e. the Permo-Triassic boundary)
- a less pervasive but more abrupt mass extinction (including the dinosaurs) was taken to mark the end of the Mesozoic Era (i.e. the **K/T boundary**).

K/T boundary = Cretaceous–Tertiary boundary (C was already used for Carboniferous, hence the use of K).

Some of the geological periods within the eras were likewise terminated with mass extinctions of varying severity, although not all were. Obviously, some of these extinctions radically reset the biological agenda for subsequent evolution, with the richness of the Phanerozoic fossil record and the accompanying wealth of geological data placing constraints upon hypotheses seeking to explain such events and their aftermaths. These are explored in greater detail in subsequent chapters, but for the present you need to consider Precambrian times. An understanding of this history is essential to investigate the origins of eukaryotes and to make sense of the part they have played in the Earth–life system.

5.2 Mat world

Life on Earth at 2.5 Ga (which was conjured up in Question 5.1) comprised greenish-grey mats along the coastline. The laminated calcareous structures they formed (called **stromatolites**) dominated the macroscopic evidence of life in all but the last two hundred million years or so of the Proterozoic fossil record. They had already become quite widespread and diverse by early Proterozoic times. In rare cases, fossil stromatolites contain tiny filamentous and spheroidal fossil remains (Figure 5.1a). But how representative are these microscopic fossils (**microfossils**) of the life forms of that time? One of the best-known examples of such preservation is the 2 Ga old Gunflint Iron Formation that straddles the boundary between Canada and the USA around Lake Superior (Figure 5.1a). The exquisite preservation of the fossil structures in this instance is due to mineralisation in chert (a form of silica, SiO_2), precipitated from mineral-rich waters seeping through the microbial mats. The mineralisation must have taken place soon after deposition and before significant decomposition of the organic materials could occur, and was probably due to molecules of silicic acid (H_4SiO_4) formed by the earlier dissolution of silica, bonding with hydroxyl ions (OH^-) on the surface of the organic matter.

The preservation of similar associations in Proterozoic rocks found elsewhere in North America, Siberia and western Australia suggests that the organisms might have been widely distributed. The minute size of the fossil structures is consistent with a microbial origin, the filamentous sheaths resembling those of living cyanobacteria (although the cyanobacterial interpretation of some Archaean fossils has recently become a matter of some controversy), and the spheroidal structures (2–15 μm across) also resembling cyanobacteria, although many other bacteria also have a spheroidal (coccoid) morphology. The filaments do not display the preferred orientations (alternations of horizontally and vertically orientated clusters) characteristic of mat-forming microbes in younger stromatolites, so the interpretation of their relationships and ecology must be tentative.

There are also some non-stromatolitic assemblages of microfossils in the Gunflint Iron Formation and elsewhere, preserved in silicified mudrocks. These include star-shaped forms interpreted as iron- and manganese-oxidising bacteria, as well as simple, rather larger (6–30 μm across) spheroidal forms of uncertain evolutionary relationships, which were probably planktonic.

A yet wider variety of stromatolites (ranging from flat-layered to domed, conical, columnar and even branched forms) is known from younger Proterozoic rocks, reaching a peak in those deposited around 1 Ga. Microfossils also testify to a rich diversity of microbes existing at that time. The association of various fossils with differing sediments suggests some diversification among habitats. Many of the microfossils are quite similar to living cyanobacteria or, more rarely, to bacterial **heterotrophs**. Larger eukaryotic forms also contributed to the diversity of that time (Figure 5.2) and Figure 5.3 shows a diagrammatic transect across a range of shelf-sea environments that can be interpreted from sedimentary rocks of mid- to late Proterozoic age, illustrating the variety of life forms associated with them. The restricted inshore to supratidal zone (marked (1) in Figure 5.3) was

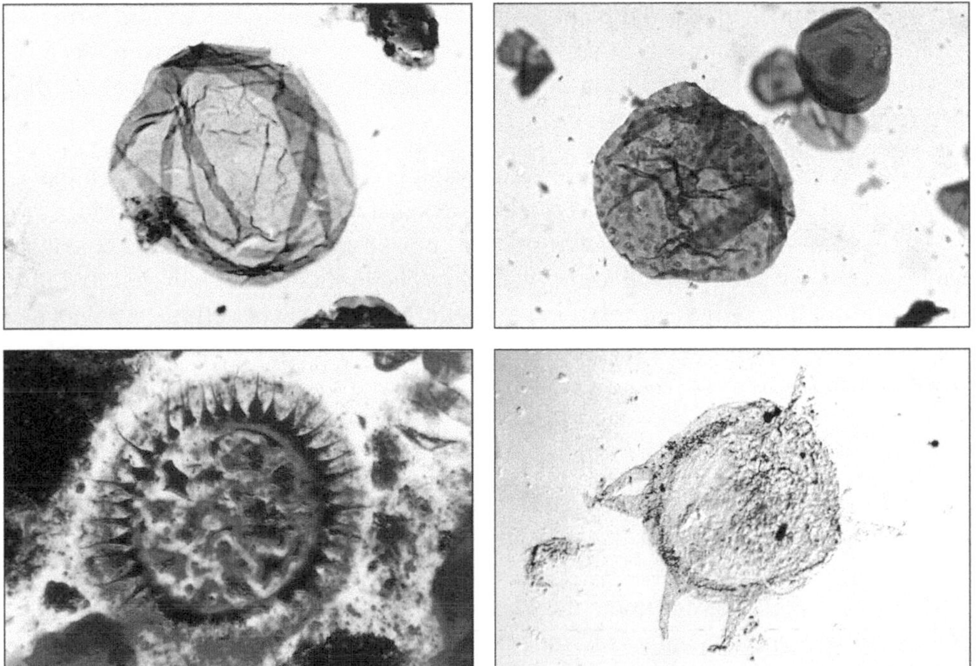

Figure 5.2 Fossils of early eukaryotes: a variety of cysts (resting stages), known as **acritarchs**, from rocks of late Proterozoic age. (Each box in the figure is about 300 μm high.) (Andrew Knoll)

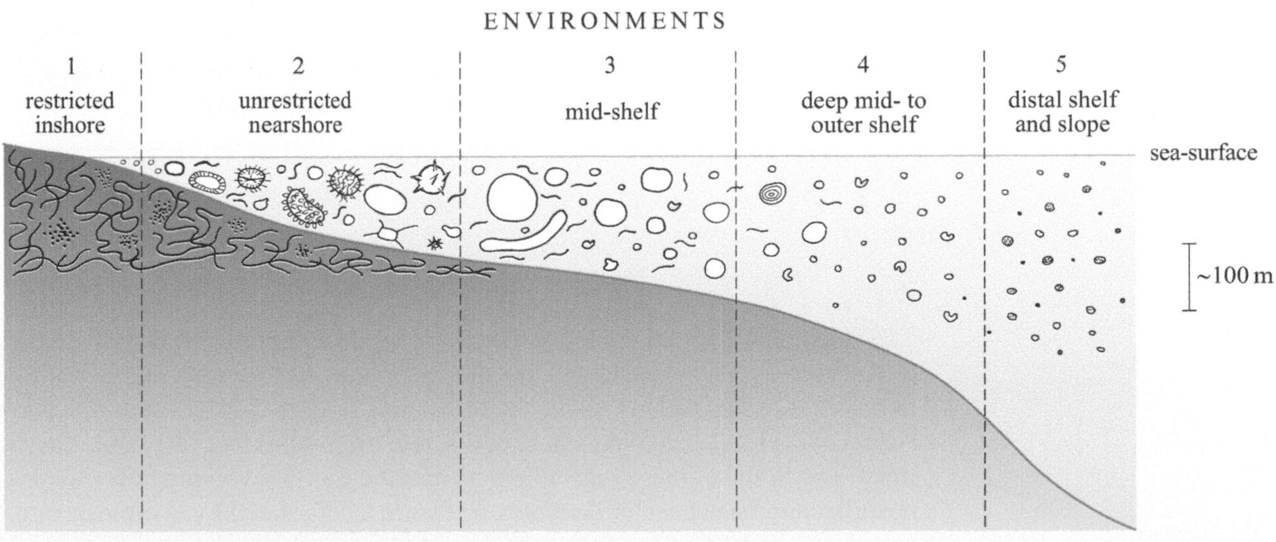

Figure 5.3 Reconstructed distribution of life forms in a range of shelf to basin environments in mid- to late Proterozoic times. The five numbered environments and their inhabitants are described in the text. *Note*: figure not drawn to scale. (Butterfield and Chandler, 1992)

dominated by tiny filamentous and spheroidal **prokaryotes** that produced stromatolites (like the earlier Gunflint examples). In unrestricted nearshore zones (2), prokaryotic filaments were joined by larger (>50 μm), elaborately ornamented eukaryotic cysts called acritarchs (discussed below), which were probably planktonic. In somewhat deeper waters on the mid-shelf (3), less ornamented, spheroidal acritarchs predominated. Over the deep mid- to outer shelf (4) and distal shelf and slope (5), there was a low diversity of small (<50 μm), simple, spheroidal plankton, with larger, probably eukaryotic forms yielding entirely to clusters of smaller prokaryotic cells out in the basins. As a result of the lack of preservation of ancient ocean floor, little is known about the life forms that lived there.

By mid-Proterozoic times, microbial ecosystems had already established most of the fundamental biogeochemical feedbacks that shape the modern world. Foremost among these was the evolution of plankton, which supplied a constant rain of dead organic material to the sea floor. There, it could be incorporated in the accumulating sediment, relatively undisturbed by current activity, and not see the light of day again until geological processes returned the sedimentary rocks to the surface.

Question 5.2

What do you think the effects of organic removal might be on the atmosphere?

The stromatolites that dominate the Proterozoic fossil record are not without their own biogeochemical significance. Their widespread development within the photic zone (from open shelf to supratidal settings) trapped large quantities of calcium carbonate ($CaCO_3$), to form beds of limestone. This represented an important geological sink for carbon (in the carbonate), although it was not accompanied by the release of molecular oxygen, unlike photosynthesised organic matter. Nevertheless, the carbonate story was, in time, to be taken much further by the eukaryotes, as you will see later in this chapter.

5.3 Empire of the eukaryotes

5.3.1 What is a eukaryote?

The cells of eukaryotes possess orders of magnitude more DNA than do those of prokaryotes, which is packaged together with proteins to form a number of thread-like bodies in the nucleus called **chromosomes** (Figure 5.4a). Besides the nucleus, eukaryote cells also contain a host of discrete **organelles** that carry out specific functions. Some of these organelles, such as the energy-evolving mitochondria, are postulated to have been prokaryotes that were incorporated by **symbiosis** into early eukaryotic cells. Another feature of the eukaryote cell is an internal framework of protein rods (microfilaments and microtubules) that plays an important role in its internal organisation. In contrast, most prokaryote cells are surrounded by a rigid outer wall, and it is to the inner surface of this that their simple loop of DNA is anchored (Figure 5.4b). In addition they also lack organelles.

Question 5.3

What is the main difference in the organisation of DNA in these two major groups of organisms, i.e. eukaryotes and prokaryotes?

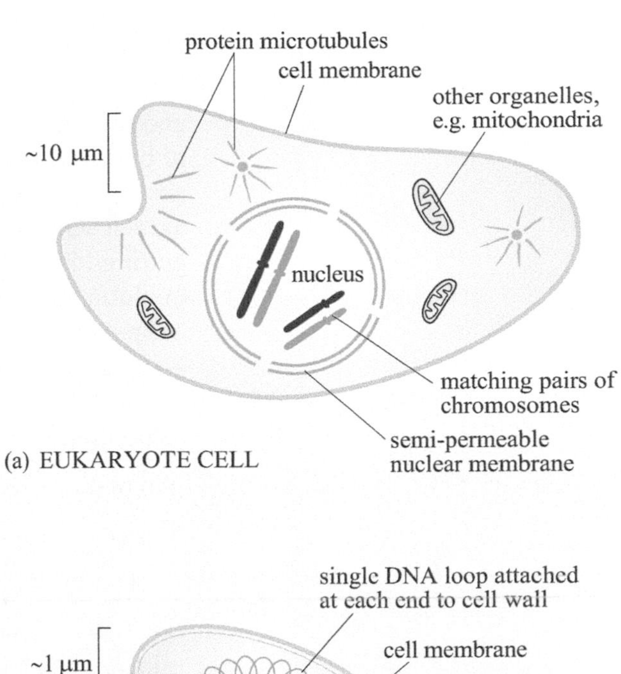

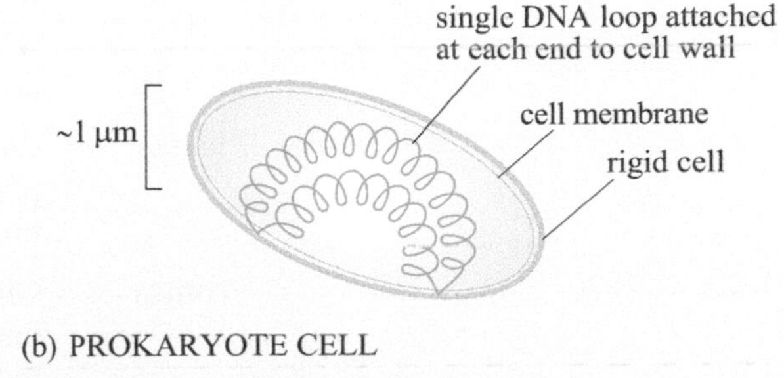

Figure 5.4 The arrangement of DNA in (a) a eukaryote cell and (b) a prokaryote cell.

For reasons that will be explained shortly, these differences in cell architecture, particularly in the arrangement of the DNA, have vastly expanded the scope of eukaryote evolution relative to that of the prokaryotes, and the contrast is readily apparent from living examples (Box 5.2). Most organisms that can be distinguished by the unaided eye are made up of eukaryotic cells. Most are multicellular, including plants, fungi and animals, although there are also many single-celled forms that reach tens to hundreds of micrometres in maximum dimension. Single-celled eukaryotes include amoebae, foraminifers, *Plasmodium* (the parasite that causes malaria) and many kinds of unicellular algae. The vast majority of prokaryotes (e.g. bacteria), by contrast, are unicellular, although a few may group together to form multicellular filaments. Their cells are usually at least an order of magnitude smaller than eukaryote cells (they are rarely more than 10 μm in maximum dimension and, in most forms, are less than 1 μm). The innovations ushered in by the eukaryotes were therefore key events in ecological history.

Box 5.2 The kingdoms of organisms

Until well into the 20th century, it seemed suitable to divide organisms into only two broad kingdoms: plants and animals. Plants are mostly **autotrophic** (from *auto* meaning self and *trophic* meaning feeding) and **sessile** (fixed), while animals are mostly heterotrophic and mobile.

As techniques of study improved, particularly with the advent of electron microscopes, it became clear that bacteria, which had traditionally been lumped with the plants, differed profoundly in structure from all other organisms. Hence, the fundamental division between prokaryotes and eukaryotes was established. However, many eukaryotes still posed problems for the old plant/animal division. Fungi, for example, contradict their traditional 'plant' status by being heterotrophic, as well as showing various unique chemical and structural traits. Moreover, unicellular eukaryotes present a highly confusing picture. For example, some eukaryotic forms swim by means of a whip-like structure called a flagellum, yet possess chloroplasts for photosynthesis, while other closely related flagellate forms feed heterotrophically by ingesting prey. For a time, a convenient solution seemed to be to put all the unicellular eukaryotes in another kingdom, Protista, so leaving the multicellular forms to be divided between plants, animals and fungi. Further detailed studies of cell structure and differences in DNA sequences have caused even this neat compromise to break down as further significant divisions were recognised, some cutting across the unicellular/multicellular divide. At present, there is no universally agreed scheme and evolutionary relationships are still debated. To avoid getting lost in detailed discussion, the pragmatic (if not strictly evolutionary) divisions of the eukaryotes shown in Table 5.1 will be adopted in this book.

Table 5.1 Divisions of the eukaryotes.

Kingdom	Description
Animalia	Multicellular, heterotrophic forms that ingest their food
Fungi	Largely multicellular (although with incomplete dividing walls), heterotrophic forms that feed on organic molecules, which they either absorb directly or break down externally with enzymes and then absorb
Plantae	Largely multicellular, autotrophic forms, comprising land plants (all multicellular) and algae (both seaweeds, which are multicellular, and related unicellular forms)
Protista	Other unicellular forms, which may be either autotrophic or heterotrophic. As used in this book, this grouping is still a motley assembly of several only remotely related groups, some of which share relatively more recent common ancestors with certain of the other kingdoms

5.3.2 Eukaryote beginnings

Fossil evidence of the earliest eukaryotes is extremely sparse. The first forms are likely to have been single-celled and devoid of any skeletal hard parts and, as such, their potential for preservation would have been slim. The discovery of exceptional fossils of such organisms, therefore, has a strong component of luck, but each year new records are discovered or appear that push the inferred schedule for various evolutionary accomplishments further back in time. Even when they are found, the fossils may be difficult to interpret as little critical structural information tends to be preserved. Criteria for recognising eukaryotes are themselves prone to reassessment as more is learned about the characteristics of living forms and the processes involved in the preservation of the fossils themselves. Figure 5.5 summarises the evidence available at the time of writing (2008) – you may choose to add any subsequent information that you encounter.

What features should be used to identify eukaryote fossils? Complex histories of chemical alteration of the original organic materials of the organisms, if preserved at all, effectively limit the analysis of body fossils to their structural features. The scale of resolution of these depends upon the circumstances of preservation. The finest preservation is found where organic materials have been directly coated by microscopic mineral crystals very soon (maybe only days or even hours) after death. The potential resolution of this style of preservation depends on the size of the mineral crystals, which is analogous to the 'grain' in a photograph affecting the detail of the image. Rare examples of petrified soft tissues are known from Phanerozoic deposits in which even the shapes of cell nuclei can be discerned. However, the preservation of structure is usually somewhat cruder. Where only impressions remain in the enclosing sedimentary rock, the grain size of the host rock itself limits the available resolution.

Question 5.4

From what you have learned so far about the differences between prokaryotes and eukaryotes, which diagnostic features of eukaryotes might you have some chance of detecting in the fossil record?

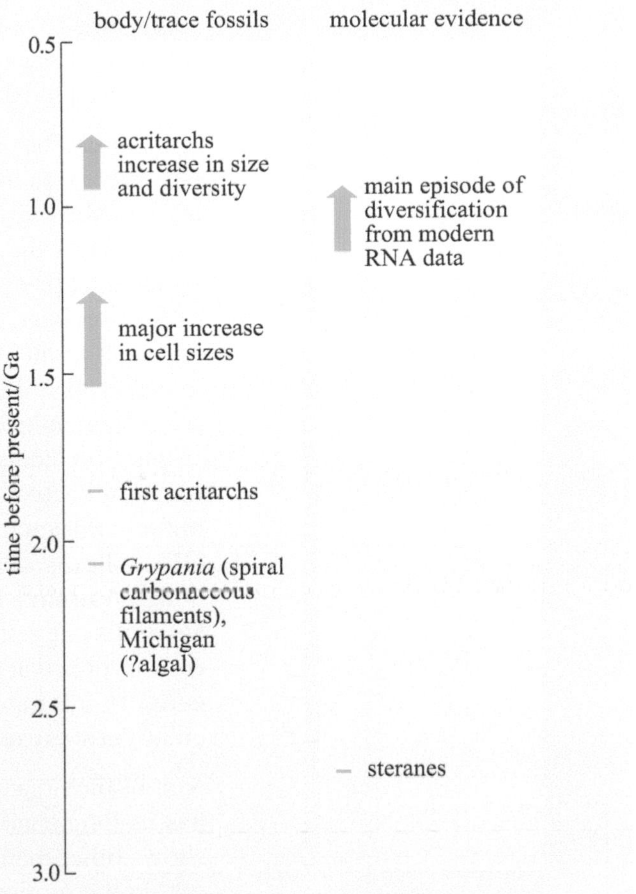

You might also have thought of the characteristic product of **meiosis** (e.g. a quartet of **haploid** cells; see Box 5.3), and indeed some Proterozoic fossil tetrahedral cell clusters have been interpreted as such. However, because such clusters might equally well have been produced through successive fissions by prokaryotes that failed to separate, this criterion is regarded as unreliable.

time before present/Ga: 0.5, 1.0, 1.5, 2.0, 2.5, 3.0

body/trace fossils: acritarchs increase in size and diversity; major increase in cell sizes; first acritarchs; *Grypania* (spiral carbonaceous filaments), Michigan (?algal)

molecular evidence: main episode of diversification from modern RNA data; steranes

Figure 5.5 Evidence for the early evolution of eukaryotes.

In practice, all three possible criteria given in the answer to Question 5.4 have proved problematical in the search for the oldest eukaryote fossils. Dark spots within some Proterozoic fossil cells preserved in cherts have been interpreted by some researchers as representing organelles, but the development of similar features produced through cell collapse in rotting prokaryotes gives cause for doubt. Evidence for cellular organisation is also not particularly pertinent in this context, as tissue differentiation only occurred much later in multicellular eukaryotes. Even cell size is not strictly diagnostic, as there is some overlap in the size ranges of prokaryote and eukaryote cells. Significantly increased size in fossil cells is, however, regarded as reflecting the spread of eukaryotes. In Proterozoic rocks deposited over the time interval from about 1.6 Ga until about 1.4–1.2 Ga, the average size of fossil cells more than doubles. Marine shales dating from about 1.4 Ga, moreover, yield resistant organic cell coats, called acritarchs (Figure 5.2), most of which probably represent the cysts (thick-walled resting stages) of eukaryotic algae that floated in the seawater.

Chemical fossils (chemofossils) push the origins of the eukaryotes even further back in time than the microfossil evidence cited above. Substances called **steranes** (Figure 5.5), formed from sterols (of which a familiar example is cholesterol), are known only from eukaryotic cell membranes that have been recorded from Australian petroleum deposits dated to at least 2.7 Ga. The interpretation of yet older fossils remains inconclusive. Carbonaceous filaments abundant in some formations dating from about 1.3 Ga onwards have also been recorded in rocks as old as 2.1 Ga (e.g. *Grypania*, Figure 5.5). These are commonly interpreted as multicellular eukaryotes that were probably algae of some sort, although a prokaryotic identity for at least some of them cannot be ruled out. Recently reported examples from China dating to about 1.7 Ga are considered to be eukaryotes.

All that can be reliably gleaned from the fossil record, then, is that the eukaryotes, as recognised on the limited criteria discussed above, had arisen at least by the middle of the mid-Proterozoic (by about 2.1 Ga), if not some time earlier. How they evolved remains open to conjecture, although a fascinating hypothesis first proposed some time ago, and which is consistent with a number of features seen in living eukaryotes, has been championed by the American biologist Lynn Margulis. Some eukaryote organelles bear striking resemblances to bacterial cells; for example, **mitochondria** (singular mitochondrion), which are responsible for the energy supply derived from aerobic (oxygen-consuming) respiration, contain small amounts of DNA in a simple loop, similar to that found in bacteria, although some of the DNA responsible for a functional mitochondrion is housed in the cell nucleus. **Chloroplasts**, the organelles that effect photosynthesis in green plant cells, show a similar pattern, and can even undergo **binary fission**. From this and the evidence of other resemblances, Margulis suggested that these organelles were indeed once independent prokaryotes that took up symbiotic residence inside ancestral eukaryotic host cells. This is known as the **endosymbiotic hypothesis**, proposed for the origin of eukaryotic organelles.

Not all the organelles can be so readily explained; the nucleus in particular has its long chains of DNA in protein-bound chromosomes (Section 5.3.1) and shows other marked differences from bacteria. It probably originated from within the original cell structure of a distinct kind of prokaryotic ancestor (which then became the primordial eukaryote host cell type for the later acquisitions mentioned above). One effect of replacing a single DNA loop with multiple chromosomes is that the process of DNA replication during cell division can be carried out at the same time on different chromosomes. It is thought that this parallel processing approach may have been an adaptation for speeding up the replication of increasing amounts of DNA.

If at least some eukaryote organelles were thus secondarily acquired through endosymbiosis, the original ancestral host may have differed from other prokaryotes only in the possession of a nucleus. This primordial eukaryote line may well have extended back into the Archaean as a product of the early evolutionary diversification of the prokaryotes.

Question 5.5

Casting your mind back to the earlier discussion of the structural differences between eukaryote and prokaryote cells, what difference from the usual prokaryote condition would have been a prerequisite for endosymbiosis?

The living bacterium *Thermoplasma* (one of the Archaebacteria), which is adapted to extreme conditions of high temperature and acidity, lacks a rigid cell wall, and in this respect provides a plausible model for the eukaryote ancestral host cell. The ancestral form is thought likely to have been an anaerobically respiring heterotroph (Figure 5.6) that in due course adapted to rising oxygen levels through symbiosis with aerobic bacteria capable of mopping up the oxygen. These eventually became the mitochondria.

- What immediate and later advantages would have been gained by the host from such an acquisition?
- The aerobic bacteria would have helped to remove molecular oxygen, which would have been toxic to the anaerobic host. Later, the host could exploit the energy yield of its aerobic guests.

Photosynthesis may also have arisen in eukaryotes through the acquisition of autotrophic **symbionts** that eventually became chloroplasts. These various acquisitions of organelle precursors were not necessarily unique events and the diversity of chloroplast types in different groups of algae implies independent acquisition in several different lines. In some cases it appears that the chloroplast precursor was itself a chloroplast-bearing eukaryote because a nuclear relic is present within the chloroplast structure; the brown algae (e.g. kelp), common along many shorelines, are one such group. There is as yet no evidence for mitochondria having been acquired more than once, but then again nothing is known of the mitochondria in the myriads of organisms that have become extinct.

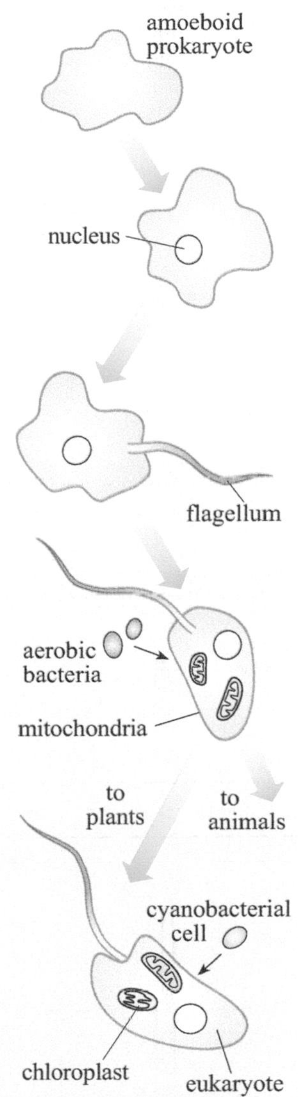

Figure 5.6 The sequence of possible acquisitions of organelles by eukaryotes according to a modern version of the endosymbiotic hypothesis. The first step (initiating the primordial eukaryote line) was the formation of a nucleus. Thereafter, some of the organelles at least may have been acquired through endosymbiosis of various prokaryotes, as shown.

The timing of this hypothesised evolutionary assembly of the variously 'kitted-out' eukaryotes can only currently be guessed at, as direct fossil evidence for the appearance of the organelles in question has not yet been found. Indeed, the succession of endosymbiotic unions may well have extended over a long interval, although in the case of the spread of mitochondria, any time after the appearance of molecular oxygen in the atmosphere and oceans would seem plausible.

Question 5.6

On this basis, very approximately, when might mitochondria-bearing eukaryotes first have begun to proliferate?

Such early evolutionary advances appear to have had only a limited initial impact on the Earth's ecosystems – or at least this appears to be the case for those ecosystems represented in the fossil record. As noted earlier, the largely prokaryote-dominated mats that yielded stromatolites continued to dominate shallow sea-floor communities for another 1000 million years, until their rapid decline from around 850 Ma. The main evolutionary divergence of eukaryote stocks, including most of the various chloroplast-bearing groups, seems to have occurred only after a prolonged period of time.

5.3.3 Eukaryote architecture and evolution

Two attributes of eukaryote cells have been of paramount importance in the evolution of their diversity and complexity. First, their flexible cell membrane is supported internally by the molecular framework of protein rods. These microfilaments and microtubules can grow or shorten and so act rather like a system of extendible tent poles inside a tent. The eukaryote cell, unlike the prokaryote cell, can therefore change shape and engulf external objects. This difference has permitted the addition of consumption by physical ingestion to the repertoire of feeding modes, and ultimately, to the build-up of multi-tiered **trophic pyramids**. The microtubules also play other crucial roles, for example in cell division by controlling internal structure (described below).

The second attribute is the vastly greater number of genes in eukaryotic cells that permits more specialised development. Multicellular forms, in particular, may contain a huge variety of cell types, making up tissues with widely different functions. Just consider those that make up your own body, for example. Specialised cells produce bones, skin and connective tissue for support and protection, while muscle cells allow movement, itself coordinated by nerve cells. A variety of secretory cells produce everything from hormones, controlling growth and behaviour, to digestive juices and mucus, while blood cells are involved in gas exchange and the immune system. Reproduction is effected by sex cells. And so the list goes on.

All these functions are underwritten by genes and nearly all cells possess the genetic information to make potentially any type of cell in the body. Development from a single initial cell means that the genes have not only to be reliably replicated as cells repeatedly divide, but their activity must also be regulated through interactions between genes to produce different types of cell. Such a complex set of instructions involves a lot more DNA than that found in the simple loop within a prokaryote. Whereas a typical bacterium may have some 4 million **base pairs** in its DNA, that of a human, for example, has about 3.5 billion base pairs (although probably only up to a quarter of this DNA actually codes for proteins). The management of so much DNA poses its own problems. For example, the total number of mutations is unavoidably increased, despite the existence of complex enzyme-controlled DNA repair mechanisms. For although the probability of mutation at any one DNA base site is extremely low (around one in a billion per replication), it is now compounded over billions of sites, so mutations are likely to arise at almost every cell division. Where these affect the functions of the proteins coded for, most will be of detrimental effect, and only a few, by chance, will be beneficial. Without some means of compensating for such errors, viable development of multicellular forms would be virtually prohibited.

One way the problem of harmful mutations is mitigated in eukaryotes involves a doubling up of chromosomes in the cell nucleus to yield what is termed a **diploid** complement of chromosomes: each chromosome thus has a homologous partner, containing corresponding genes. If a gene in one set becomes corrupted by a mutation then, to some extent, its function can be covered by the matching gene from the other set. Given the low probability of mutation of any one gene, the chances of both homologous genes being coincidentally affected is almost negligible. A twin-engined aeroplane provides an analogy: it may be able to continue flying even if one engine fails, whereas a single-engined craft will crash if its engine fails. Nevertheless, chromosome doubling by itself is an insufficient insurance against the risk of disadvantageous mutation when multiplied by the colossal span of geological time.

Question 5.7

Can you think why this should be so?

A further eukaryote innovation, **sexual reproduction**, gets around the problem of disadvantageous mutation by regularly mixing together genes from different individuals. Two steps, **mitosis** and meiosis, are entailed in sexual reproduction (Figure 5.7). Fusion of special parental cells called **gametes** (e.g. sperm and egg cells), each containing only a single or haploid set of chromosomes, produces a new diploid cell (the fertilised egg). For the process to be repeated in subsequent generations without continually doubling chromosome numbers, there clearly has to be a corresponding process to halve the number of chromosomes again between generations, i.e. to form haploid gametes. Normal cell division, known as mitosis (see Box 5.3), which is involved in body growth, conserves chromosome numbers in the nucleus, whereas meiosis, a special kind of cell division, halves the number of chromosomes, while also exchanging segments of DNA between the matching chromosome pairs of the diploid parent (see Box 5.3). Together, meiosis to yield haploid gametes and fusion of the gametes to generate new diploid combinations, shuffle and distribute the genes of parents so as to deal a unique genetic hand to each of their offspring (except in the case of identical twins).

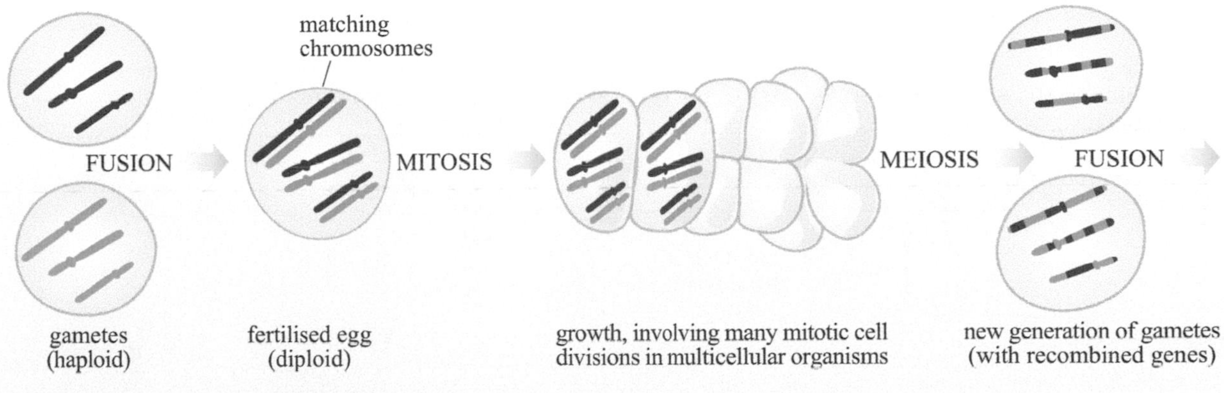

Figure 5.7 The cycle of sexual reproduction. See text for explanation.

Box 5.3 Cell division in eukaryotes: mitosis and meiosis

Mitosis is an elaborate mechanism for cell division that has evolved in eukaryotes and ensures the rapid duplication of DNA in the daughter cells with a high degree of accuracy. Prior to cell division, the DNA sequence of each chromosome (Figure 5.8a) replicates and the pairs of replicate strands for the time-being stay together (Figure 5.8b). A spindle-shaped structure of protein microtubules then forms within the cell, with the chromosomes aligned along its equator (Figure 5.8c). At about the same time, the nuclear membrane breaks down. The replicates of each chromosome separate and are pulled by the microtubules to the opposite poles of the spindle (Figure 5.8d). There, each set of chromosomes becomes re-enclosed by a new nuclear membrane, while the spindle disintegrates (Figure 5.8e). Inward pinching of the outer cell membrane completes the process of mitotic cell division. This mode of cell division ensures that the huge amounts of replicated DNA are correctly divided between daughter cells, such that each receives an identical, full set of chromosomes.

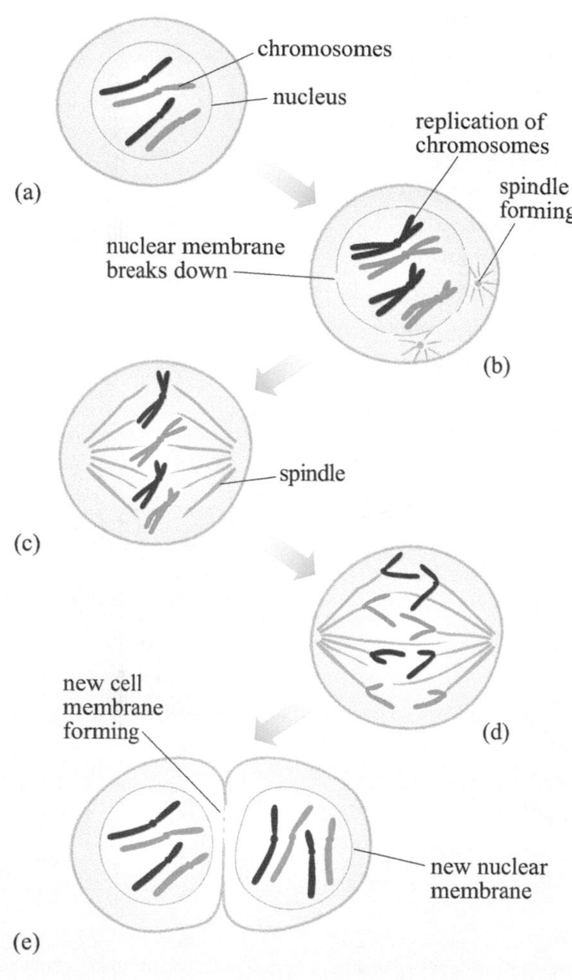

Figure 5.8 The process of mitosis. The chromosomes in (a) are shaded differently to indicate that they were originally from different individuals. See text for explanation.

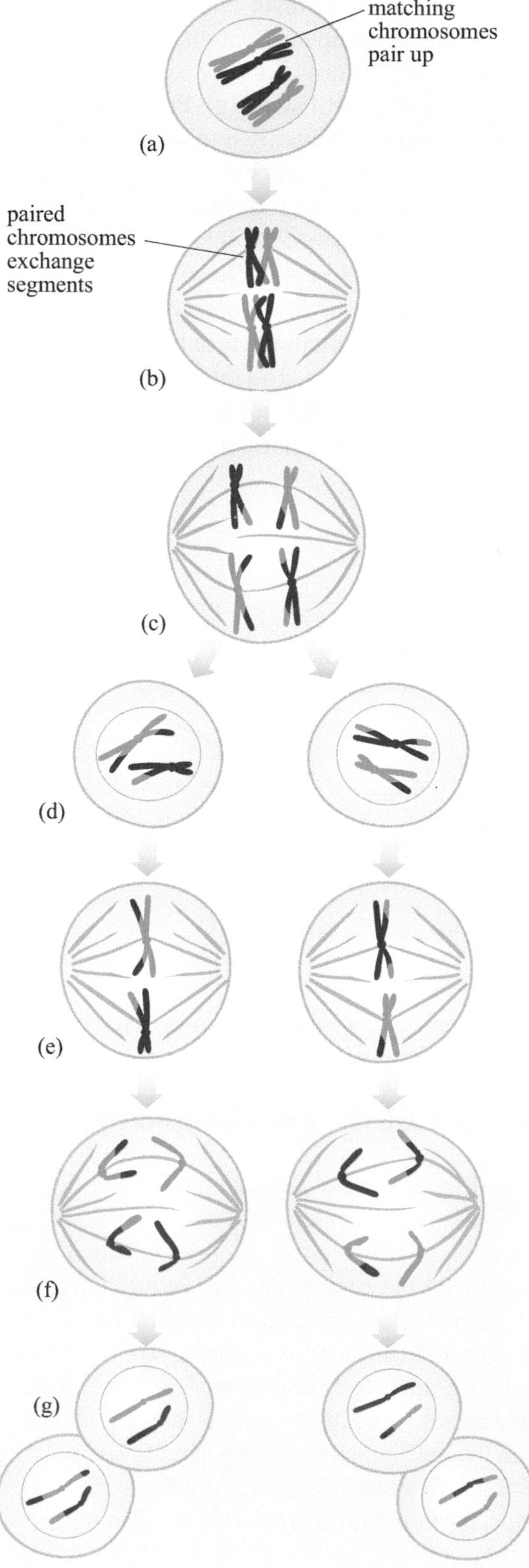

Figure 5.9 The process of meiosis. See text for explanation.

A second type of division called meiosis produces haploid daughter cells (i.e. each daughter cell has half the original number of chromosomes) and forms the counterpart to fusion in the cycle of sexual reproduction (Figure 5.7). Meiosis again uses spindles, although in two successive divisions. As before, the already replicated chromosomes (Figure 5.9a) become aligned along the equator of a spindle for the first division (Figure 5.9b). However, in contrast to mitosis, homologous chromosomes, derived from the fusion of parental gametes, are now also paired up alongside each other. A very curious process then ensues. Each homologous pair (there are now four strands in all in each bundle, as each chromosome is already replicated) proceeds to twist along its length, in the manner of a towel being wrung out (Figure 5.9b). The replicated strands repeatedly snap, but their loose ends reattach, although not necessarily to those from which they have just separated. Consequently, the four entwined strands randomly exchange segments of their gene sequences with each other. Eventually, the chromosome pairs separate again (Figure 5.9c) and migrate to opposite poles of the spindle to form two daughter cells (Figure 5.9d). The second division of the daughter cells is essentially mitotic in character (Figure 5.9d–g), separating the two strands of each chromosome (with their shuffled sequences). A quartet of haploid daughter cells results, with each cell containing only a single unpaired set of chromosomes (Figure 5.9g). Together, the crossing over of bits of the homologous chromosomes and the independent assortment of the paired chromosomes into the daughter cells bring about a thorough shuffling, or recombination, of the parental genes. No two daughter cells (e.g. sperm or egg cells) receive the same half share of genes. Harmful mutations can therefore be effectively screened out by natural selection.

The evolution of sexual reproduction in eukaryotes has had an enormous impact on the history of life, for it has altered the very rules by which evolution through natural selection proceeds (Figure 5.10 overleaf). To a large extent, prokaryote evolution is a straight contest between clones (races of genetically identical individuals). Their reproduction by simple fission (involving straightforward replication of their DNA loop) produces virtually identical offspring, varying through (rare) mutation only, and many individuals must be produced to bring about a favourable combination of genes by coincidental mutation (Figure 5.8a). In contrast, the small size and rapid reproduction of prokaryotes means that competition can certainly be fast and furious as clones with new advantageous mutations supplant others. Sexual reproduction in eukaryotes provides a means for regularly varying the genetic complement that an individual passes on to its offspring through recombination (Figure 5.8b). However, it is not quite as simple as that, for some exchange of DNA between individuals can occur, providing an additional, if irregular, source of genetic variation. Potentially, all possible permutations of the genes available in a population can be assembled in different individuals and played off against one another in the Darwinian struggle for existence. Those genes that generally combine to yield the best adapted individuals will be selectively preferred through the generations. A beneficial mutation can be rapidly promoted (i.e. its frequency in the population increased) by natural selection, unhampered by the other genes present in the original individual in whom it arose. The great leap in the efficacy of natural selection that this represents means that even large and complex multicellular organisms, with relatively slow rates of reproduction (compared with those of prokaryotes), can nevertheless evolve rapidly.

There are, however, several reasons why, at the individual level, asexual reproduction can offer advantages over sexual reproduction. These are:

- energy (hence resources) is not wasted on finding a suitable mate and achieving fertilisation
- a female in a sexual population who switched to asexual reproduction would endow each of her offspring with 100% rather than only 50% of her genes, and thus stand to pass on twice as many of her genes to subsequent generations
- proven favourable combinations of genes risk being broken up by sexual reproduction.

In view of these and other advantages of asexual reproduction, it is difficult to see how sexual reproduction could have gained a foothold and spread through populations at all; natural selection should surely suppress sexual reproduction because of the disadvantages at the individual and population levels. Several hypotheses have, however, been proposed by evolutionary biologists to counteract such arguments, referring to the possible advantages of sexual reproduction at the individual level. One model, for example, considers the effect of parasites; rapidly reproducing parasites can soon adapt to a given combination

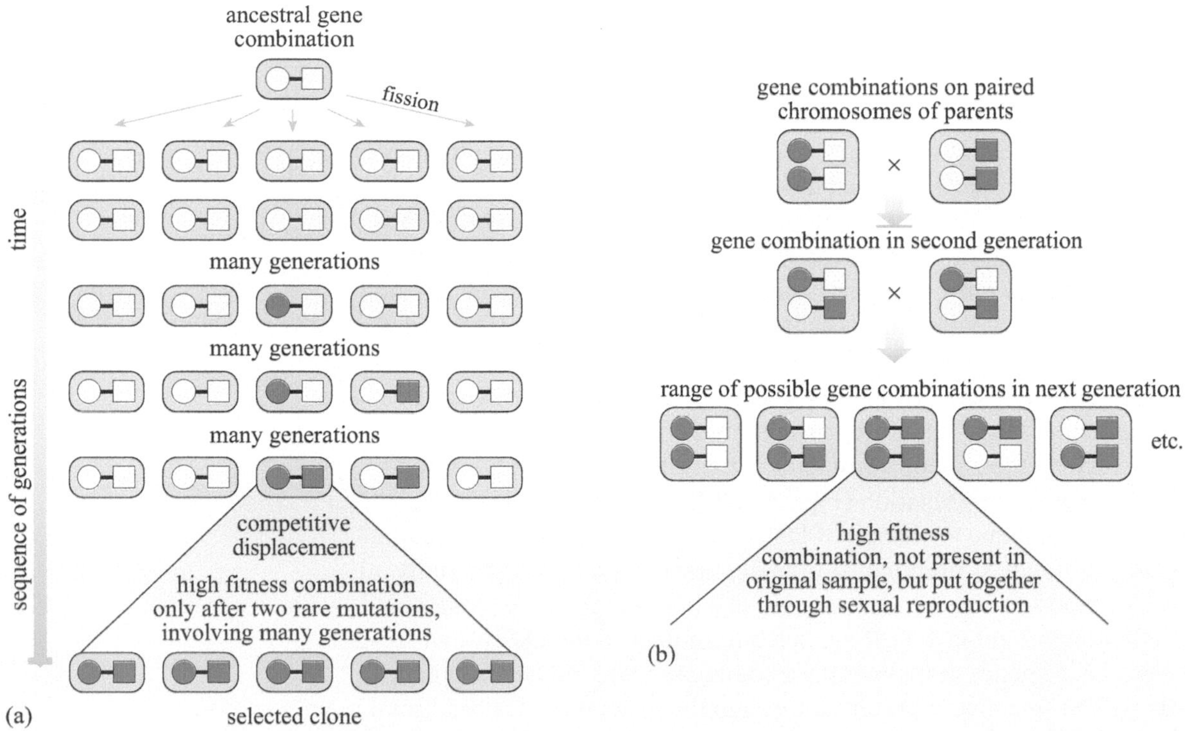

Figure 5.10 The effect of sexual reproduction on natural selection: (a) competition in asexually reproducing organisms is between clones, with variant offspring produced by rare mutations (represented by the dark symbols); (b) sexual reproduction furnishes each individual with its own unique permutation of genes, offering greater genetic variety for natural selection to act upon. The light and dark symbols represent alternative gene types. Only two genes on each chromosome are shown for the sake of simplicity. A high fitness combination of genes is one that, through preferential survival and/or reproduction of individuals bearing it, becomes relatively more frequent in subsequent generations.

of genes in their host organism and can thus threaten the survival and reproductive success of all potential host individuals with that combination of genes. Asexual reproduction provides virtually no escape from this threat; the offspring will automatically carry the same combination of genes. Sexual reproduction, on the other hand, does offer an escape by varying the genetic complements of offspring, such that some at least may prove robust if not immune to infection by the parasites. So there may well be advantages to sexual reproduction at the individual level offsetting the disadvantages, which could explain how it became established and maintained in populations. Once there, it can provide the fortuitous gift of evolutionary flexibility to those species that have adopted it, and it is to this effect that the eukaryotes owe much of their evolutionary diversity.

5.3.4 Diversification of the eukaryotes

The fossil record of eukaryotes points to a marked proliferation of types in the late Proterozoic era. Although known from older rocks, acritarchs for example become abundant and diverse in rocks dating from around 1 Ga onwards. Andrew Knoll, from Harvard University in the USA, has led an intensive study of a sedimentary succession of marine origin, ranging in age from 850 Ma to 600 Ma, on the Arctic island of Spitsbergen. Fossil records reveal the presence of both prokaryotes and eukaryotes from a variety of habitats. However, while the prokaryotes look very much like those living in corresponding environments today and show little evidence for change, the eukaryote fossils present a different story. They reveal a marked diversification, with a relatively rapid turnover of species up through a **sedimentary succession**. Alongside single-celled forms similar to living unicellular green algae and dinoflagellates are various multicellular algae (seaweeds). Conspicuous by its absence, however, is any fossil evidence for animal life. Subsequent studies elsewhere appear to corroborate this story.

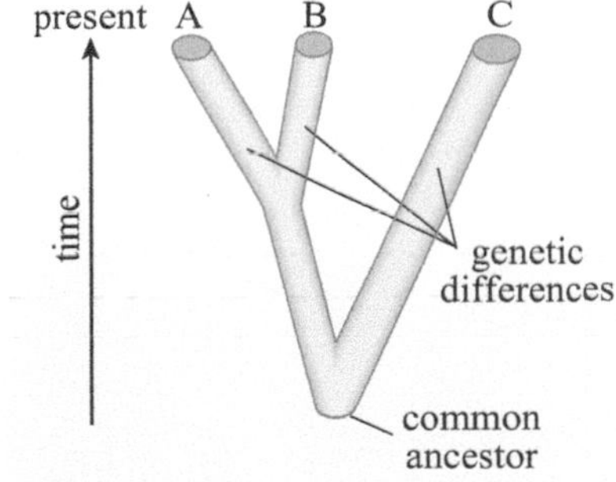

Figure 5.11 The relationship of genetic differences between organisms and their times of divergence from common ancestors, according to the molecular clock theory. As the lineage leading to C split first from a common ancestor to A and B, its genetic sequence would have had more time to change independently than the lineages leading to A and B. Provided that the average rate of change in all lineages stayed the same, the genetic sequences of A and B would remain more similar to each other than to that of C. Such reconstructions of evolutionary trees are only partial because fragments of fossilised DNA sequences are exceedingly rare and analysis of them is fraught with problems.

Another kind of evidence comes from comparing the DNA and/or RNA sequences of living eukaryotes (i.e. the amino acid sequences of proteins derived from DNA and RNA). The hypothesis is that as evolutionary lineages diverge from a common ancestor, so too do the amino acid sequences, as a consequence of cumulative (advantageous) mutational change in each lineage. There are theoretical reasons backed up by empirical evidence for presupposing that the average rate of such divergence remains approximately constant, at least for particular parts of the sequences. This proposal is known as the **molecular clock theory**, and refers to the supposedly clock-like gradual change of the sequences. In so far as this theory is correct (and there is plenty of debate about that), it implies that, for any three organisms, the pair that shows the smallest mutual divergence of sequences shares the most recent common ancestor (Figure 5.11). On this basis, a hierarchy of evolutionary relationships can be established for large numbers of living species, from which the shape of their evolutionary family tree can be partially reconstructed.

Absolute dating of some of the evolutionary branching points reconstructed from molecular data may be established from fossil evidence where the record is sufficiently complete. Such calibration allows the average rate of change of the molecular clock in question to be estimated and the whole evolutionary tree accordingly set against an absolute timescale. At the very least, the method provides a means of producing hypotheses concerning the pattern and chronology of evolutionary history that can be further tested against the direct fossil record.

Comparisons of certain RNA sequences from living organisms suggest that the majority of the major eukaryote groups separated from one another in a flurry of evolutionary diversification from common ancestors between about 1.2 Ga and 1.0 Ga. What sparked this burst of evolutionary activity is again open to conjecture. As with the origin of the eukaryotes, the fossil record bears witness to the resultant evolutionary explosion, but is mute on how it came about. There is no geological evidence for any significant environmental change (e.g. of atmospheric composition or climate) around the time that might be considered to have precipitated this evolutionary explosion. A pertinent biological observation, though, is that those groups of living single-celled organisms that appear to represent the earliest branches of the eukaryote evolutionary tree are all effectively asexual.

Question 5.8

What hypothesis might be deduced from this observation to explain the main burst of eukaryote diversification (i.e. that the earliest branches were effectively asexual)?

Although only a hypothesis, the answer to Question 5.8 highlights the important point that not all major evolutionary events need be explained by reference to external environmental triggers. Internal modifications may fortuitously open up major new opportunities for further evolutionary change, and the evolution of sexual reproduction in eukaryotes is a likely example, offering an explanation for what might be considered one of the biggest gear changes in evolutionary history.

Whatever the circumstances, from 1.2–1.0 Ga onwards, the takeover of the Earth's ecosystems by eukaryotes seems to have become almost unstoppable. Meanwhile, the diversity of stromatolites, which had reached a peak at ~1.0 Ga, began to decline, especially from about 850 Ma, although they remained widespread until Phanerozoic times, when they became largely restricted to stressful environments (e.g. highly salty bays or tidal flats) or cryptic habitats (such as cavities within reefs). Whether these changes reflect direct interactions between the prokaryotes and the eukaryotes, or whether stromatolite-building microbes declined in response to other environmental stresses, with various eukaryotes moving in on their vacated habitats, remains unclear. This is not to say that prokaryotes became unimportant – far from it! Rather, they adapted to the eukaryote world in a variety of ways. Some entered into symbiotic relationships (including disease), and others adopted specialised roles, often in extreme conditions. Indeed, all life continues to rely upon the fundamental biochemical activity of prokaryotes for its continued existence.

5.4 The carnival of the animals

5.4.1 Ediacaran faunas

Section 5.3 described how fossil evidence for animals in the first major episode of eukaryote diversification is absent. Fossils widely interpreted as 'animals' make a late but spectacular entry in the stratigraphic record of the **Ediacaran Period** (635–542 Ma), representing the last 90 Ma or so of the Proterozoic Era. From many parts of the world, assemblages of characteristic types of body fossils (Figure 5.12) have been recovered, which are collectively known as **Ediacaran faunas** after the Ediacara Hills in South Australia, from where many such fossils were first described in detail.

With the exception of a few small, simple forms in northwestern Canada, dated to ~600 Ma, diverse Ediacaran assemblages range in age from at least 565 Ma until the end of the Ediacaran Period. A few distinctive types are known to have survived until mid-Cambrian times, especially in deeper-water habitats.

5.4.2 Interpreting the Ediacaran fossils

What were these Ediacaran organisms like? Perhaps their most striking feature is their size, with some of the later forms reaching up to a metre or so in length. Yet there is no evidence for them possessing any supporting skeletal hard parts. The fossils are found in sedimentary rocks originating from a variety of environments, ranging from deep marine areas to inshore, even tidal flat, settings (Figure 5.12). The fossils are preserved only as flattened impressions at the base of sandy to silty sedimentary layers. Palaeontologists, therefore, have to interpret the Ediacaran fossils on the basis of their gross anatomical features. Needless to say, opinions have differed quite sharply. Earlier descriptions usually saw them being allocated to those phyla (major groups) still surviving today with which they seemed to have most in common (even if only superficially). Thus, various kinds of segmented annelid worms (similar to ragworms), primitive elongate arthropods (shrimp-like animals) and bun-like echinoderms (relatives of sea-urchins) were all recognised in the Ediacaran fauna, along with assorted cnidarians (jellyfish and colonial polyps). However, many of these identifications remained problematical. For example, in only one form, *Dickinsonia* (Figure 5.12a), have structures interpreted as guts been detected. Moreover, the intact condition of the fossils, even in sandstones that were deposited in shallow, agitated and presumably well-aerated water, raises questions as to how they could have been preserved. Today, a combination of predation, scavenging and decay, not to mention disturbance by burrowing organisms, virtually rules out such a style of preservation of soft-bodied organisms in equivalent environments.

(a)

(b)

(c)

(d)

Figure 5.12 Examples of some Ediacaran fauna fossils: (a) *Dickinsonia*, an elongate pancake-shaped worm; (b) *Cyclomedusa*, a jellyfish; (c) *Tribrachidium*, a bun-shaped organism with three spiral tracts on its upper surface; (d) *Inkrylovia*, an elongate bag-like form with transverse partitions. All photographs are to the same scale. ((a and b) Simon Conway Morris; (c and d) Peter Crimes)

One explanation given for these exceptional assemblages is that large-scale predators, scavengers and deep burrowers simply did not exist at that time, making the Vendian a sort of privileged window of opportunity for preservation. Another proposal views the majority of Ediacaran organisms as a distinct eukaryote group separate from the existing kingdoms, and of very simple construction. They are portrayed as having had a quilted mattress-like body filled with fluid enclosed in a tough cuticle (explaining their good preservation), and to have not yet evolved a digestive system consisting of guts. It is suggested that they either absorbed dissolved food molecules over their broad body surfaces or that they may have contained autotrophic symbionts within their tissues, avoiding the need for a digestive system. At least in the case of *Dickinsonia* (Figure 5.12a), however, this interpretation can be rejected as, apart from the possible gut structure, one specimen has been described as having been preserved with its segments torn across prior to burial. The impression it created in the underlying sediment is just as deep as that made by intact specimens, showing that the body was firm and not, after all, fluid-filled. Another researcher has suggested that the Ediacaran organisms may perhaps have been lichens (consisting of symbiotic associations of algae and fungi). Most authorities, however, would regard the pendulum as having swung rather too far from the conventional view in such interpretations.

The most widely accepted hypothesis at present is that most of the Ediacaran fossils represent primitive animals, with the majority having a grade of organisation equivalent to that of the cnidarians. Living cnidarians have a sac-like body wall directly surrounding a digestive cavity that is reached via a single orifice as, for example, found in a sea anemone. The tough body wall consists of outer and inner cellular layers that are separated by a gelatinous layer and, in a few forms, segmentation and bilateral symmetry of the body seem to be well developed (Figure 5.12a). This may perhaps be regarded as the primitive relative of more advanced animals. Certain problematical forms, however, may still turn out to be algae rather than animals.

In addition to these originally soft-bodied forms, a few kinds of skeletal components are also known from the Late Vendian Period, including some networks of sponge spicules (mineralised needles that provide a skeletal framework for many sponges).

Haemoglobin is the red pigment that transports respiratory gas molecules in the human bloodstream.

The recognition of Ediacaran animals is consistent with molecular evidence that suggests an even earlier time of origin, using the molecular clock theory. Figure 5.13 shows the estimated divergence of the amino acid sequences of the protein haemoglobin between various pairs of living animal species plotted against the minimum possible ages for the original divergences from common ancestors based on fossil evidence. The minimum age for divergence in each case is taken as that of the oldest fossil that can be attributed to one, but not to the other, of the evolutionary branches in question: the existence of such a fossil shows that the split from a common ancestor must already have taken place. As there may have been successive amino acid substitutions at any given site on a haemoglobin molecule, values for the *cumulative* amounts of divergence may exceed 100% and the maximum is 190%.

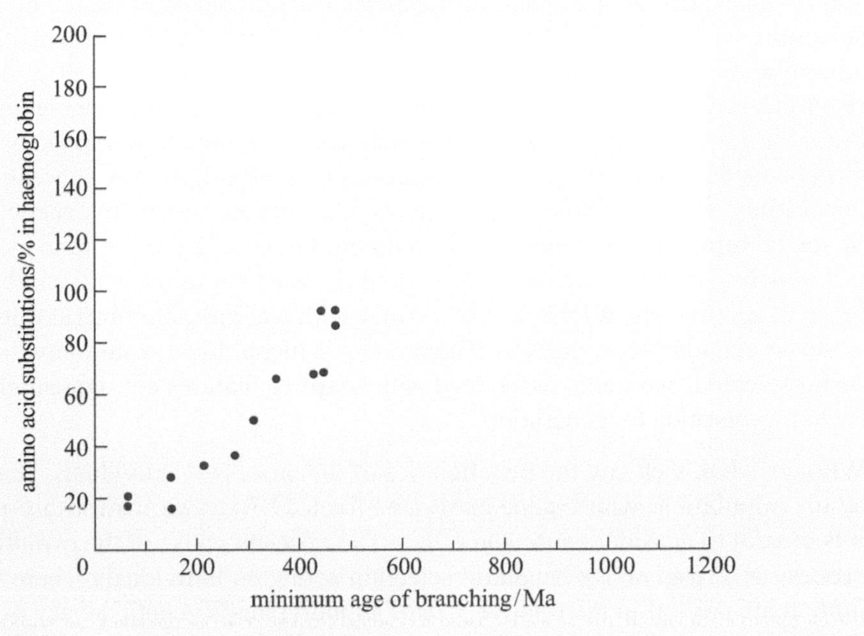

Figure 5.13 Measures of the relative amounts of divergence in the amino acid sequences of haemoglobin between pairs of living animal species plotted against the time by which divergence from a common ancestor must have occurred, based on fossil evidence. (Runnegar © Norwegian University Press)

Question 5.9

Using the molecular clock theory as a guide, extrapolate from the data shown in Figure 5.13 to deduce the inferred latest time of origin of the animals, based on the maximum value of haemoglobin divergence given above. (*Hint*: note that the origin of the animals can be assumed to have preceded the first divergence to have taken place within the group.)

The value derived in Question 5.9 is, of course, only a minimum estimate due to the possibility of yet more divergent forms having become extinct. On the other hand, the molecular clock may not itself have been regular, so the conclusion is by no means infallible.

The Ediacaran animals thus appear to represent a major evolutionary proliferation of large-bodied animals some time after the appearance of the first animals, which may date back to around 1.0 Ga.

5.5 The rules of the new evolutionary game

What was it about the evolution of the eukaryotes and, ultimately, the Ediacarans that was so innovative? One of the most important developments was sexual reproduction (Section 5.3.3). In the final part of this chapter it is worth following through the theoretical implications of the innovation of sexual reproduction for the evolution of adaptations in organisms. Analysing who or what benefits from adaptations will provide a useful perspective for ongoing considerations of the relationship between the Earth and life.

The characteristic adaptedness of organisms is a product of evolution by natural selection, which may be summarised as follows. The reproductive potential of a population usually well exceeds the numbers that can be sustained, giving rise to Darwin's theory of the struggle for existence. Individuals possessing heritable variations that promote their survival and reproduction in the prevailing circumstances relative to other individuals, preferentially endow subsequent generations with their offspring and hence their genes. Given the genetic mixing of sexual reproduction, genes associated with favourable traits tend to increase in frequency in the population. Through time, selection of these favourable traits leads to adaptations of form and behaviour with recognisable functions in relation to survival and/or reproduction. The process is hierarchical, with individuals being selected, and genes associated with adaptive features consequentially sorted, generation by generation.

Who, or what, then, are the beneficiaries of the process – individuals, genes or whole populations whose gene pools are affected? To try to answer this question, it is crucial to consider cause and effect. The efficient cause of the evolutionary process described above is natural selection acting on individuals. There is an immediate pay-off in the relative size of an individual's genetic legacy in the next generation. This is expressed as an individual's production of offspring, themselves surviving to be capable of reproduction, relative to other individuals – referred to as its **fitness**. The sorting of the genes, and any consequences for populations, however, are the incidental effects of such fitness differences. Hence, the only logically necessary beneficiaries of the adaptations that evolve are the individuals possessing them, who are rewarded with higher average values of fitness relative to other individuals. (The 'selfish gene' concept of Richard Dawkins, set out in his book with that title in 1976, argues that it is the selected genes that get the ultimate benefit. Yet that is, nevertheless, still contingent on the successes of the well-adapted individuals carrying them.)

At first sight, two evolutionary phenomena may seem to contradict the conclusion above concerning the benefits of adaptations. The first is what biologists call 'altruism' within species (although with no implication of intentionality), and it involves one individual behaving in such a way as to benefit the reproductive success of another, although at a cost to itself. The second is the coevolution of different species to yield mutually beneficial features, such as that between flowers and their insect pollinators. Closer analysis of both phenomena shows them to conform, albeit in subtle ways, to the conclusion above. A discussion of this is beyond the scope of this book.

A classic example of altruism, for example, is the worker honey-bee, which feeds the queen and her offspring, and may die defending them, while herself failing to reproduce. How could such apparently selfless behaviour have evolved? The answer to this conundrum was provided in the 1960s by the British evolutionary biologist W. D. Hamilton. In most sexual organisms, the mean amount of genetic overlap between siblings (i.e. genes received in common from the parents) is 50%, because each individual gets a random 50% of each parent's genes (Box 5.3). Social insects such as honey-bees, however, have a rather curious pattern of genetic inheritance because males are haploid and females are diploid. Hence, females (including the workers) receive all of their father's genetic complement, together with 50% of their mother's genetic complement. Consequently, the majority (75% on average) of the genes received by each worker are shared with its siblings – all the queen's progeny, including new queen larvae. By helping the queen and her daughter queens who are especially adapted for copious reproduction, a worker ensures a far greater representation by proxy of her (shared) genes in subsequent generations than she might achieve by her own reproduction. So her effective genetic legacy or 'inclusive fitness', which takes those shared genes into account, is increased by her apparently altruistic behaviour. Such traits are said to have arisen by a special form of natural selection termed **kin selection**.

- Why do you suppose the cells making up your own body do not compete with your sex cells to generate new individuals?
- Since all the cells in your body are derived from a single fertilised egg, they all share the same genes. The sex cells, therefore, transmit the genes of all the other cells by proxy, leaving the other cells dedicated to their various specialised functions.

Evolved adaptations are effectively of selfish, transient benefit to the individual genetic entities concerned, and their rewards are strictly dependent on the prevailing circumstances. The adaptations provide no guarantee for the future welfare of the population (or even species) as a whole. A change in circumstances can precipitate extinction, as the testimony of the fossil record abundantly illustrates. Some adaptations may tend to prove disadvantageous to populations in the longer term, notwithstanding individual benefits in the shorter term. For example, in some environmental circumstances the acquisition of an asexual mode of reproduction by normally sexually reproducing organisms can lead to greater reproductive success (Section 5.3.3). However, those species that become fully asexual often appear to become extinct more rapidly than related sexually reproducing species. This is probably because of the associated decline in evolutionary flexibility of the asexual species. The short-term gain of individuals may eventually be at the longer-term expense of the population.

What is the implication of these arguments for the way in which living organisms interact with the Earth? Of great importance among the adaptations of organisms are systems of self-regulation, or **homeostasis**, which maintain stable conditions within the body in the face of a range of environmental perturbations. A familiar example of such a system is that which maintains your normal body temperature. In response to feeling cold, your body produces heat by various means (such as shivering), while it prevents overheating by, for example, sweating. Thus you are equipped with an integrated system of sensors (temperature-sensitive nerves in this case) and compensatory devices that maintains a constant core temperature in the body. Such internal constancy is advantageous because enzymes, which are themselves adapted to function optimally in the body's normal state, may be highly sensitive to fluctuating conditions. It is tempting to try to draw a parallel between such self-regulation within organisms and the system of feedbacks that appear to regulate conditions on Earth. Such a parallel has been explicitly proposed in the Gaia hypothesis (Box 5.4).

Box 5.4 The relationship between life and Earth, according to the Gaia hypothesis

The Gaia hypothesis, first formulated by James Lovelock in 1972, asserts that 'the climate and chemical composition of the Earth's surface environment is, and has been, regulated at a state tolerable for the biota' (Lovelock, 2000). The Earth and its biota are regarded as having evolved together as a tightly coupled system, with self-regulation of important properties, such as climate and chemical composition, arising as emergent properties. The whole system is thus explicitly likened in this respect to a 'super-organism'. The kind of process Lovelock had in mind when proposing this hypothesis was the long-term maintenance of levels of molecular oxygen in the atmosphere sufficient to keep us alive, despite the short residence time of this highly reactive gas.

No theoretical explanation has been advanced as to why the feedbacks involved should necessarily serve to regulate conditions in the interests of the biota. Instead, a computer model to suggest how such a system might operate has been developed. Called 'Daisyworld', it envisages an idealised world similar to our own, receiving energy from a gradually warming sun. The planet is imagined to have been seeded by two sorts of daisies, one black and one white. The essence of the model is that, to start with, when solar heating is modest and the surface temperature is below the optimum for the daisies, the black daisies warm up more quickly than the white ones and their growth is advantaged. As the black daisies spread, they reduce the planetary albedo, allowing more heat to be retained, and so help warm the atmosphere to an optimum level for the growth of daisies. Later, with increasing solar flux, optimum atmospheric temperatures may be surpassed in places, and it is the turn of the white daisies, whose greater reflectance helps keep them cool, to be at an advantage. As they now spread, displacing the black daisies, they increase the planetary albedo, and so serve to counteract the overheating. Thus, the planet and its biota together furnish a self-regulating system that, for a time at least, maintains optimum conditions for the daisies. Further refinements were later added to this model (with the addition of, for example, grey daisies and a fauna of rabbits and foxes), none of which was found to disrupt the self-regulatory nature of the system. The Daisyworld concept came to prominence in the influential 1980s nuclear thriller *Edge of Darkness* written by Troy Kennedy Martin.

To test whether Daisyworld is a valid model of the real world, it is necessary to see if the feedbacks between life and Earth have consistently operated over geological time in such a way as to maintain optimal conditions for life. (This question is revisited in the final chapter of this book.)

Question 5.10

Would you expect homeostatic systems like those found in living organisms to evolve by means of natural selection at a higher level than the individual, operating, say, for the benefit of whole populations, ecosystems or even life as a whole?

It is important to distinguish between evolved homeostasis in organisms of the kind explained above and simple equilibration brought about by feedbacks. A mixture of ice and water will equilibrate at the freezing temperature of water, for example, because the bonds that hold water molecules together in the ice crystals provide a means of negative feedback. Any heat added to the system is taken up by some of the ice melting (a process that absorbs heat as the bonds break). Loss of heat is compensated for by the heat released through ice formation. So, for as long as both pure water and ice are present, the temperature stays constant. This system has no checks and balances built into it, however, to guarantee constancy in the face of other perturbations. The equilibrium temperature in this case can, for example, be altered simply by adding salt, which modifies the energy balance and lowers the freezing temperature. Without intervention, the system will not rid itself of the salt to retrieve the previous equilibrium temperature. By contrast, core body temperature is not expected to fluctuate (except temporarily, when ill), for example according to changes of diet. Evolution by natural selection has furnished each human being with homeostatic devices that compensate for all the kinds of environmental changes that were faced by our ancestors (who did survive them and reproduce, as our own existence shows). Equilibria in systems that do not reproduce in the manner of organisms, and have not therefore been honed by natural selection, lack such built-in safeguards to their stability.

Certain aspects of ecosystems, and indeed the Earth and life as a whole, may certainly equilibrate because of the balancing effects of feedbacks (e.g. interactions between levels in the trophic pyramid) within them. But there is no reason to expect the essentially selfish adaptations of organisms to underwrite a given environmental equilibrium. Should conditions change, the winners in the 'struggle for existence' will simply be those that adapt to the new circumstances. Hence, no support is forthcoming from the natural selection theory, at least, for the claim of the Gaia hypothesis that the Earth and its life together regulate conditions *in the manner of a living organism*.

Nevertheless, the feedbacks between the Earth and its life have certainly yielded conditions quite unlike those that would have prevailed had the Earth been lifeless. The innovations introduced by the eukaryotes discussed above have allowed them in their turn to add their own significant twists to the tale.

Summary of Chapter 5

1 While some conditions at the Earth's surface (e.g. mean temperature) may have remained within modest limits for most of its history, others (e.g. oxygen concentrations) have undergone radical changes. Yet life has apparently been present almost throughout, although the main diversification of eukaryotes only occurred over the last 1000 Ma.

2 Stromatolites dominate all but the last part of the Proterozoic fossil record, reaching a peak around 1.0 Ga and declining thereafter. Exceptionally well-preserved early examples contain microfossils of tiny filamentous and spheroidal prokaryotes. By mid-Proterozoic times, there was a clear differentiation of communities in different habitats, including both prokaryotes and simple eukaryotes. Planktonic forms supplied organic material to the deep sea floor.

3 Fossils of probable eukaryotes show that they had evolved at least by the mid-Proterozoic (around 2.1 Ga). Organelles such as mitochondria and chloroplasts probably evolved from endosymbiotic prokaryotes that took up residence in the ancestral eukaryotic cells. The nucleus, by contrast, probably evolved within the original cell structure. The acquisition of mitochondria might have been in response to the first appearance of molecular oxygen.

4 Both fossils and molecular data point to an evolutionary explosion of eukaryotes commencing between about 1.2 Ga and 1.0 Ga. This may reflect the evolution of sexual reproduction: no particular environmental trigger has been implicated.

5 The last 70 Ma of the Proterozoic (the Ediacaran Period) was marked by the appearance of assemblages of large, enigmatic fossils collectively referred to as Ediacaran faunas. They have been subject to a variety of alternative interpretations, but most are likely to have been primitive animals. Molecular data confirm the existence of animals from about 1.0 Ga.

6 Eukaryote cell nuclei contain orders of magnitude more DNA in chromosomes than the simple loop attached to the inside of the rigid outer wall of prokaryote cells. The eukaryote cell also possesses an internal framework of protein rods, which can alter cell shape and control its internal structure. These differences have vastly expanded the relative scope of eukaryote evolution to include multicellularity with cell differentiation, allowing them to build up multi-tiered trophic pyramids. Prokaryotes, meanwhile, have been confined to production and decomposition, although under a wide range of conditions.

7 There is an increased probability of mutation associated with the increased amounts of DNA in eukaryotes. Compensation for this is provided by doubling of the chromosomes, together with the mixing of genes from different individuals through sexual reproduction.

8 Sexual reproduction has increased the efficacy of natural selection, potentially allowing all possible permutations of the genes available in a population to be tested in the struggle for existence. This allows even complex multicellular organisms with slow rates of reproduction to evolve rapidly.

9 Adaptations that evolve through natural selection remain of selfish benefit, in terms of fitness, to the individuals (or genetic entities) possessing them. Adaptation may influence higher levels, e.g. populations and ecosystems, but only by way of incidental effect, which may be good or bad. Hence, while natural selection can explain adaptations for homeostasis in individuals, the theory does not predict the emergence of analogous systems at higher levels.

Learning outcomes for Chapter 5

You should now be able to demonstrate a knowledge and understanding of:

5.1 How changing physical conditions on Earth (and in particular increasing oxygen concentrations) can be linked with the development of early life on Earth through biogeochemical feedback mechanisms.

5.2 The main differences between prokaryotes (single-celled organisms) and eukaryotes (multicellular organisms) in terms of their cell size, structure, function and methods of reproduction.

5.3 The important role played by eukaryotic cells in the evolution of increasingly diverse and complex of life on Earth, and how all life can be classified into four main kingdoms.

5.4 The scarcity of information on the earliest life forms and how the explosion in diversification of the Ediacaran fauna has been interpreted and compared to modern organisms.

5.5 The evolutionary process of natural selection and how aspects such as altruism and adaptiveness can favour an individual within a population, as well as alter the characteristics of the population over time.

Life in the Phanerozoic

Reconstructing the pre-Phanerozoic history of life, as discussed in previous chapters, depends upon rather patchily distributed, often problematical, fossils and on somewhat speculative interpretations of other biological, geological and geochemical data. In contrast, the more complete Phanerozoic marine fossil record allows quantitative estimates of the turnover in groups of organisms to be made. Thus, it is possible to identify the timing and scale of **evolutionary radiations** (phases of significant increase in numbers of species within groups of organisms), as well as mass extinctions, when abnormally large numbers of species became extinct together. Such information helps to narrow the search for cause and effect in the evolution of life.

The story starts close to the Precambrian–Cambrian boundary. This was arguably the most important interval in the evolution of life on Earth, when complex life emerged. The story of this transition is not, however, completely straightforward. The prokaryote world did not suddenly disappear and the new world order of modern animals and plants did not suddenly appear. Rather, there was a gradual development of highly 'experimental' animals, i.e. the Ediacarans, which persisted for a brief interval of geological time (less than 80 Ma) in the late Precambrian, followed by the diversification of modern **bilaterian** life (i.e. animals with bilateral symmetry) at the beginning of the Phanerozoic, an event that occurred very rapidly in geological terms and which has been called the **Cambrian Explosion**.

The start of the Phanerozoic (and, therefore, the end of the Precambrian) is set at 542 Ma. The boundary that defines this age – the **Global Stratotype Section and Point (GSSP)** – has been officially placed in an obscure rock outcrop in southeastern Newfoundland. The placement of such an important boundary here, however, is not haphazard. The trace fossils of southeastern Newfoundland are the first record of the Cambrian Explosion.

- ■ What are used to define boundaries in the sedimentary record?
- □ Normally fossils, i.e. the evidence of life. The Precambrian–Cambrian boundary, however, is unusual because the fossils used to define it are trace fossils rather than body fossils.

Recalling that Phanerozoic means 'visible life', to the 19th century geologists who formalised geological time, the situation was very clear-cut: there was a time without life – the Precambrian, and a time with visible life – the Phanerozoic. In fact, Charles Darwin was so worried about the lack of fossils in the Precambrian that he said people would be entitled to reject his theory of evolution by means of natural selection on those grounds alone. As discussed in Chapter 5, however, during the Precambrian the Earth teemed with prokaryotic life and saw the rise of the cellular innovations that would, in time, give rise to the animals and plants. This process of complex animal evolution *started in the late Precambrian* with the evolution of the Ediacarans. The good news, of course, is that Charles Darwin, if he were alive today, could relax about the issue of the missing Precambrian fossils: they are no longer missing. All that was required to clear up that mystery was another century of investigation!

The transition from the Proterozoic to the Phanerozoic was a time of momentous change, both among organisms and in the conditions at the Earth's surface. To understand how these changes related to one another, it is first necessary to try to establish what took place, and when. Many of the time periods when these changes occurred have been established, largely during the past decade, through the use of high precision isotopic dating techniques.

Figure 6.1 shows the order and timing of events schematically expressed in the form of a 24-hour clock, with midnight being the formation of the Earth at 4.55 Ga and midnight 24 hours later being the present day (0 Ga). This approach is a useful way of understanding the immensity of geological time and the distribution of geological and biological events through that time. Crucial events in the history of life are marked on this diagram.

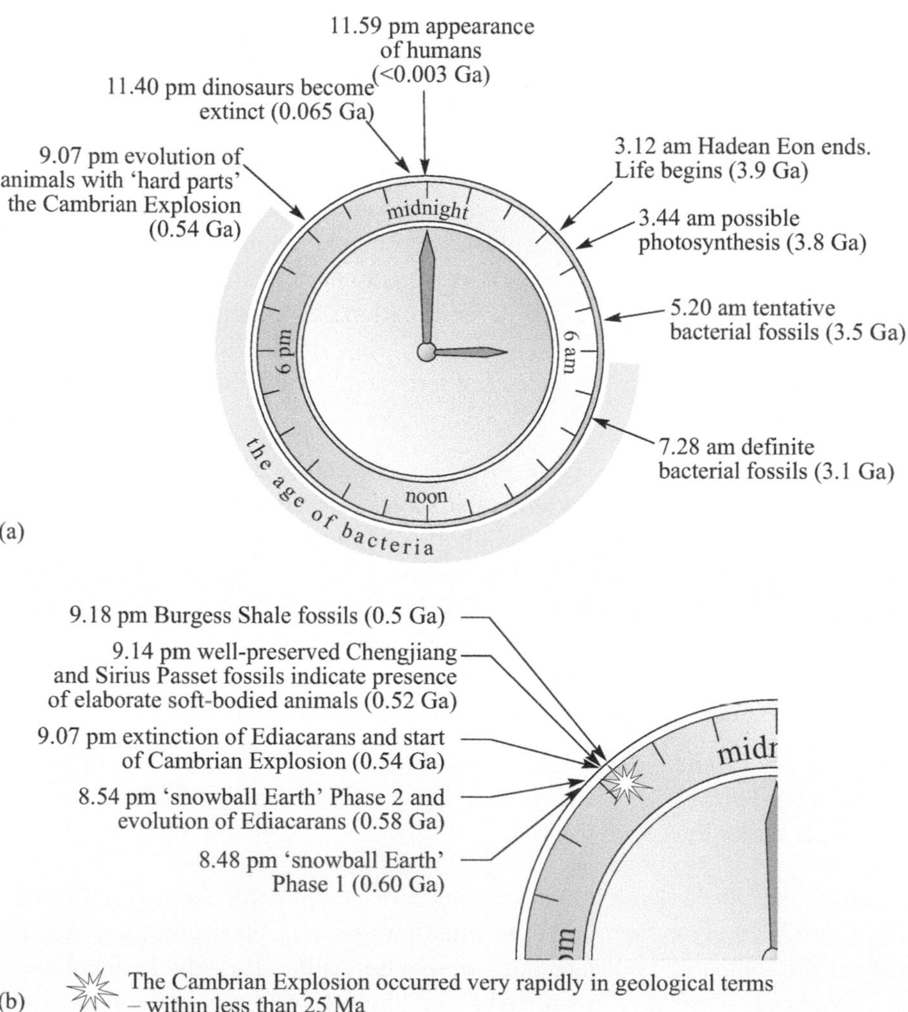

Figure 6.1 The history of life expressed as a 24-hour clock.

Nothing illustrates the importance of the Precambrian–Cambrian boundary more than appreciating how quickly the events associated with it happened when compared with the very long time that the world stayed in the 'age of bacteria' (where 'bacteria' is used loosely here to describe all prokaryotic and single-celled eukaryotic life). Life evolved at about 3.8 Ga (3.12 am on the clock) and persisted at the single-celled level of organisation until about 0.54 Ga (9.07 pm on the clock). At this time, the Ediacarans evolved and were followed by the rapid evolution of the animals of the Cambrian Explosion. Note, however, as discussed in Chapter 5, that the age of bacteria was not without its own crucial evolutionary innovations. Perhaps the most important of these was the evolution of the eukaryotic cell, with sub-cellular organelles such as mitochondria and chloroplasts evolving from more primitive prokaryotic cells.

Interpreting the geological record of the Precambrian–Cambrian transition has been hampered by two major problems. First, correlation of the sedimentary successions in different parts of the world (i.e. recognising beds of the same age from their fossil content or other age-diagnostic features) is extremely difficult; second, no one has yet discovered a complete and unbroken sedimentary succession through this interval.

The last 10 years has seen extraordinary advances in the synthesis of knowledge about this crucial time, and not all of this work has involved palaeontology or geology. Much information has come from genetics, particularly the process by which genetic sequences of different organisms are compared with each other to gauge their level of similarity. These comparisons are known as **molecular phylogenies** and provide an invaluable test of the findings of the fossil record.

An effect of this comparison process is that it is possible to estimate the times at which major groups of organisms diverged from each other, assuming there is a more-or-less constant rate of mutation (a conclusion that, to a first approximation, appears to be experimentally justified). This 'molecular clock' technique is another way in which the synergy of genetics, palaeontology and geology is revolutionising our understanding of the events around the Precambrian–Cambrian boundary.

One of the major events around the Precambrian–Cambrian boundary was the rapid evolutionary proliferation of animals with readily preservable skeletal hard parts or shells, which allowed better preservation in the fossil record. This innovation occurred while the Ediacarans dominated the oceans just before the Cambrian Explosion.

The oldest examples of these so-called small shelly fossils are some tiny calcareous tubes named *Cloudina* (Figure 6.2 overleaf), which are common in certain limestones dating from the late Ediacaran (at 550 Ma). This raises a number of key questions, which will form the basis of discussions in Section 6.1, namely: what sparked this explosion? What became of the earlier Ediacaran organisms? What were the ecological and environmental consequences of this rapid evolution? What is striking about the dawn of the Phanerozoic, though, is the sheer diversity of skeletalised organisms that began to appear. Section 6.2 will explore the rich fossil record that followed the explosion to determine any patterns of relationship in time that can be detected between evolutionary radiations, mass extinctions and major environmental changes. This in turn will emphasise the key theme of the nature of the relationship between life and the Earth, as seen from the context of the Phanerozoic record.

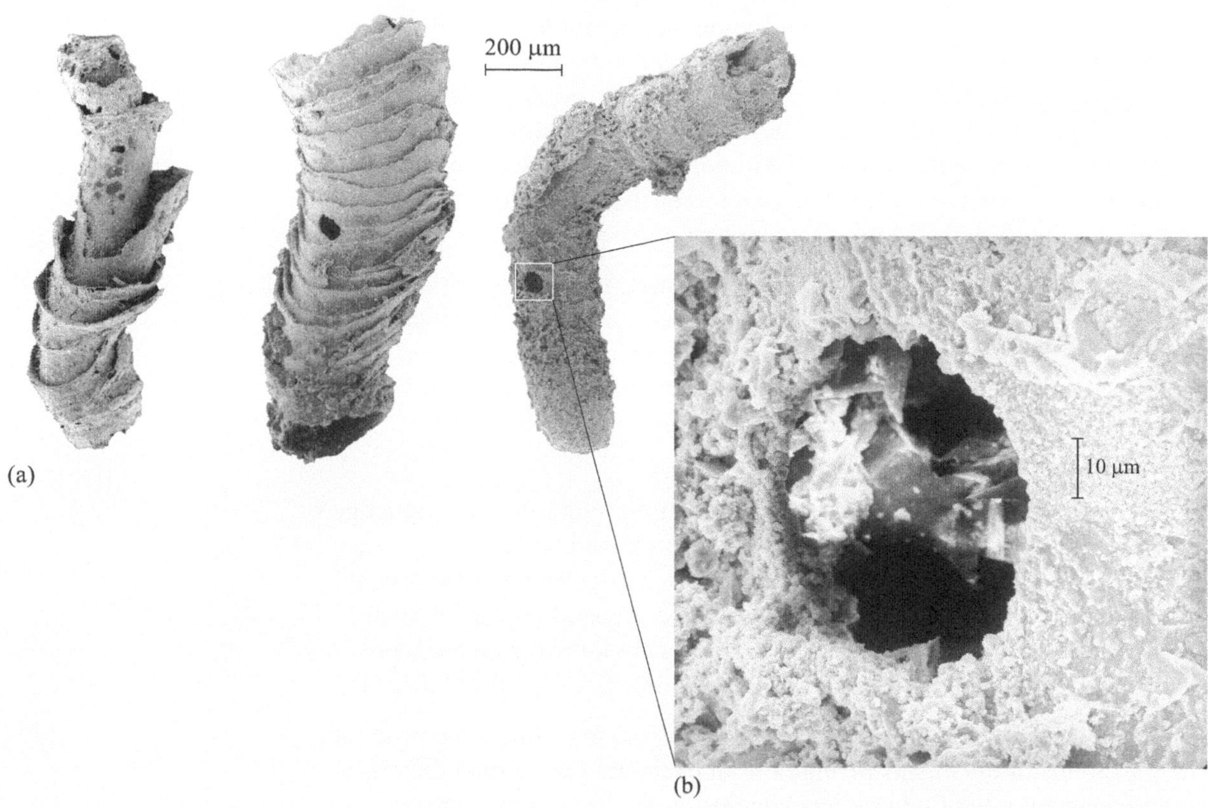

Figure 6.2 *Cloudina* from late Ediacaran age limestones of China. (a) Three isolated specimens. These have been replaced by calcium phosphate following burial and so could be extracted by dissolving the enclosing limestone in acid. (b) Detail showing a hole bored in the wall of one of the shells by a presumed predator. (Bengston and Yue, 1992)

6.1 The Proterozoic–Phanerozoic transition

6.1.1 The snowball (or slushball) Earth

One of the more remarkable ideas to emerge from the study of ancient climates in recent years has been the suggestion by Harvard geologists Paul Hoffman and Dan Schrag that the world was covered from the Equator to the poles by a sheet of ice several kilometres thick in the late Proterozoic termed **snowball Earth**. They base this extraordinary conclusion on a combination of geochemistry (specifically the isotopes of carbon, as discussed in Box 4.1) and classical geology (understanding the climatological significance of different rock types and how their distribution in time can be used to decode ancient climates).

Evidence for these extreme ice ages has been found in rocks of late Early Proterozoic age from every continent. Evidence from fossil magnetism in the rocks shows that some of these glacial sediments were deposited at low latitudes; so, it is clear that these ice ages were global in extent.

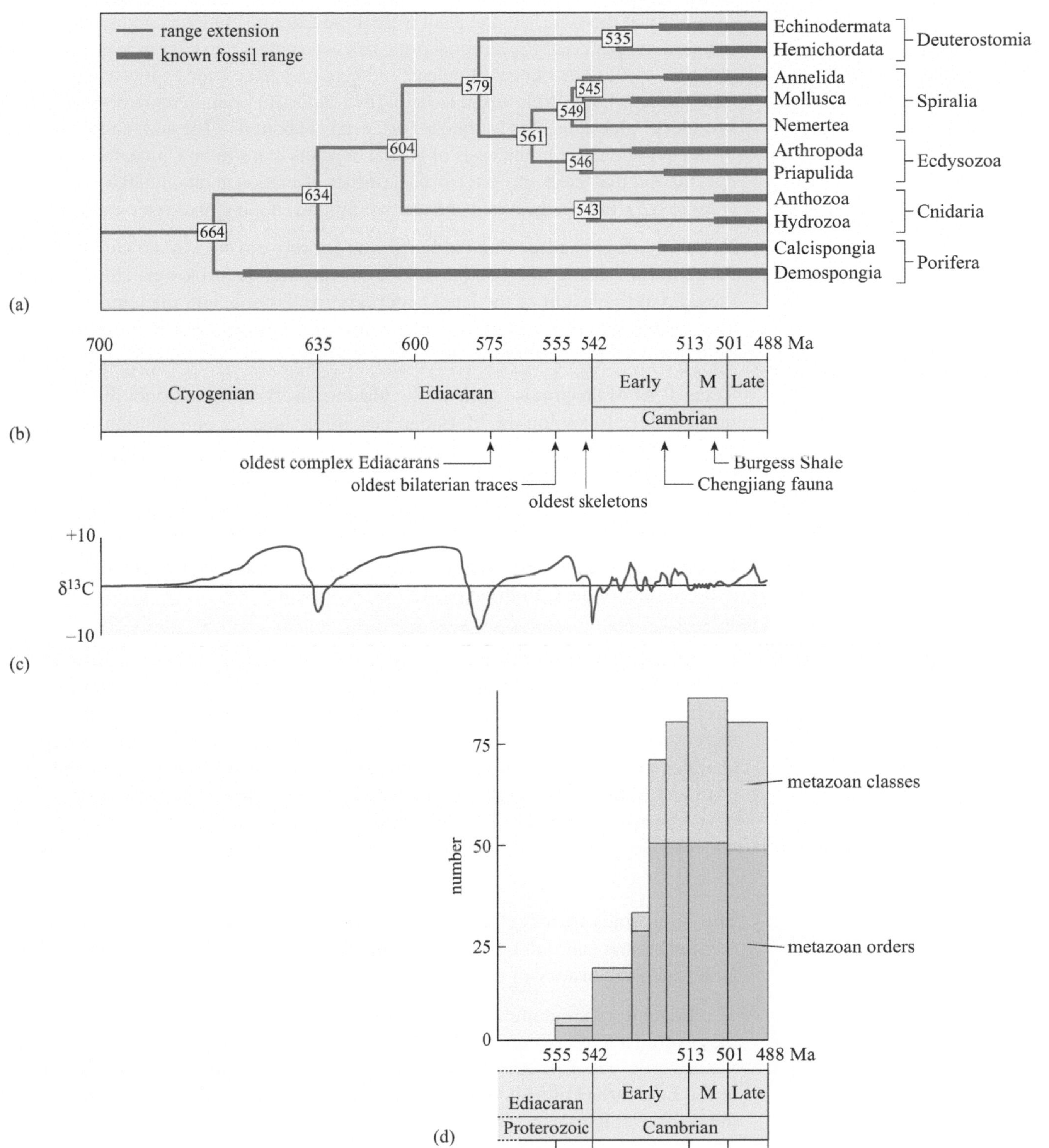

Figure 6.3 Early animal evolution placed in the geological context of the Early Proterozoic–Cambrian transition: (a) the best estimates of the times of divergence of the major groups of animals; (b) timing of major events in the early history of life; (c) the carbon isotope curve of inorganic carbon indicating times of inferred major glaciation (negative inflections); (d) numbers of metazoan classes (blue, top of bars) and metazoan orders (red, bottom of bars) over the Precambrian to Cambrian transition. Classes and orders are discussed in Box 6.1.

The fluctuations in carbon isotope values imply at least considerable climatic instability at this time. So prevalent were these extreme climatic conditions that the penultimate period of Precambrian time has been named the Cryogenian. Geological and geochemical evidence is unclear, but there may have been as few as two or as many as five Early Proterozoic ice ages; the important point to remember is that the last global glaciation (the Marinoan) occurred at about 635 Ma and marked the end of the Cryogenian. On the basis of glacial deposits in northern China, there is some speculation that there may have been a further glaciation at about 580 Ma (Figure 6.3), but it is generally agreed that this was not a global-scale event.

Not all geologists agree that the Earth was entirely covered in ice during the late Precambrian. Some suggest that the ice sheets extended to lower latitudes than they did at the height of the late Quaternary glaciations, into the temperate zones and, as this idea is not as radical as the idea of a snowball Earth, it is known as the 'slushball Earth' compromise.

Regardless of the precise scale of the glaciations, there is no doubt that immediately following the Marinoan glaciation the pace of evolution accelerated, generating the Ediacaran organisms first, followed by the bilaterian animals of the Cambrian Explosion.

Question 6.1

Referring to Figure 6.3c, what does the negative excursion between the Ediacaran and the start of the Cambrian (~542 Ma) represent?

It is tempting to think that the rapid diversification of the Ediacarans and the bilaterian animals of the Cambrian Explosion were direct responses to the environmental changes at the time. However, it seems likely that, in addition to the environmental instability, there may have been an internal trigger related to changes in the genetics of these organisms. These questions are considered in more detail below, but first it is important to understand what the first tier of post-Marinoan organisms (i.e. the Ediacarans) actually were.

6.1.2 The Ediacarans

The Ediacarans are an extremely important part of the story for the emergence of the metazoans (animals), so much so that they are briefly discussed here again, along with the history of their discovery.

The Ediacarans are named after the fossil locality in Australia (Ediacara) where they were first discovered in 1946 by Reginald Sprigg, an assistant geologist to the government of South Australia, who was examining old silver and lead mines in the Ediacaran Hills about 600 km north of Adelaide with a view to prospecting them for uranium.

Apart from his professional interest in uranium, Sprigg was also interested in the possibility that there might be fossils in these Precambrian sequences – an idea that at the time flew in the face of received palaeontological reason. As he approached the old mines he noticed numerous impressions of what he took to be jellyfish outcropping in the Pound Quartzite. He took the fossils home, wrote them up in a scientific paper and then spent years trying to convince the palaeontological community that they were indeed evidence of Precambrian life.

By the late 1950s, additional material had been found in the Flinders Range (also in Australia) and, crucially, from another locality on the other side of the world. In Charnwood Forest, Leicestershire, England, a schoolboy named Roger Mason came across a large fern-like impression embedded in the sandstone. Mason contacted a palaeontologist at the University of Leicester, Trevor Ford, who returned with him to the quarry. Over the next few months they discovered several other fossils, and the Ediacarans were firmly established as a real page in the book of life.

But what are the Ediacarans? The crucial pointer is something that Sprigg noticed instantly: they looked like jellyfish. Jellyfish are a member of a group of animals, namely the cnidarians and the ctenophores, that are united in their common possession of radial symmetry and only two layers of body tissue, which are separated by a layer of jelly. Such creatures are quite different from the rest of the animal world, the bilateria, which are characterised by having bilateral symmetry and three layers of body tissue. Thus, it seemed quite reasonable to the palaeontologist who carried out the first in-depth study of the Ediacarans, Martin Glaessner, to consider them as an evolutionary step on the road to the more complex bilaterian animals of the Burgess Shale. This view was, however, rejected as some of the strangest Ediacarans that do not look like jellyfish appear to have no living relatives. For example, *Spriggina* (named after Sprigg) may be a primitive arthropod (although there is as yet no consensus) whereas *Dickinsonia* (Figure 5.12a) certainly has no living relative. This strange creature was probably entirely sessile, living on the seabed and perhaps deriving nutrition through symbiosis with photosynthetic algae. Unlike the majority of the Ediacarans, both *Spriggina* and *Dickinsonia* are bilaterally symmetrical.

Swayed by their apparent similarity to the cnidarians and ctenophores, Glaessner tried to fit all members of the Ediacaran fauna into conventional Kingdoms or Orders. It was only in 1983 that Adolf Seilacher of the University of Tübingen, Germany came up with a different and radical notion. Seilacher noted that the animals of Ediacaran times do not fit into the same basic body plan categories as the animals around today (or indeed those for the Burgess Shale fauna). In his view, the Ediacara consisted of separate segments, similar superficially to a quilted airbed, a plan that is not used in oceanic organisms today. To him, these creatures seemed to show continuous gradations in form between animals that, if they were alive today, would be placed in separate phyla. For example, *Spriggina* shares certain features with the Ediacaran sea pen *Charniadiscus* found in Leicestershire. As such, Glaessner called the Ediacarans the 'Vendobionta'.

This view of the Ediacarans as being morphologically unique has recently been acknowledged and amplified by Martin Brasier and Jonathan Antcliffe of Oxford University. They have suggested that many of the Ediacaran fossils are different life stages of the same creature and that the radical differences between them can be explained by a process called **heterochrony**, where the characteristics of part of the life cycle are retained into another part of the life cycle of the organism. This, of course, does not help with understanding what the Ediacarans were exactly. There are some who say that they were not multicellular animals at all; recent speculations have suggested that they may have been related to the group that gave rise to the fungi or were perhaps even lichens (a symbiont – an organism that is associated with another in a mutually beneficial relationship – consisting of fungi and algae) in the modern sense of the word. This may seem absurd, but in the **Universal Tree of Life** (Figure 6.4) you can see that animals, plants and fungi are very closely related.

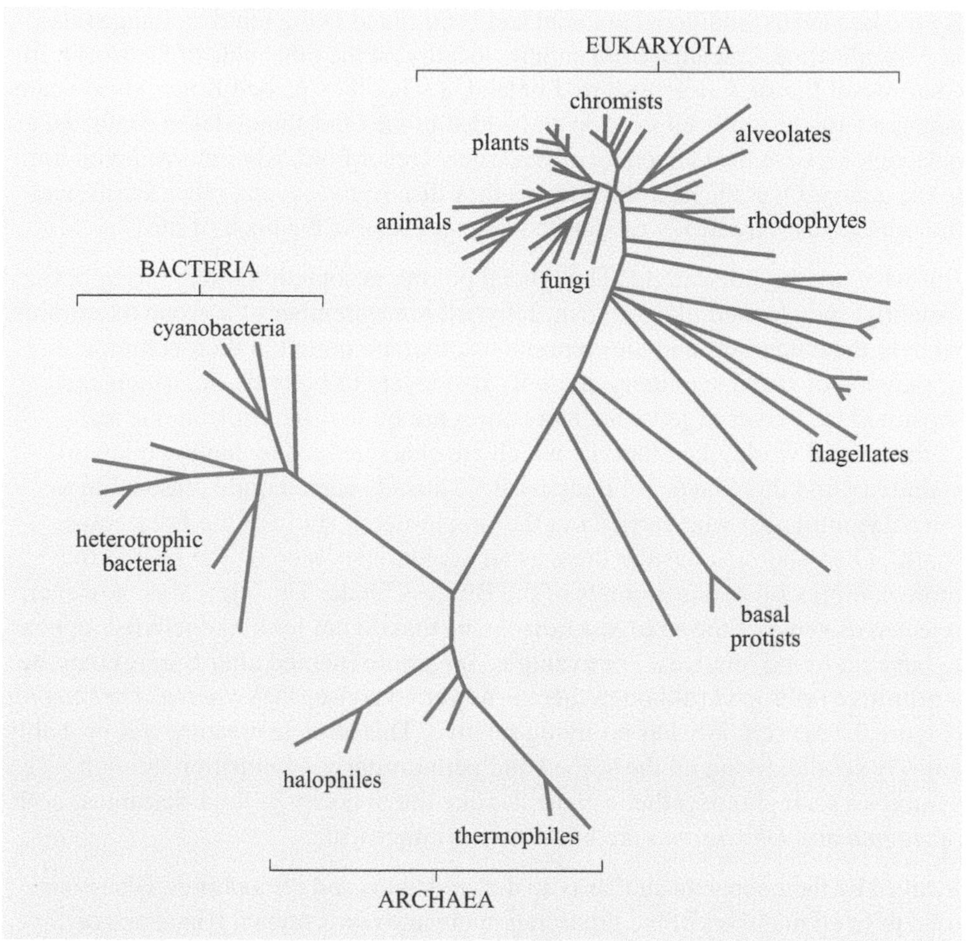

Figure 6.4 The Universal Tree of Life showing the relationship between animals, plants and fungi in one branch and the great disparity between this branch and the Archaea and Bacteria branches. The longer the line, the greater the evolutionary distance.

Whatever the truth, the consensus is that the Ediacarans are a group of wholly extinct multicellular organisms that were constructed around a different type of tissue organisation from the rest of the known animal world. To many, these creatures, lacking both head and gut, are a failed experiment in the history of life.

Nothing illustrates the importance of the Ediacarans more than the fact that, in the last few years, the terminal period of the Proterozoic has been renamed after them. The Vendian (from which the Vendobionta took their name) has been renamed the Ediacaran and the end of the Ediacaran marks the end of the Proterozoic and the time when bilaterian animals exploded onto the world stage.

6.1.3 The Cambrian Explosion

At the same time as the Ediacarans were alive, the ancestors of modern bilaterian animals were diversifying with unprecedented rapidity; this is called the Cambrian Explosion.

The Cambrian Explosion represents a time when a large range of different bilaterian body plans appeared in the fossil record in a relatively short period of time. So major were these body plans that they equate to different phyla – one of the highest taxonomic groupings. At this point, a distinction needs to be made between two aspects of the Cambrian Explosion that are often confused: the diversification of complex bilaterians known from a handful of exceptionally well-preserved faunas dotted around the world as opposed to the sudden proliferation of shelly fossils in the sedimentary record.

From the palaeontological perspective, there are only a few localities where the world of the earliest Cambrian fossils may be observed:

- the famous Burgess Shale fauna of British Columbia (of middle Cambrian age, dated to ~505 Ma)
- the Chengjiang (~520 Ma) and Sirius Passet (~530 Ma) faunas of China and Greenland respectively.

The reason why the **radiation** of bilaterians in the early Cambrian is considered an 'explosion' is because no examples of these bilaterian animals have been found at the time of the Ediacarans while by ~530 Ma (Sirius Passet fauna) there were abundant examples of animals that can be recognised as ancestral to those alive today. These animals, therefore, appear in the fossil record within a relatively short period of ~12 Ma duration – hence the term 'explosion' (Figure 6.1).

The exact timing of this event has been narrowed down by examining fossils from Siberia where, on the banks of the Kotuikan River, one of the most spectacular Cambrian outcrops in the world can be observed. At river level (where there are only a few unspectacular trace fossils), the rocks have been dated at about 545 Ma. A little higher in the section is a sandstone unit that has been dated as 544 Ma – a difference of only one million years. Starting immediately above this unit, small shelly fossils begin to increase in frequency up the rock section. It is the first evidence of the shelly part of the Cambrian Explosion. Further up the cliff, sediments dated at 530 Ma contain more than 80 recognisable **taxa** (singular taxon), including small shelly fossils, tracks, burrows and trails.

The origins of the Cambrian Explosion

The first 10 Ma or so of the Cambrian was a time when small shelly fossils developed in diversity and complexity. It seems likely that the ancestors of molluscs and an extinct mollusc-like taxon known as the halkierids evolved at this time (Figure 6.5 overleaf). After this 10 Ma period and close to the end of the early Cambrian, **crown groups** (a group of closely related organisms that includes the common ancestor plus all its descendants) of all the major phyla alive today (e.g. annelids, arthropods and brachiopods, and the chordates that ultimately gave rise to the vertebrates and humans) suddenly appeared on the scene.

Obviously, these fossils cannot have appeared fully formed out of nowhere, and advances in molecular phylogenetics and molecular clocks have shed crucial new light on the timing of their evolution. Until quite recently, the molecular evidence seemed to point to an origination time of the crown group bilateria of about 1000 Ma. It has since become clear that the vertebrate genes that were used to calibrate the molecular clock mutate much more slowly than invertebrate genes, giving much older divergence times than are implied by the fossil record. The latest studies indicate that the bilaterians arose between 630 and 600 million years, i.e. between 88 and 58 million years before the Precambrian–Cambrian boundary.

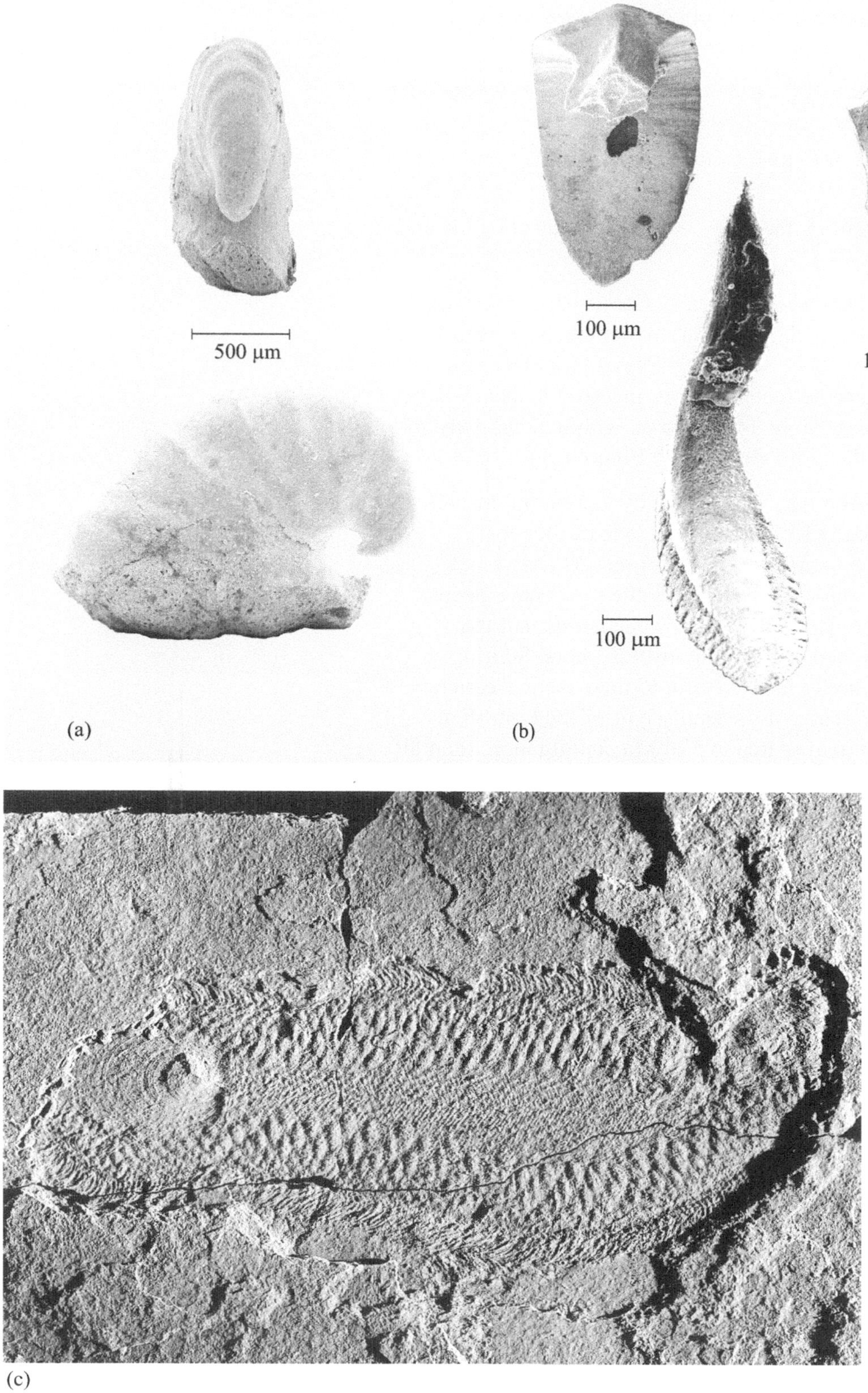

500 μm

(a)

100 μm

100 μm

100 μm

100 μm

(b)

(c)

Figure 6.5 Examples of Lower Cambrian small shelly fossils: (a) a mollusc shell; (b) isolated sclerites, an extinct mollusc-like group; (c) chain-mail-like covering of sclerites on a complete organism (*Halkieria evangelista*). The fossil in (c) is about 3.5 cm long. ((c) Simon Conway Morris)

This clearly implies that the bilaterians were contemporary with the Ediacarans; but, as previously discussed, there is virtually no evidence for this from the fossil record. The reason for this lack of evidence is believed to be because all the late Proterozoic bilaterians were microscopic and, hence, were not preserved in the fossil record. Whether or not the bilaterian animals did originate in the Ediacaran (as seems overwhelmingly likely), it is clear that they evolved very rapidly in the early Cambrian, possibly due to an extinction event that decimated the Ediacarans.

The causes of the Cambrian Explosion

The causes of the Cambrian Explosion fit into two categories: internal (i.e. biological) and external (i.e. environmental) causes. Historically, much effort has been focused on relating the diversification of organisms in the early Cambrian to environmental change. This approach follows on from the work of Darwin, who believed that external influences drove the evolution of organisms.

One of the most exciting developments has been the insight that developmental processes within organisms can, and do, have much responsibility for these changes. One of the most important examples of this has been the recognition of **Hox genes** as an important part of animal development. Hox genes are responsible for *pattern formation*, i.e. the overall arrangement of appendages on the body of an organism. Comparison of the distribution of Hox genes in diploblastic animals (animals having a body made of two cellular layers) and their distribution in the bilateria shows that there are many more varieties of Hox gene in the latter group than in the former. The much higher number of Hox genes in the bilaterians was probably caused by a process known as **gene duplication**. These duplicated Hox genes were subsequently assigned to control the development of other body parts. Hence, it seems likely that Hox genes are implicated in the rapid diversification of the bilaterians in the early Cambrian. Hox genes may be thought of as the 'coarse' (as opposed to the 'fine') control lever of evolution, and sudden access to it during the early Cambrian drove the Cambrian Explosion.

Some scientists have suggested that the regulation of many new body patterns by Hox gene control allowed the evolution of **macrophagy**, i.e. the ability to ingest and digest food extracellularly by allowing the development of the gut. By allowing the bilaterians to eat and digest food particles that are larger than single cells, organisms were able to grow bigger. This might explain the sudden appearance of the small shelly fossils in the early Cambrian. It would also support the notion that the stem-group bilaterians were contemporary with the Ediacarans. Thus, the bilaterian increase in body size may be ascribed to the development of a novel method of feeding that developed due to the sudden evolutionary deployment of a powerful set of genes. The evolution of macrophagy in turn set up a whole new world of competition, as predators and prey suddenly indulged in the biggest evolutionary arms race of all time.

From the evidence of molecular phylogenetics, it seems likely that the Hox gene revolution occurred sometime between 650 Ma and 550 Ma, precisely coincident with the Ediacarans and the implied hidden beginnings of the Cambrian Explosion.

So far, this section has examined the internal triggers for the Cambrian Explosion. However, there is no denying that the late Proterozoic and the early Phanerozoic were times of enormous environmental change as well. The terminal Proterozoic snowball event (the Marinoan glaciation) finished at about 635 Ma, the same time as when the Hox gene revolution may have been occurring. A heavily glaciated Earth would have forced the early biota into **refugia** (i.e. geographically isolated populations) with limited gene flow between them that could have stimulated rapid evolution and the easy dissemination of novel gene plans like the Hox complex. In addition, it has been suggested that there may have been an increase in deep-ocean oxygen concentration at this time, which would have assisted organisms increasing in size in the rapidly evolving bilateria.

6.2 Radiations and extinctions

6.2.1 Estimating the turnover of life

The Cambrian Explosion is the first of the major evolutionary radiations that can be charted in some detail from the fossil record. Figure 6.6 shows the total numbers of marine animal families (most containing several species; see Box 6.1) based on fossil evidence and estimated to have coexisted at different times through the Phanerozoic. This was initially limited to the marine fossil record, simply because this was relatively more complete than that of land-dwelling organisms.

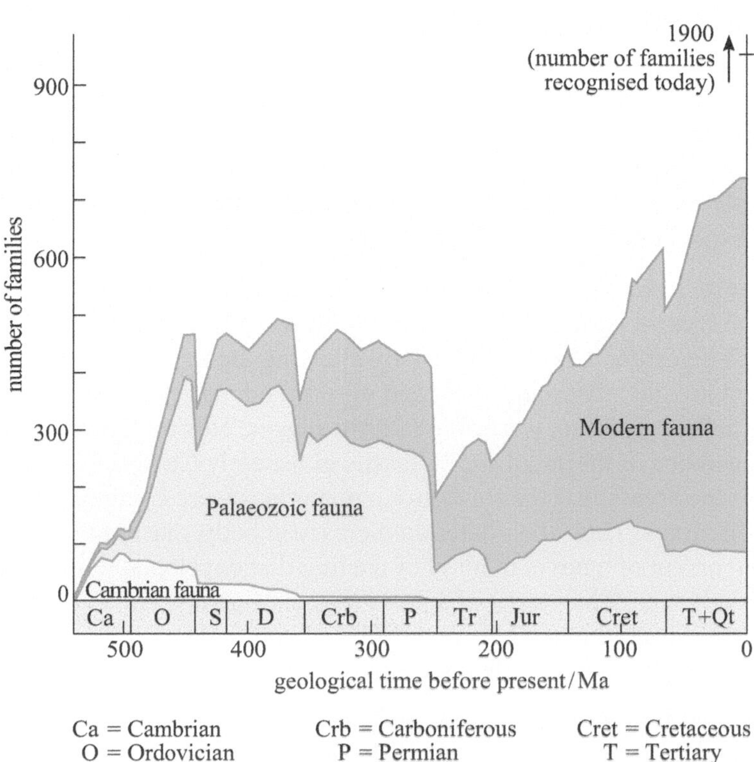

Figure 6.6 Changes in the numbers of families of marine animals through the Phanerozoic, based on fossil evidence. See Section 6.2.3 for explanation. (Sepkoski, 1990)

■ What might explain the contrast in relative completeness between the marine and land-based fossil records?

■ Most sediment ultimately ends up in the sea, so sedimentary deposition there is relatively more widespread, and less interrupted, than on land.

A later compilation of families (by Mike Benton of the University of Bristol, UK, 1995), covering all marine and land organisms, is shown in Figure 6.8a (overleaf); Figure 6.8b and c show the respective land-dwelling and marine components of that compilation. For all three graphs, minimum and maximum estimates of family

Box 6.1 The hierarchy of classification

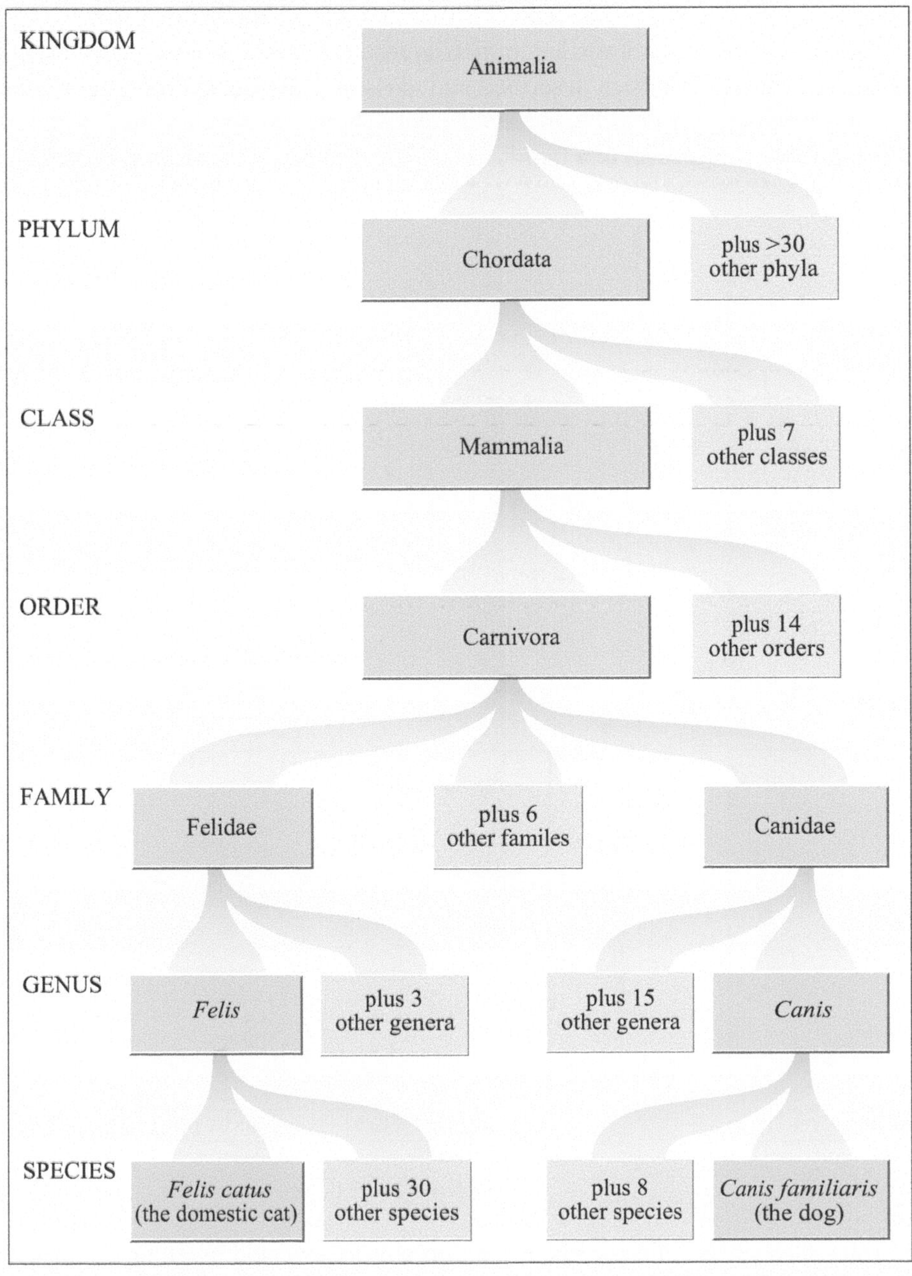

Organisms, both living and fossil, are classified into a nested series of increasingly inclusive groups, known as taxa, all the way up to the level of the kingdoms discussed in Box 2.2.

This system, known as the **taxonomic hierarchy**, is ideally intended to reflect evolutionary relationships. Where evolutionary relationships are unclear, a pragmatic scheme has to be adopted instead, which can be subject to later correction in the light of new findings. Taxonomic categories at different levels in the hierarchy are given standard names (e.g. family, species, order), each taxon being assigned its own Latinised name (e.g. Canidae, the dog family). The taxonomic categories used for animals are shown in Figure 6.7, along with the classification in this system of the domestic cat and dog.

Figure 6.7 The taxonomic hierarchy showing the classification of the domestic cat and dog.

numbers are shown to allow for uncertainties in the number of extinct families and the representation of some fossils in both marine and continental deposits. Note that Figure 6.8c differs in detail from Figure 6.6, reflecting a mass of slight differences in the family classifications used, and in the estimates of their stratigraphical ranges. These differences provide a useful reminder that such information is not a fixed set of facts, but the result of interpretation both in the classification of fossils and in the correlation of sedimentary sequences. The record is always subject to revision both in terms of new finds and through new analyses of pre-existing data. The similarity of the two marine compilations nevertheless suggests an underlying pattern can be identified and made available for interpretation.

You may wonder why the numbers of families have been plotted instead of the number of species. After all, families are artefacts of classification, whereas species are natural biological units. The reason is that this is the simplest way to cope with the incompleteness of our knowledge; fossils record a fraction of past species, and only some of these have been described and classified. By being more inclusive and using families, it is possible to have a relatively more complete record. A single specimen from one species in a family is sufficient to record the family's presence.

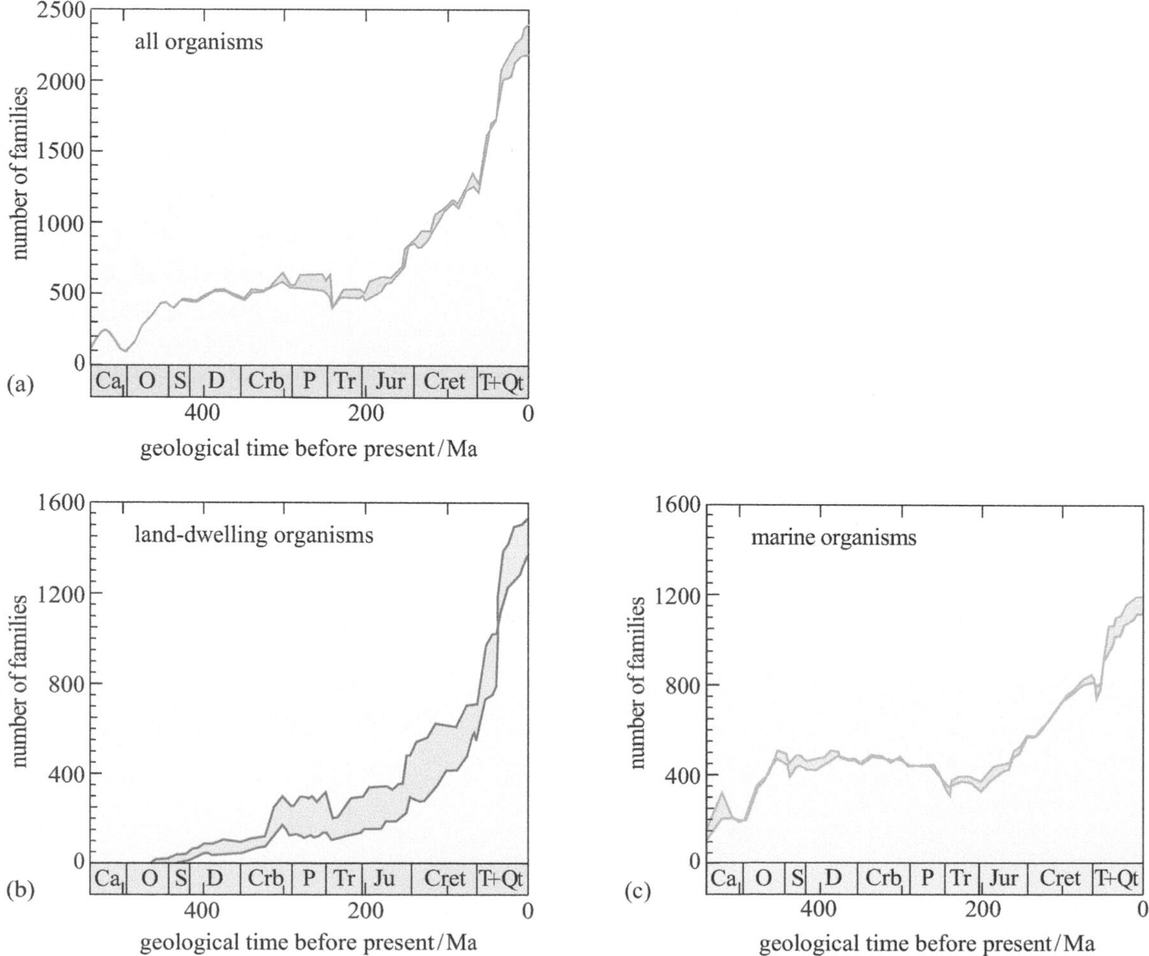

Figure 6.8 Changes in the numbers of families through the Phanerozoic: (a) all organisms; (b) land-dwelling organisms only; (c) marine organisms only. Maximum and minimum estimates are shown. (Benton, 1995)

(By analogy, a single telephone directory would provide you with a fairly comprehensive list of the households in a town, but a great deal more effort would be required to account for each individual.) Families are, therefore, a useful proxy for numbers of species. Of course there is a risk of biased representation because families are of very different sizes. The working assumption, however, is that such inequalities average out, or at least still allow broad patterns at the species level to be revealed, when large enough samples are considered. The skeletal hard-part fossil record represents only a fraction of life, but it at least provides a consistent basis for monitoring the relative variations in family numbers by allowing similar types from different times to be compared.

The steep rise in family numbers in Figure 6.6 throughout the early Cambrian represents the Cambrian Explosion. Throughout this episode, the evolution of new families (i.e. first appearances of species, recognisable by diagnostic characteristics) clearly exceeded the number of extinctions over this time. This illustrates the importance of the evolutionary innovations that occurred in the bilateria during the Ediacaran period and how they fuelled rapid diversification in the early Cambrian.

The pattern has been likened to what happens to numbers of individuals in a population when the birth rate exceeds the death rate.

Question 6.2

Briefly describe an example of the effects of the birth rate exceeding the death rate.

6.2.2 Mass extinctions

Diversity did not increase indefinitely, however, and the data in Figure 6.6 suggest a temporary lull in the late Cambrian before diversification rapidly increased during the Ordovician. The reasons for the late Cambrian lull are still unclear. However, more striking is the sharp drop in family diversity (number of families) accompanying the close of the Ordovician, which marks the first of five large-scale mass extinctions during the Phanerozoic as shown by the fossil record.

- Referring to Figure 6.6, what was the approximate percentage decrease in total numbers of families recorded from fossil hard parts at the end of the Ordovician?
- The total family numbers fell from around 470 to about 340. Hence, the percentage decrease in family diversity was:

$$\frac{470-340}{470} \times 100 = 28\% \text{ to 2 sig. figs.}$$

Two points need to be raised about this result: first, it only represents the net decline in family numbers; given the appearance of new families over the same interval, the percentage *extinction* of pre-existing families would have been somewhat higher. Second, the values of percentage extinction at lower taxonomic levels (i.e. of genera and species), if calculated, would have been higher still. For example, consider a family that lost all but one of its species: it is still counted as having survived, although at the species level there may have been a considerable loss, which must be added to the number of species extinctions in families with no survivors. By sampling well-documented groups at lower taxonomic levels it is possible to estimate these losses, and it has been estimated that ~85% of all species may have become extinct at the end of the Ordovician.

Question 6.3

Referring to Figure 6.6, state the times of the other four great mass extinctions that occurred within the Phanerozoic Era.

Greatly increased rates of extinction over limited periods of time are implicated in all of these mass extinctions, as well as in a number of other smaller scale examples. A temporary suppression in the rate of speciation, and hence origination of new families, is also evident in some cases. By far the most severe mass extinction was that at the end of the Permian, when more than half (57%) of marine animal families became extinct, while the estimated toll at the species level was a staggering 96%. The possible causes of this disaster are examined later in this chapter, from the context of global change that was occurring at that time.

A postulated link between the mass extinction (including extinction of the dinosaurs) at the close of the Cretaceous and a major asteroid impact (recognisable in the sedimentary record by a sharp peak in abundance of iridium), has generated much debate and popular interest in the causal link between mass extinctions and planetary events. In 1984, Jack Sepkoski and David Raup undertook a detailed analysis of the distribution in time of all detectable mass extinctions. They concluded, for the Mesozoic examples at least, that they all appeared to show a periodicity of approximately 26 Ma, and suggested that this could reflect a regular cycle of extraterrestrial bombardment. A vigorous debate ensued, with opponents criticising aspects of the database and of the analyses, particularly with respect to the recognition of a periodic signal.

There is no dispute that there have been asteroid and cometary impacts (some of considerable size and effect) on Earth as several forms of evidence exist, such as preserved impact craters, deposits of beads known as tectites, **shocked quartz** and other distinctive minerals formed under very high pressures, and tell-tale geochemical signatures (e.g. sharp increases in iridium) in sedimentary sequences. What is open to question, however, is the extent to which impacts have been responsible for mass extinctions and, in particular, whether mass extinctions really show a periodicity that might reflect astronomical 'forcing'. At present, the jury remains out on the latter question, with critics arguing that the observed pattern does not deviate significantly from what can be produced by random coincidence in a model involving numerous unrelated kinds of environmental perturbation (including occasional impacts).

Evidence linking some of the other mass extinctions of the Phanerozoic to impacts is subject to debate, although such evidence has been reported at a couple of modest Tertiary mass extinctions overall. The apparent durations of the various mass extinctions, and their relative effects on different groups of organisms, are by no means consistent from one example to another. Hence, it is likely that a variety of causes may be needed to explain them, and the geological record offers several compelling Earth-bound alternative models (Box 6.2).

One point that is nevertheless clear is that all five major mass extinctions were brought about by environmental crises of one sort or another. There is no evidence that the rapid diversification of any groups of organisms directly provoked any of these past mass extinctions. Each mass extinction seems to have come as a shock, cutting sharply across any pre-existing pattern of change in diversity, with each one resetting the agenda for subsequent evolution.

Box 6.2 The 'big five' Phanerozoic mass extinctions

In detail, many of the mass extinctions show a more complex pattern than revealed in Figure 6.6.

The main episodes of mass extinction in the Phanerozoic fossil record, together with their main 'casualties' and likely causes, were as follows.

1 Late Ordovician

Two main peaks of extinction towards the end of the period were separated by hundreds of thousands of years. Both plankton (e.g. graptolites) and bottom-dwelling life (e.g. trilobites and reef-building organisms) were affected. Associated events were the growth and decay of a vast ice sheet on the southern supercontinent of Gondwana as it moved over the South Pole, drastically changing the sea level, climate and ocean chemistry. However, the patchy geological record has yet to reveal any record of a clear causal event.

2 Late Devonian

This pattern of extinction is as yet unresolved, but possibly consists of a series of extinctions extending over at least 3 Ma, with the most severe effects ~5 Ma before the end of the period. Shallow marine ecosystems were most affected, with tropical reef-dwellers particularly hard hit. Declines in temperature seem to be associated with widespread anoxia in shallow seas. Although there is no direct evidence for glaciation at this time, sea levels fluctuated and fell overall. A positive shift in $\delta^{13}C$ values in the C_{carb} record would be consistent with burial of organic carbon associated with increased photosynthesis, removing CO_2 from the atmosphere and sequestering it in sea-floor sediments. Alternatively, there is some evidence for meteor impacts provided by glassy fragments found in Belgium, as well as impact swarms in Chad and North America.

3 Late Permian

Increased extinction rates occurred over ~3–8 Ma at the end of the period, although recent findings suggest that these fell largely in two distinct episodes. Marine organisms were particularly devastated (Figure 6.6). At this time the continents had amalgamated into the supercontinent Pangaea and the Earth was emerging from the Carboniferous icehouse (see Chapter 7). Consequently a complex, probably synergistic, array of causal factors, including biological feedbacks, has been postulated for this extinction.

■ Can you recall from Chapter 4 what other major event coincides with the Late Permian extinction?

■ The eruption of the Siberian flood basalt province (Table 4.3).

In addition to environmental stresses with largely unknown causes, the eruption of over 2 million km^3 of basalt could have had a significant impact on the global climate, causing periods of cooling, through the emission of SO_2, periods of warming, through the emission of CO_2, and the resultant acidification of the environment. In a global system that was already under environmental stress, the eruption of such a large volume of basalt over a relatively short period of time could have provided the *coup de grâce* of this the greatest of all mass extinctions.

4 Late Triassic

There were at least two, maybe three, extinction peaks during the last 18 Ma or so of this period. In the sea, both free-swimming animals (especially ammonoids and marine reptiles) and bottom-dwelling forms (including many reef-building organisms) declined. On land, many reptilian groups, including mammal-like forms, were lost, along with large amphibians and many insect families, although there was no marked global extinction of land plants. The marine extinctions, at least, coincided with marked changes in sediment type, strongly suggestive of major climatic change and, in particular, more extensive rainfall. Nevertheless, a huge impact crater in Quebec, about 65 km across, dates from around the Triassic–Jurassic boundary. The late Triassic also coincides with the eruption of the Central Atlantic Magmatic Province (Table 4.3).

5 Late Cretaceous

Two patterns of extinction appear to be superimposed: a gradual decline followed by a rapid collapse. Several groups of marine animals, both free swimming (e.g. belemnites and some marine reptiles) and bottom dwelling (e.g. certain kinds of bivalves) appear to have declined over the last 9 million years of their period. At the end of the Cretaceous an abrupt collapse occurred, especially among the plankton. This collapse happened over ~100 000 years or less, and is marked by fluctuations in the carbon isotope record of marine limestones. Ammonites combined both patterns of extinction with a slow decline terminated by the rapid extinction of a sizeable number of the remaining species at the end of the Cretaceous.

On land, the most famous extinction is that of the dinosaurs – although the pattern of their decline is still debated. They were joined in their demise by the pterosaurs (flying reptiles), while flowering plants also suffered major losses at the end of this period, especially at mid-latitudes in North America (as shown by carbon isotope data from organic residues indicating mass mortality) and at high latitudes in Asia. In the Southern Hemisphere, changes in speciation were gradual or non-existent.

A number of other groups were little affected by extinction, including the crocodiles, snakes and placental mammals. Although the final collapse is widely interpreted as impact related (with evidence mounting for the impact crater being buried in the subsurface of the Yucatan Peninsula, Mexico), it affected ecosystems that were already perturbed by other environmental causes, as shown by the gradual background decline. This can be related to substantial changes shown by the continental configuration, global climate change and oceanic circulation patterns at the time. Once again, however, the K–T boundary is marked by the eruption of a major flood basalt province, in this case the Deccan lavas of India.

6.2.3 Evolutionary radiations

Look at Figure 6.6, and in particular the periods immediately following each of the 'big five' mass extinctions.

Question 6.4

What happened to diversity in the immediate aftermath of each of these extinctions? On the basis of what you have read so far in this chapter, suggest a reason for the pattern shown.

This represents the immediate consequences of the mass extinctions, but what longer-term trends can be recognised?

- What happened to overall levels of diversity over the longer term?
- Overall, diversity rose – although not in a continuous fashion. After the Ordovician, an approximate plateau of diversity was established, remaining until the late Permian mass extinctions. Thereafter, diversity started to rise once again and continues to do so at the present time.

After the Ordovician (and throughout the rest of the Palaeozoic), the recovery of diversity following the mass extinctions appears to have tapered off to similar levels each time, equivalent to between 400 and 500 families. There has been much debate about whether the Palaeozoic plateau of diversity levels reflects some kind of evolutionary equilibrium, with the post-Palaeozoic trend signifying growth to a new, higher, equilibrium level. The implicit assumption of this idea is that increased diversity leads to increased rates of extinction and decreased rates of origination, as a consequence of increased competition between species.

Question 6.5

Describe how the relationship between diversity, rates of extinction and origination could yield an equilibrium in diversity.

It remains open to question whether the rate of extinction caught up with that of origination (other than during mass extinctions) to yield the postulated equilibrium levels. An alternative interpretation is that the Palaeozoic plateau of diversity might be the net effect of several major and minor mass extinctions (the latter being below the level of resolution of Figure 6.6), superimposed on a background level of sustained diversification. This is the current preferred explanation to describe the patterns of change in Figure 6.8. Possible controls on the rates of origination and extinction of taxa, and hence on global diversity, are still, however, vigorously debated.

Whatever the long-term influence of the extinctions on the numbers of families over time, the outcome left an impact on the composition of marine life. This effect is evident in the blue bands in Figure 6.6. These represent sets of major groups (mainly classes) of marine animals, with each set showing a characteristic pattern of family turnover. These sets are referred to as **evolutionary faunas**; their separate diversity histories, along with some of the representatives of the major groups, are illustrated in Figure 6.9 (overleaf). The first set, referred to as the Cambrian fauna, diversified more rapidly during the initial radiations of the Cambrian than the others. However, it soon experienced a decline in numbers of families as the next Palaeozoic fauna began to diversify. The Palaeozoic fauna continued diversifying until the late Ordovician, after which it began a slow decline in numbers, while the modern fauna diversified rapidly, eventually dominating the scene in post-Palaeozoic times.

Note that the names of these evolutionary faunas refer only to their times of dominance: all three existed throughout the Phanerozoic (although very few members of the Cambrian evolutionary fauna survived beyond the Palaeozoic). Moreover, the three faunas do not represent discrete sets of animal groups, as in some cases different classes from a single phylum have been allocated to all three faunas, e.g. the molluscs, which consist of bivalves (clams) and gastropods (snails) in the modern fauna and cephalopods (including the ammonoids) in the Palaeozoic fauna.

These broad patterns of change can be explained in terms of a complex version of the equilibrium diversity model outlined above, where each successive fauna (for whatever unknown reasons) had its own characteristic diversity-dependent rates of origination and of extinction of families, and thus its own intrinsic equilibrium diversity level. The details of how this model could explain the histories of the three faunas shown here, and the arguments for and against it, are beyond the scope of this book. Nevertheless, the response of these evolutionary faunas to mass extinctions is worth considering.

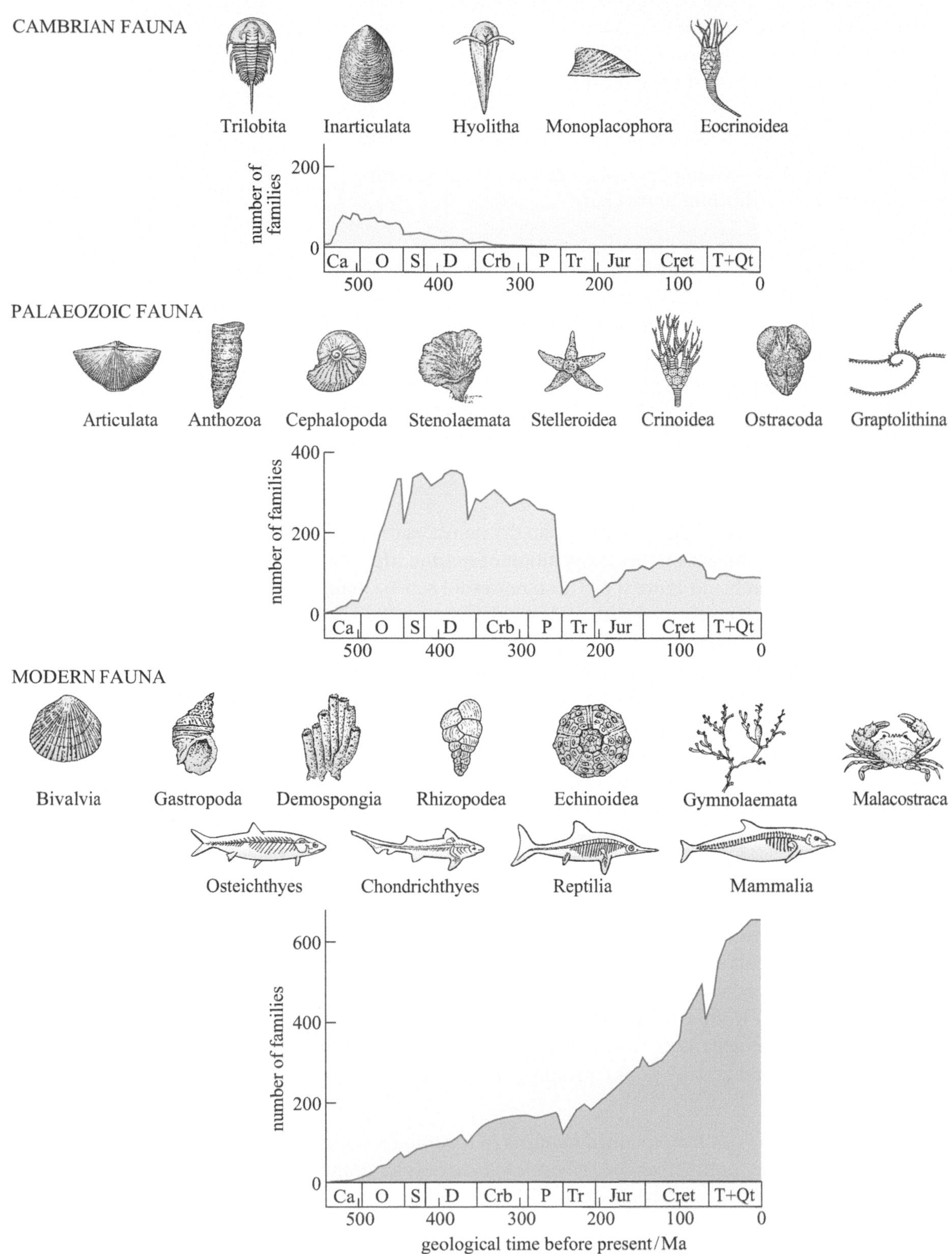

Figure 6.9 Examples of representatives and diversity histories of the three evolutionary faunas of skeletalised marine animals indicated in Figure 6.6. (Sepkoski, 1990)

Question 6.6

Did mass extinction affect all three evolutionary faunas equally? Explain your answer with a reasoned argument.

The ensuing radiations built on the characteristics of the survivors. The complexion of animal groups coexisting at any given time has, therefore, been influenced as much by the devastations of extinction as by the achievements of adaptive evolution. But what were the ecological consequences of these major faunal changes?

6.2.4 Ecological relationships and their consequences

Perhaps the most remarkable of the faunal changes occurred after the late Permian mass extinction and involved predation. Although predation was common in the Cambrian, with some predators even ingesting entire small shelly prey, throughout the Palaeozoic there was an abundance of exposed shelly animals that lived rooted to the sea floor (Figure 6.9) with evidence for predation (e.g. in the form of diagnostic damage preserved in the fossil shells, Figure 6.2) rare. Throughout the Mesozoic, however, the fossil record shows several lines of evidence for a marked increase in the intensity of predation on shelly prey. These include an increase in the frequency of predator damage, the appearance of many new kinds of predators with specialised adaptations for tackling shelly prey, and the emergence of new defensive adaptations by shelly organisms (prey) (Figure 6.10 overleaf).

Among the newly evolving predators in early Mesozoic seas were crabs and lobsters, along with several groups of fish and marine reptiles, variously equipped to crush, smash or pierce shells. New kinds of starfish evolved the ability to pull open bivalves, insert their stomachs into the opened shell and digest the occupant. In the Cretaceous, these adapted predators were joined by gastropods capable of drilling through shells to reach their prey, while in the Tertiary, various shell-breaking birds and mammals evolved.

Although the major groups of shelly prey animals developed a wide variety of protective adaptations, the one notable evolutionary trend was a boom in burrowing, especially by bivalves and echinoids (sea urchins) to increasing depths in the sediment to gain refuge from predators. The extent of **bioturbation** of the surface sediment was therefore both intensified and deepened. This effect, coupled with an increase in the amount of disturbance at the surface by grazers, detritus feeders, and the number of predators excavating the sediment, made the larval settlement of shelly animals permanently anchored to the surface more hazardous. As a result, these animals, so successful in the Palaeozoic, waned in relative diversity.

These linked ecological changes can be collectively referred to as the Mesozoic marine revolution, transforming the character of marine life to that which can still be recognised today. It could also be said that this revolution helped to provoke the unprecedented rise in marine animal diversity observed in the fossil record from the end of the Palaeozoic onwards. The most profound feedback to biogeochemical cycles, however, came from associated changes in plankton. Starting in the late Triassic, groups of microscopic plankton with calcareous skeletons began to appear, including both single-celled algae (e.g. coccolithophores) and protists (e.g. planktonic foraminifera). One potential

Figure 6.10 Reconstruction of the predators and prey of a shallow Cretaceous sea floor in southern England. In the section of burrowed sediment in the foreground, a burrowing clam (bivalve) is being attacked by a shell-drilling snail (gastropod), while a sea urchin (echinoid) has burrowed into the sediment for protection. In the middle distance, a starfish tackles a mussel attached to some kelp, while a regular echinoid grazes on some small, sessile, colonial animals. To the right, mussels and oysters (both bivalves) are being attacked by another drilling gastropod, a crab and a lobster (both arthropods). Various shell-crushing fish hover in the background, and the jaw of one (a shark), with its pavement of flat teeth, lies on the surface in the centre foreground. In the foreground, the remains of some bivalve shells with tell-tale gastropod drill holes can be seen.

interpretation of this simultaneous adoption of calcareous skeletons (or toughened organic walls in some other planktonic groups) by both groups is as a defensive adaptation against increased grazing pressure.

Question 6.7

What effect might the evolution of calcareous plankton have had on the distribution of carbonate sediments?

Going back again in time, however, some quite different evolutionary changes during the Palaeozoic had even more considerable environmental consequences – perhaps the most profound since the oxygenation of the atmosphere in the Proterozoic. These were the changes that allowed plants and (ultimately) animals to invade the land.

6.3 The greening of the land

6.3.1 Beginnings

So far, this chapter has investigated the emergence of metazoans in general terms. In this section, the focus turns to a specific evolutionary change that occurred during the Palaeozoic that was of major importance for the distribution of life on Earth and its eventual effects on climate – namely the colonisation of the land by plants. It is generally accepted that the main colonisation of the land by green plants took place in the Silurian or late Ordovician. However, there is evidence that bacterial mats and possibly fungi were prolific enough on damp land surfaces to have formed primitive soils as early as the late Proterozoic. This is not hard to imagine because there is a present-day analogy: the sandy desert floors of the southwest of the USA are partially stabilised against erosion by crusts of cyanobacteria. Either way, although these early soil-forming microbial communities would have altered the colour of the local landscape and incorporated some organic matter into the land substrates, they would have had a minimal to negligible effect on global climate, unlike the changes that took place during the Silurian and Devonian, as will shortly be discussed.

The first convincing evidence that plants had begun to adapt to a dry-land environment comes in the form of microscopic fragments of tube-like structures, cuticle and **spores** (Box 6.3).

Box 6.3 What is a spore?

'Spore' is a generic term; a commonly known example is pollen. Reproduction in some plants is achieved by the formation of spores instead of the production of separate male and female gametes directly (Figure 6.11).

A spore consists of a living cell surrounded by a tough water- and chemical-resistant outer coat. The cell is haploid and develops into a haploid individual which, in turn, produces gametes by mitosis. Fusion of two gametes yields a new diploid individual that goes on to produce a new generation of spores by meiosis.

The spore coat protects the haploid generation during dispersal and, because the coat is so tough, the spore has a high probability of being preserved in the fossil record. Many sedimentary rocks contain thousands of fossilised spores in each cubic centimetre, and the diversity of spore contents of rocks allows stratigraphic correlations to be made.

Pollen grains are specialised male spores in which the development of the haploid generation is highly abbreviated: the sperm cells are released directly and then fuse with egg cells to produce seeds. In this way, seed plants have effectively curtailed the haploid generation in favour of the development of diploid individuals.

Figure 6.11 Scanning electron micrograph of a specialised male spore, or pollen grain, of chicory (*Cichorum intybus*). The complex ornamentation (i.e. the spikes on the surface) aids in dispersal by insects and can be used for spore identification. (Bob Spicer/Open University)

6.3.2 Interpreting the structure of early land plants

The move from an aquatic to a terrestrial environment was accompanied by fundamental changes in plant architecture. Initially, land plants were low-growing, sheet-like structures that existed almost entirely within a relatively still and humid layer of air next to the ground: the **boundary layer**. In other words, if a wind is blowing (or air currents are in motion) and the free air (i.e. air that is well above the ground surface and unimpeded by, for example, trees and buildings) is moving relative to the ground at a certain speed, then close to the ground the air will be moving much more slowly because of the effect of friction (viscous drag) caused by the ground surface. The wind speed will, therefore, decrease progressively towards the ground surface; immediately above the ground there will be a relatively still layer, i.e. the boundary layer (Figure 6.12).

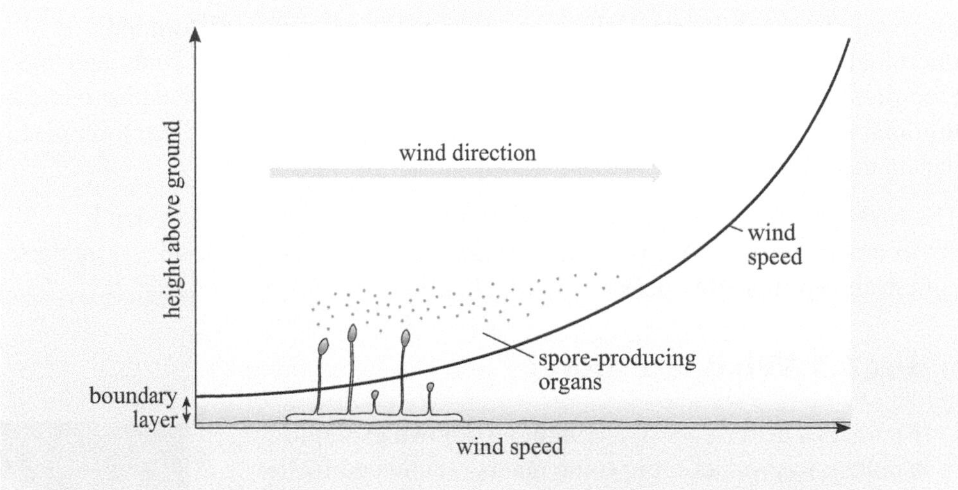

Figure 6.12 Variation in wind speed with height above a surface over which air is flowing. Superimposed on this plot is a diagram showing the height and characteristics of a typical early low-growing land plant, such as *Sporogenites*, which has only its spore-producing organs projecting above the boundary layer.

The thickness of the boundary layer will depend on the speed at which the free air is moving and on the roughness of the surface. Typically, the thickness of the boundary layer over a sand or gravel surface will be of the order of a few centimetres. If the ground surface is moist, then the boundary layer can become saturated with water vapour. This in turn means that it will have a higher relative humidity than the free air above it (unless, of course, all the air is water saturated).

A plant growing within the moist boundary layer would tend not to dry out because of this high humidity. Moreover, it would not need to expend energy and resources developing internal methods of storing moisture, enhancing its mechanical strength to cope with changing water contents and hence internal strength, or making cuticles to prevent loss of H_2O through evaporation. Yet, the lack of flowing air or water around the plant body in the boundary layer would be disadvantageous for dispersing progeny. By contrast, any plant capable of growing up into the moving air above the boundary layer would not have this problem of spore dispersal (Figure 6.12) and so taller early land plants would have had an advantage in the race to colonise the land surface.

Accordingly, in early land plants the ground-hugging body of the plant remained in the boundary layer, while stalks with special pouches containing the spores grew upwards (Figure 6.13). These spore-producing organs exploited the moving air

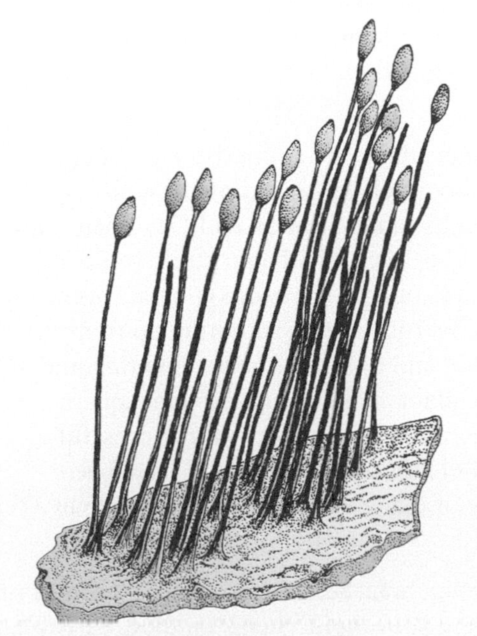
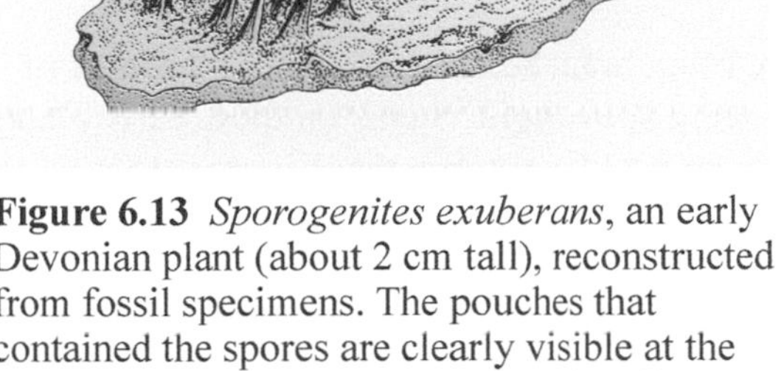

Figure 6.13 *Sporogenites exuberans*, an early Devonian plant (about 2 cm tall), reconstructed from fossil specimens. The pouches that contained the spores are clearly visible at the tops of the stalks. (Stewart and Rothwell, 1993)

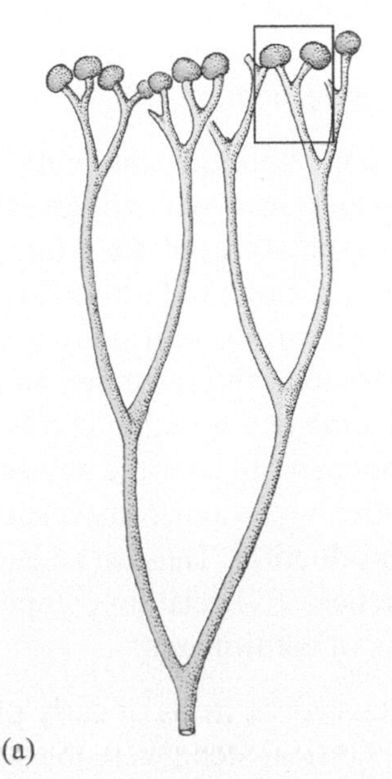

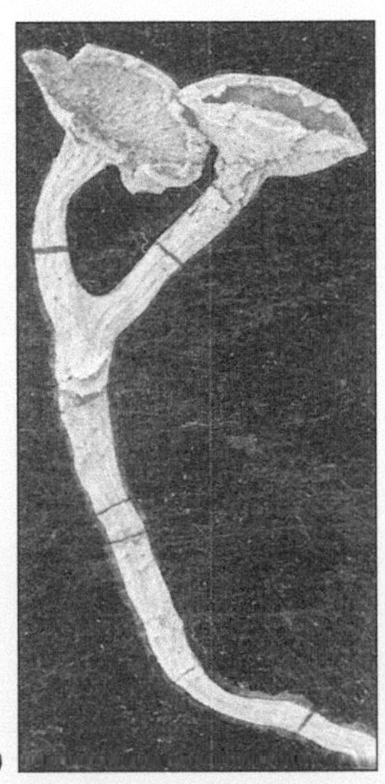

Figure 6.14 (a) Reproductive units (about 7 cm long) of *Cooksonia caledonica*, a late Silurian plant, reconstructed from fossil specimens such as that shown in (b). The specimen in part (b) is 4 mm in length. No specimen of the rest of the plant has been found. ((a) Stewart and Rothwell, 1993; (b) Dianne Edwards)

currents above the boundary layer to carry the spores away on the wind to colonise new territories and, if the spores were small enough, they could travel great distances quickly.

Reproductive units are preferentially preserved in the rock record because the presence of a waxy cuticle enhances their preservation potential. As such, many land-plant fossils consist of isolated reproductive units only, whereas the rest of the plant remains unknown (Figure 6.14).

■ How might the increasing vertical height of early plants have affected the thickness of the boundary layer?

□ The presence of taller plants would have, in effect, increased the roughness of the ground surface and hence the thickness of the boundary layer.

This positive feedback loop would have promoted vertical growth because a higher boundary layer would have extended favourable moist growing conditions upwards, permitting greater plant growth and requiring the greater projection of spore-producing organs above the boundary layer. With vertical growth would have come an increase in the cost of construction because greater structural strength would have been required. In addition, a mechanism by which to move fluids up (water) and organic molecules in solution down (sugars produced by photosynthesis) through the plant body would have been needed. This increase in construction cost would, in turn, have necessitated more food production – in other words, more photosynthesis. More photosynthesis would have required more gas exchange with the atmosphere, more plant mass and, therefore, a larger (and possibly higher) surface area to intercept light. The result of this positive feedback was the construction of branching and shading plants, and the need for more vertical growth.

To cope with the competition, many plants adapted to minimise their energy expenditure while still growing tall. All common early land plants and representatives of the groups of taller plants in the Silurian and Devonian shared a number of similarities (Figures 6.15 and 6.16); all had a prostrate stem, known as a **rhizome**, with small root-hair-like appendages (**rhizoids**) that anchored the rhizome to the substrate, and all also had vertical stems sprouting up from the rhizome at varying intervals (Figure 6.15a and b). As the horizontal rhizome branched out over the substrate, a single plant could occupy several square metres with numerous vertical stems, increasing the chances for successful reproduction. Thus, in an early Devonian landscape there would have been patches of vegetation composed entirely of extensive thickets no taller than a few tens of centimetres.

The vertical stems of early plants were often branched and, at or near the apex of the vertical stem, were the spore-producing **sporangia**. Stems on these early plants bore no leaves, but often had hooked or spine-like outgrowths (Figure 6.15c; Figure 6.16b and d). These outgrowths were quite small and poorly supplied with fluid transport tissue and, as such, their role in increasing the plant's photosynthetic area was probably minimal, unlike leaves on modern plants.

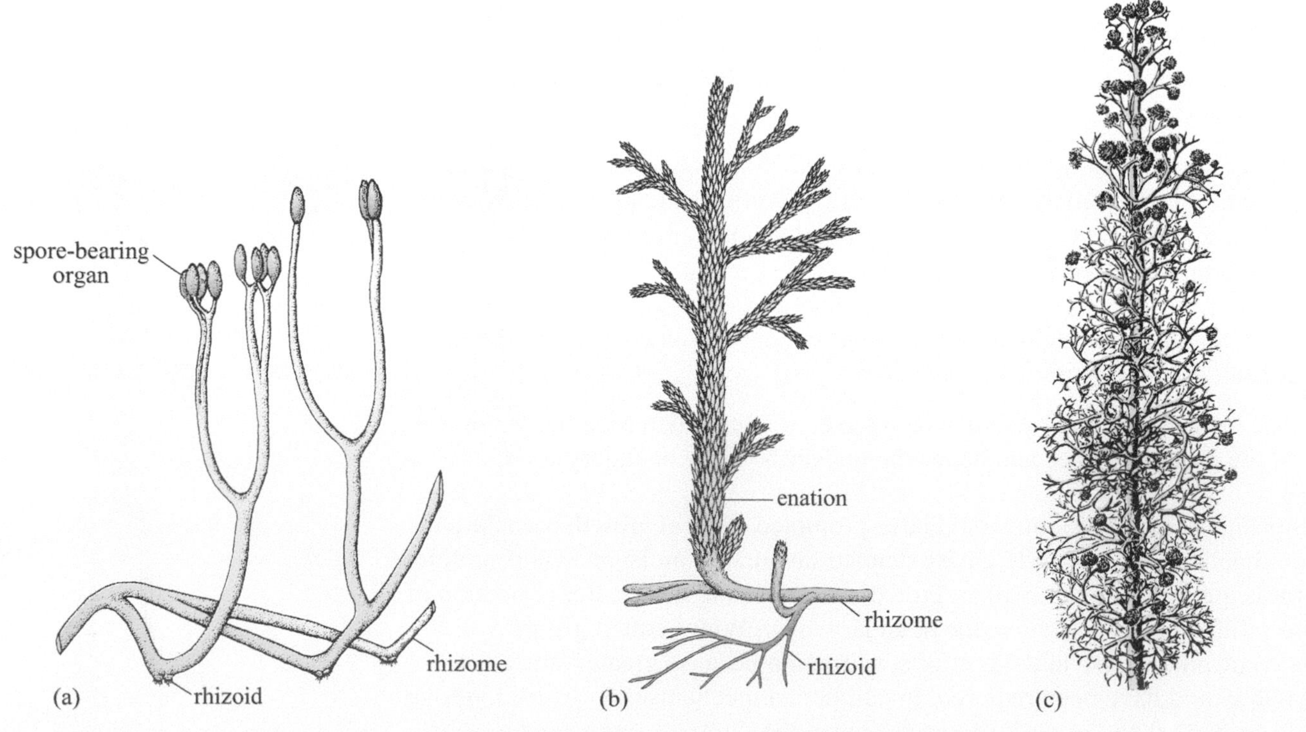

Figure 6.15 Reconstructions of some common early land plants, showing various structural features. (a) *Aglaeophyton major*, Early Devonian (about 50 cm tall). (b) *Asteroxylon mackiei*, late Early Devonian (about 30 cm tall); enation was a special kind of outgrowth from the stem that functioned as a small leaf. (c) *Pertica quadrifaria*, Early Devonian (about 1 m tall overall).

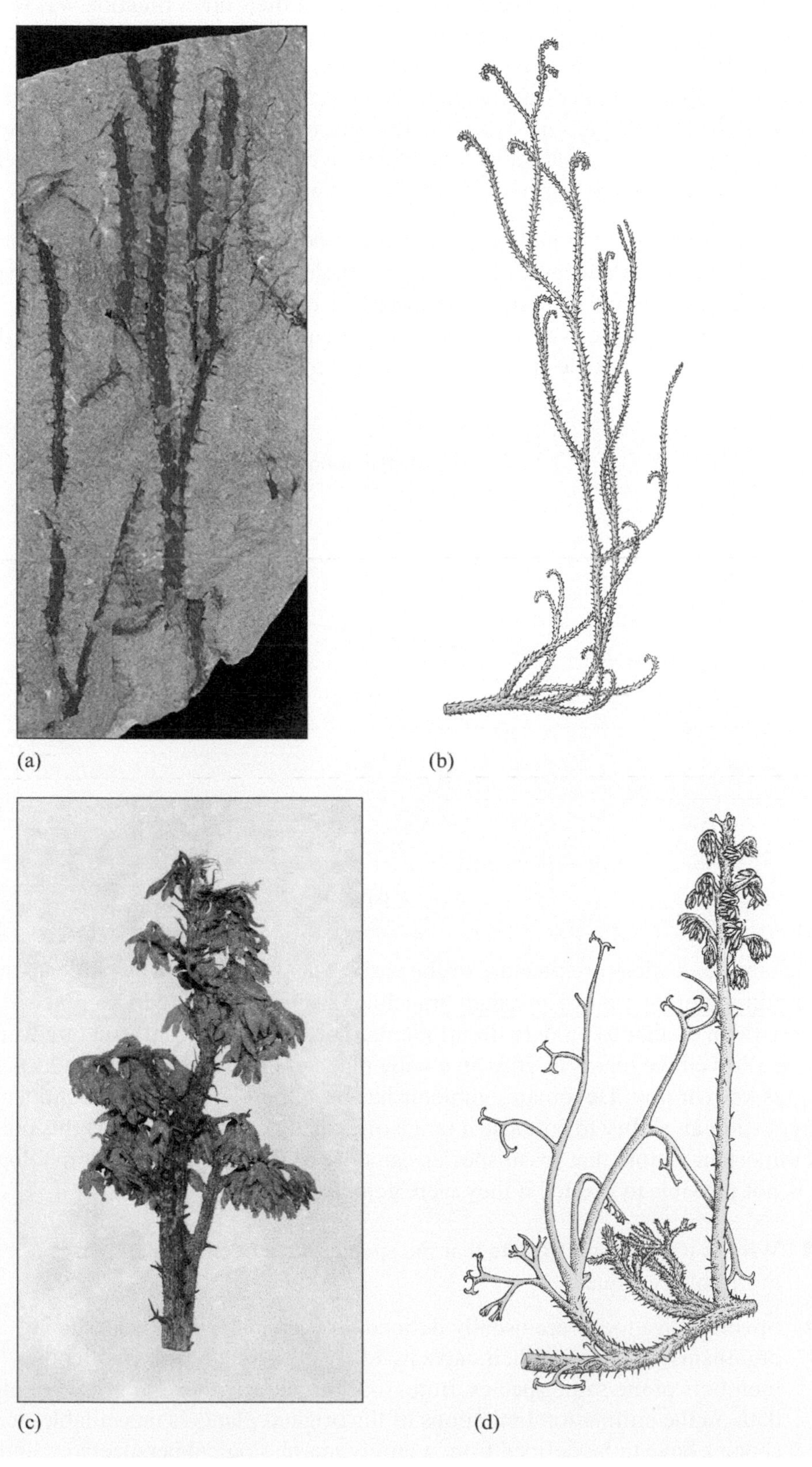

Figure 6.16 Fossils and reconstructions of: (a) and (b) *Sawdonia ornata*, Early Devonian (about 50–90 cm tall); (c) and (d) *Psilophyton dawsonii*, Early Devonian (about 50 cm tall). (Stewart and Rothwell, 1993)

Instead, their curled or hooked form suggests that their main function was to enmesh neighbouring stems, thereby providing mutual support. Upright growth was aided further by the crowding of the stems, which increased the height of the boundary layer. A modern-day analogy of this crowding effect can be seen in a field of wheat: a single stalk has very little mechanical strength and, if exposed to wind on its own, would quickly be blown over; many stalks growing closely together, however, can survive upright throughout an entire growing season.

To summarise, surface roughness and, hence, the height of the boundary layer increased as land plants evolved, their growth driven upwards by reproductive advantage and mutual shading (minimises loss of moisture) (Figure 6.17). Beneath this canopy of vegetation a humid environment was maintained, so that the need for thick cuticles to help retain water content was reduced.

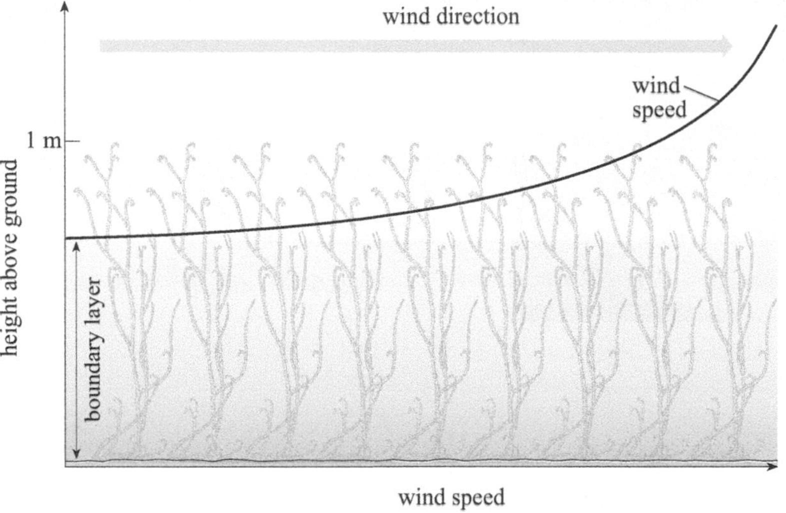

Figure 6.17 *Sawdonia* heath superimposed on an increased boundary-layer profile. As in Figure 6.12, two diagrams are superimposed here: one shows the height and characteristics of the plant and the thickness of the boundary layer, and the other shows the wind speed in relation to the height above the ground.

The simple, leafless architecture of the earliest land plants conveys little about the climate of the time. The naked branching stems with limited **vascularisation** make them similar to modern desert plants. This simple internal structure would have allowed the plants to grow in a wide range of environments; indeed, each genus known from Devonian sediments has been found over several continents, suggesting an ability to colonise a range of climatic environments. At this point it is important to note that fossil species can only be defined by their morphology; it is not possible to deduce if they were genetically related.

■ Why is it important to state that the species have been defined by their morphology alone?

□ Species alive today are usually defined (at least in sexually reproducing organisms) in terms of their capacity to cross-pollinate freely with other members of the same species. In fossils, particularly those of primitive plants, data on the pollination limitations of the original plants is unavailable, so species have to be defined from a purely morphological perspective. Similar-looking plants might, therefore, have been genetically distinct and so unable to cross-fertilise.

Box 6.4 The evolution of the leaf

By the Late Devonian, multi-layered forest ecosystems were fully developed, populated by plants with broad, flattened leaves with high leaf areas. Evaporation and transpiration, and hence the cycling of atmospheric water over land, must at last have approached present-day levels. Although earlier Devonian heaths of plants such as *Sawdonia* would have had a major impact on reducing erosion and sedimentation rates due to the binding action of the rhizoids and rhizomes, their effect on the hydrological cycle would by no means have been as great as that caused by the advent of leafed plants. Without the evolution of the leaf, our present world could not exist.

A leaf can be thought of as a flattened organ that produces food for the plant by photosynthesis. Leaves are generally green due to the colour of the light-trapping pigment, chlorophyll. Leaves come in a variety of shapes and sizes, and their architecture varies with the species and the environment. The environmental constraints on leaf architecture are discussed in Chapter 7, as fossilised leaves provide a powerful tool for determining past climates; first, it is important to consider why the leaf evolved and why it is found on the vast majority of terrestrial plants.

Take a leaf from any common plant – preferably one that has leaves that are partially transparent. What do you notice? One of the most obvious features is that most leaves are very thin, often not more than a fraction of a millimetre thick. This ensures that all the cells in the leaf are close to the atmosphere with which the plant has to exchange gases. Where leaves are thicker, this is a specialisation that has developed to increase the plant's ability to conserve water, reducing the plant's surface area to volume ratio. Some leaves may also have thicker than usual coverings of cuticle or waxes, which enhance water conservation. Overall, the size, shape and thickness of a leaf are a compromise between conflicting demands, e.g. maximising light capture and minimising water loss.

Another notable feature of leaves is their network of veins (Figure 6.18). Usually, the network consists of a **midvein**, or **primary vein**, from which a series of thinner **secondary veins** branch. These in turn have even thinner **tertiary veins** running between them. Sometimes even finer orders of veins criss-cross the leaf until very small areas of the leaf are enclosed by veins (areoles).

The veins of a leaf have two functions:

- to supply water and nutrients and distribute the carbohydrate products of photosynthesis to the rest of the plant
- to provide structural support for a web of photosynthetic tissue that makes up the leaf; it can exceed a square metre in area.

Figure 6.18 The vein system of a modern leaf. (Bob Spicer/Open University)

The branching pattern of veins is reminiscent of the branching pattern of a tree. This similarity is not altogether accidental, for it appears that almost all the leaves present today, whether from an oak tree or a fern, were derived from modified branches. The only exceptions to this are the leaf-like **enations** of some primitive plants, such as club mosses or *Asteroxylon* (Figure 6.15b), which are outgrowths of the stem.

So, how did leafed plants evolve? Examine Figure 6.19, which attempts to summarise what is known about the evolution of the plant characteristics that have been significant in increasing the gas-exchange, carbon-fixation and water-cycling processes in early land plants. The important point to extract from the figure is that there appears to be a broad correspondence between the drawdown of atmospheric CO_2 and the innovations in land-plant architecture. As time progressed, plants increased in size and their simple naked branched stems developed a single main stem with side branches. These side branches eventually became flattened and webbed with tissue to form large photosynthetic surfaces (leaves).

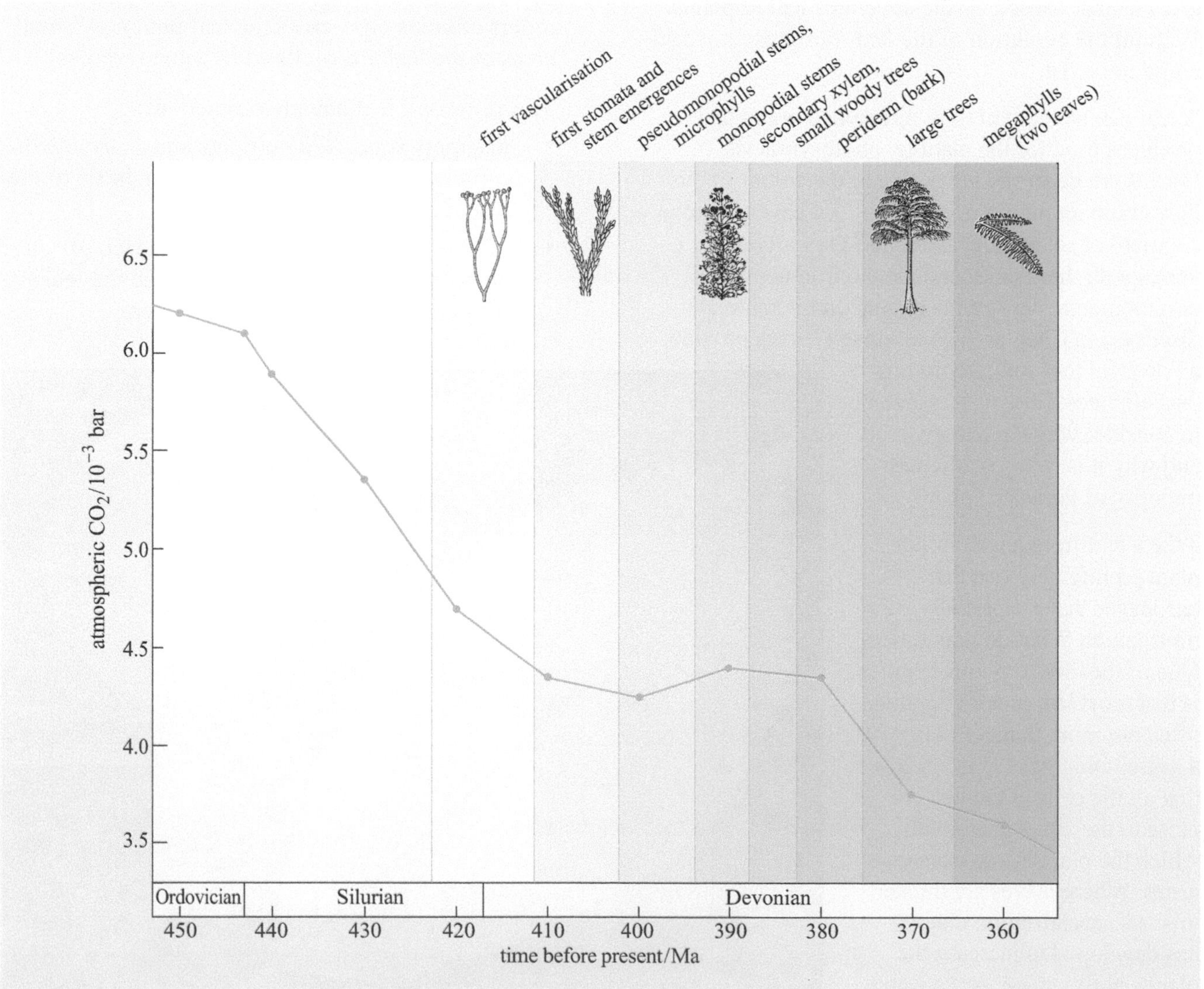

Figure 6.19 Major innovations in land-plant architecture with respect to the atmospheric CO_2 level. Pseudomonopodial stems refer to a single vertical stem made up of unequal branches, whereas monopodial stems denote a genuine main stem. Microphylls are small outgrowths of the stem, whereas megaphylls are true leaves. Secondary xylem is the tissue that makes up wood, whereas periderm is similar to bark.

This increase in leaf area demanded greater fluid movement in the plant body, so vascular systems became more complex and more efficient to deal with demand. In turn, this provided more fluids and nutrients to the leaves, which became even larger, allowing more carbon to be fixed and providing more carbohydrates with which to build larger plants. This positive feedback loop was eventually moderated (i) at the point at which atmospheric CO_2 concentrations became limiting in the context of temperature, water availability and other factors, and (ii) when the structural costs of building larger plants became prohibitive, if not in absolute terms, then in competitive terms.

Figure 6.19 shows that, in the Devonian, atmospheric CO_2 was estimated to have been about ten times higher than the present level of 0.38×10^{-3} bar. If correct, this high concentration of CO_2, coupled with the relatively small biomass of plants, would have meant that CO_2 was unlikely to have acted as a limiting factor on photosynthesis at that time. Instead, the primitive vascular systems in plants coupled with the fact that there was less water than now being cycled through the atmosphere, suggest it is likely that water rather than CO_2 may have limited photosynthesis in a large number of plants. Intriguingly, in many early Devonian plants, stomata, which in the absence of leaves would have been scattered over the stems, were rare compared with their prevalence in today's plants and those in the Carboniferous.

Question 6.8

Why do you think early Devonian plants could have functioned with only a few stomata?

The evolution of leaves in the Late Devonian greatly increased the surface area for gas exchange and also for the evaporation of water vapour into the atmosphere, thereby causing the atmosphere to cycle water at a greater rate. The evolution of leaves also contributed to the cycling of carbon because many long-lived plants discarded their leaves and replaced them numerous times during their lifetimes, enhancing carbon transfer from the atmosphere to the soil and the sediments. The innovation of leaf loss and replacement appears to have evolved in the Devonian. As the land surface became increasingly enriched with organic matter, so rates of chemical weathering increased, and rich soil profiles developed across the land. In summary, the carbon, water and other biogeochemical cycles can be inextricably linked by the photosynthesis of land plants.

6.3.3 The first forests

The Late Devonian saw a major diversification in land plants and most of the main groups of plants (with the exception of the flowering plants – the **angiosperms** – and the **cycads**) appear to have had their origins in Late Devonian times. One of these early groups, the **progymnosperms**, was especially important as it represented a further increase in carbon sequestering.

The progymnosperms were the dominant canopy formers in the earliest forests. The story of their identification is of interest because it provides an example of why it is important not to be too restricted in the concepts of ancient organisms, i.e. it can be misleading to interpret ancient plants only in terms of plants that are living today. The progymnosperms produced wood for the first time that had all the characteristics of the wood of modern conifers (which, together with other primitive seed-bearing plants, are referred to as **gymnosperms**). However, other progymnosperms were found to have foliage that looked like the modern fern and which bore reproductive structures that produced spores like a fern, rather than seeds like the modern conifers. As the wood and foliage were initially found unattached to each other, it was thought that they came from different plants and, as such, they were classified as belonging to different groups. The wood, given the generic name *Callixylon* (Figure 6.20a), was classified as belonging to the gymnosperms, whereas the foliage, assigned to the genus *Archaeopteris*, was thought to belong to the ferns (Figure 6.20b). In 1962, the American palaeobotanist Charles Beck described fossil specimens that showed the wood and foliage attached to each other, reconstructing the *Archaeopteris* 'plant' as a forest tree, about 20 m tall (Figure 6.20c), so creating the progymnosperm class.

Figure 6.20 (a) *Callixylon*, showing general characteristics of secondary xylem. (b) *Archaeopteris* foliage. (c) Reconstruction of the *Archaeopteris* tree. (b and c: Beck, 1962)

It seems unlikely that the *Archaeopteris* tree would have retained its leaves throughout its life; they were probably replaced many times, so enhancing the flux of carbon from the atmosphere to the soil. Moreover, the wood of *Callixylon* is made up of significant amounts of the complex organic polymer lignin, which decays more slowly than most non-woody plant tissues. The lignin in woody plants results in them having a higher probability of being represented in the fossil record over non-woody (**herbaceous**) plants, as well as increasing the probability of them contributing to long-term carbon sequestering.

The progymnosperms are, therefore, significant because they illustrate the innovations in land-plant biology that brought about significantly increased rates of carbon sequestering from the atmosphere. This effect may in turn have led to the global cooling in the late Carboniferous and early Permian.

Summary of Chapter 6

1 The geological record for the late Ediacaran suggests a number of environmental upheavals during this stage involving a global fall in sea level and marked fluctuations in the rate of burial of organic material (i.e. climatic instabilities). This was followed by lowered rates in the Cambrian. Nevertheless, the Ediacaran faunas persisted to the end of the Ediacaran, reaching a maximum diversity in the final 6 Ma, with a few forms surviving into the Cambrian. It is unclear whether the majority of Ediacaran animals underwent a mass extinction at the close of the Vendian or whether they were simply ecologically displaced by newly evolving animals.

2 Exceptionally preserved fossil assemblages of soft-bodied animals from the Cambrian reveal anatomical advances over the earlier Ediacaran fauna. Many of the Cambrian forms exhibit greater differentiation of body parts, including the concentration of food-trapping organs around a 'head' end, and the appearance of limbs and of discrete tubular, two-ended guts. These innovations may have been fuelled by Hox gene duplication.

3 A revolution in feeding relationships accompanied these anatomical changes. In particular, the rapid proliferation of skeletal hard parts may reflect the rise in predation, and hence multi-tiered food chains.

4 The proliferation of shells from early Cambrian times impinged upon biogeochemical cycles, leading to increased deposition of limestones in offshore open marine environments. Phanerozoic seas thus saw a shift in emphasis from the burial of carbon in organic material to the fixing of carbon in carbonate rocks. Increased burrowing in offshore sediments also helped to reduce the extent of organic carbon burial there.

5 The fossil record for families of marine animals with hard parts provides a general guide to Phanerozoic mass extinctions and radiations. The early Palaeozoic radiations are consistent with a relative ecological 'vacuum'. Subsequent diversification was interrupted by a succession of mass extinctions, of which five were notably severe, occurring in the late Ordovician, the late Devonian, the late Permian (the most devastating), the late Triassic and the late Cretaceous.

6 Although periodic extraterrestrial impacts have been proposed for these and other smaller mass extinctions, differences in the relative durations and the effects of the extinctions, together with other geological data, suggest a mixture of Earth-bound and extraterrestrial causes.

7 Each mass extinction was followed by a relatively rapid rise in family numbers. Over the longer term, an overall increase in diversity levels occurred, although whether this was the result of successively higher equilibrium levels or merely the effects of a dynamic interplay between ever-rising diversity and numerous extinction events remains unresolved.

8 Within the pattern of diversification of marine animals, three 'evolutionary faunas' may be distinguished: the Cambrian fauna that dominated the initial radiations; the Palaeozoic fauna that continued diversifying through the Ordovician, but then began a long, slow decline; the modern fauna that slowly, but relentlessly, expanded, rising to dominance after the Palaeozoic. Each successive fauna was less drastically affected by mass extinctions than its predecessor. The extinctions left a biased line-up of survivors.

9 Post-Palaeozoic faunas show an intensification of predation, especially upon shelly prey. One major defensive adaptation appearing among the prey animals was deep burrowing into the sediment, which led to yet further churning of surface layers. These, and other linked changes, are collectively referred to as the Mesozoic marine revolution. The associated rise of various planktonic groups with calcareous skeletons in the Mesozoic enhanced carbonate sedimentation in deeper water, increasing the oceanic carbonate sink.

10 Most of the Earth's non-bacterial biomass is in the form of green land plants. The evolution of terrestrial vegetation had a profound effect on the Earth's surface systems.

11 Plants adapted to this environment initially by growing close to the substrate surface and so staying within the boundary layer where they were not subject to the more intense desiccation that might occur in the faster-moving free air above.

12 The evolution of leaves improved the distribution of the carbohydrate products of photosynthesis to the rest of the plant and provided structural support to the photosynthetic apparatus.

Learning outcomes for Chapter 6

You should now be able to demonstrate a knowledge and understanding of:

6.1 The factors that may have led to the diversification of species that mark the transition from the Proterozoic to the Phanerozoic Era.

6.2 How evidence from exceptionally preserved fossil assemblages has been used to investigate the potential cause(s) of rapid expansion in the number of species during the Cambrian (referred to as the Cambrian Explosion), and the associated differentiation in body parts and functions, feeding mechanisms, level of predation and rise in hard external skeletal parts.

6.3 Why mass extinctions and radiations are best investigated by examining changes in the number of families (rather than species) over time.

6.4 How each of the five major mass extinctions was followed by a relatively rapid rise in diversity levels and the impact these extinctions had on each of the faunal successions.

6.5 How the structure and complexity of land plants developed over time in response to competition for light and nutrients as well as changing environmental conditions, and how this in turn resulted in the specialisation and adaptation of leaf structures and new reproductive mechanisms (changing from spores to seeds).

The Earth at extremes

Chapter 7

During the history of the Earth, the climate has swung to hot and cold extremes compared with today's climate. This chapter looks at the Earth in the late Carboniferous and Permian, an interval spanning over 70 Ma (323–251 Ma), focusing on the late Carboniferous–early Permian glaciation before moving on to look at the greenhouse world of the Cretaceous. Although these two periods have very different climatic regimes and the biota that inhabited them were different, they are both linked by the role of CO_2 in controlling planetary temperatures. You will look at the changes in the geography, climate and life, and proposed theories linking them. The aim of this chapter is not to describe every aspect of life during these times, but to concentrate on those organisms that contributed most and/or which were most affected during each event in order to understand how very different climatic regimes and their influence on the Earth's biota can be explained by various atmospheric and geological processes.

7.1 The icehouse world

7.1.1 Geographical perspective

Students of geology living in Europe (or in North America) might tend to think of the Carboniferous as a time of warm, humid conditions, when the Earth was covered in lush vegetation. Recent history would have been very different without the fortuitous accumulation of Carboniferous plant remains across these regions some 300 Ma ago, as these provided the coal reserves upon which the 19th-century industrial revolutions of the developed world were based. By contrast, the same students might tend to think of the Permian as a time of dry and desolate conditions, on a planet of deserts and salt lakes. An Australian or a Brazilian student of geology, however, would have a different perspective, as the evidence shows that their countries were at times covered by extensive ice sheets during the Carboniferous, while major coal reserves formed during the Permian across parts of Australia, South Africa and India.

So the perception of the Carboniferous world, seen from a European or North American perspective as warm and wet, and the Permian as hot and dry, is overly influenced by the local to regional geological record. The coal deposits of those continents were produced under tropical conditions, not because the whole planet was warm but because Europe and North America were in low latitudes at that time. In fact, the late Carboniferous and early Permian represent a cooler phase in the Phanerozoic from a global viewpoint.

To make sense of these varying perspectives, you only need to look around the world today to see the enormous variety of climate, environments and life forms, ranging from tropical rainforests through deserts, temperate forests and tundra to ice caps. So it is logical to expect to find a similar scale of heterogeneity in the Carboniferous and Permian worlds. During these times, however, the continents were arranged very differently from the way they are today (Figure 7.1).

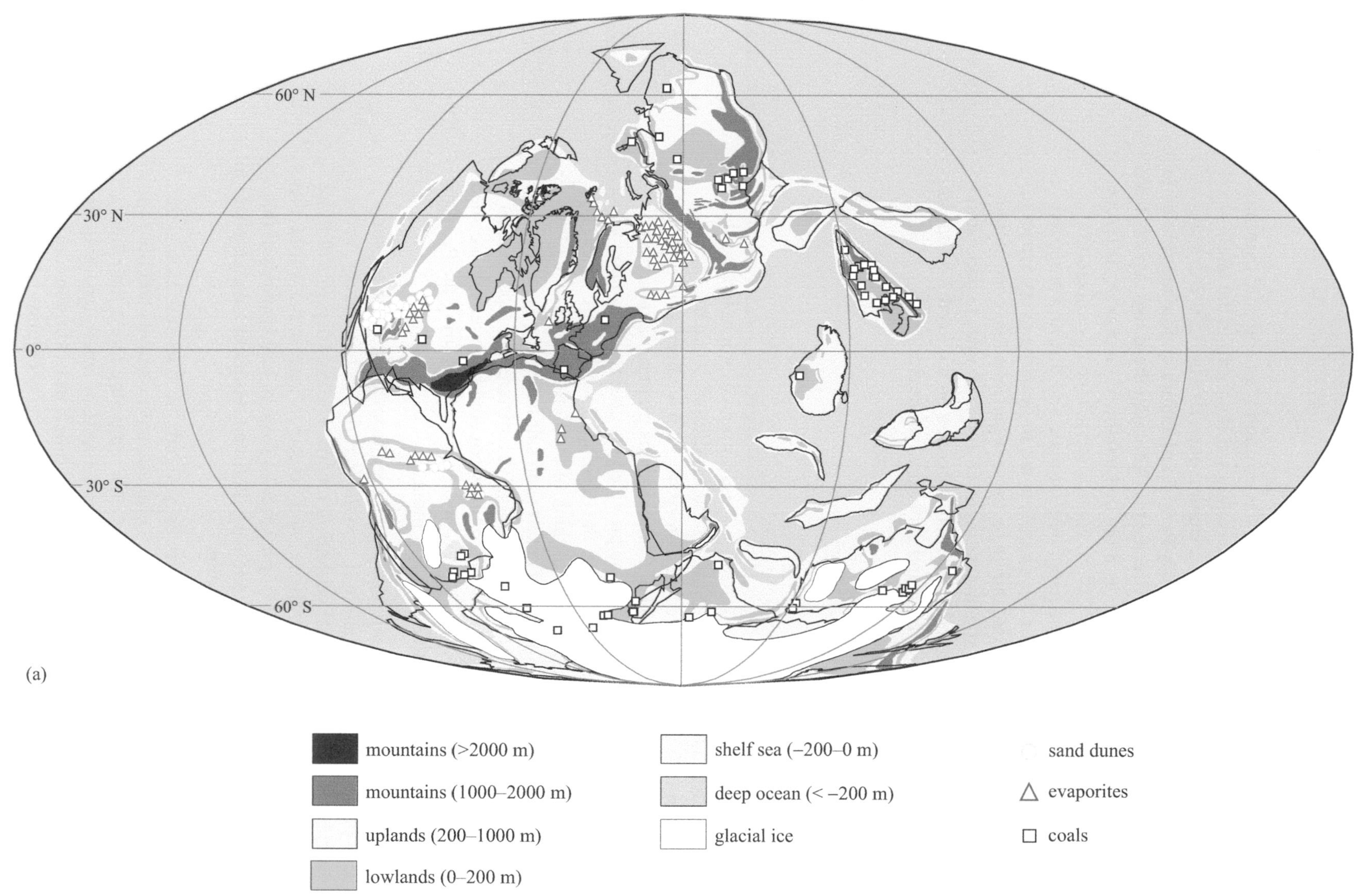

Figure 7.1(a) Pangaea in the earliest Permian (281 Ma). (Martin, 1996)

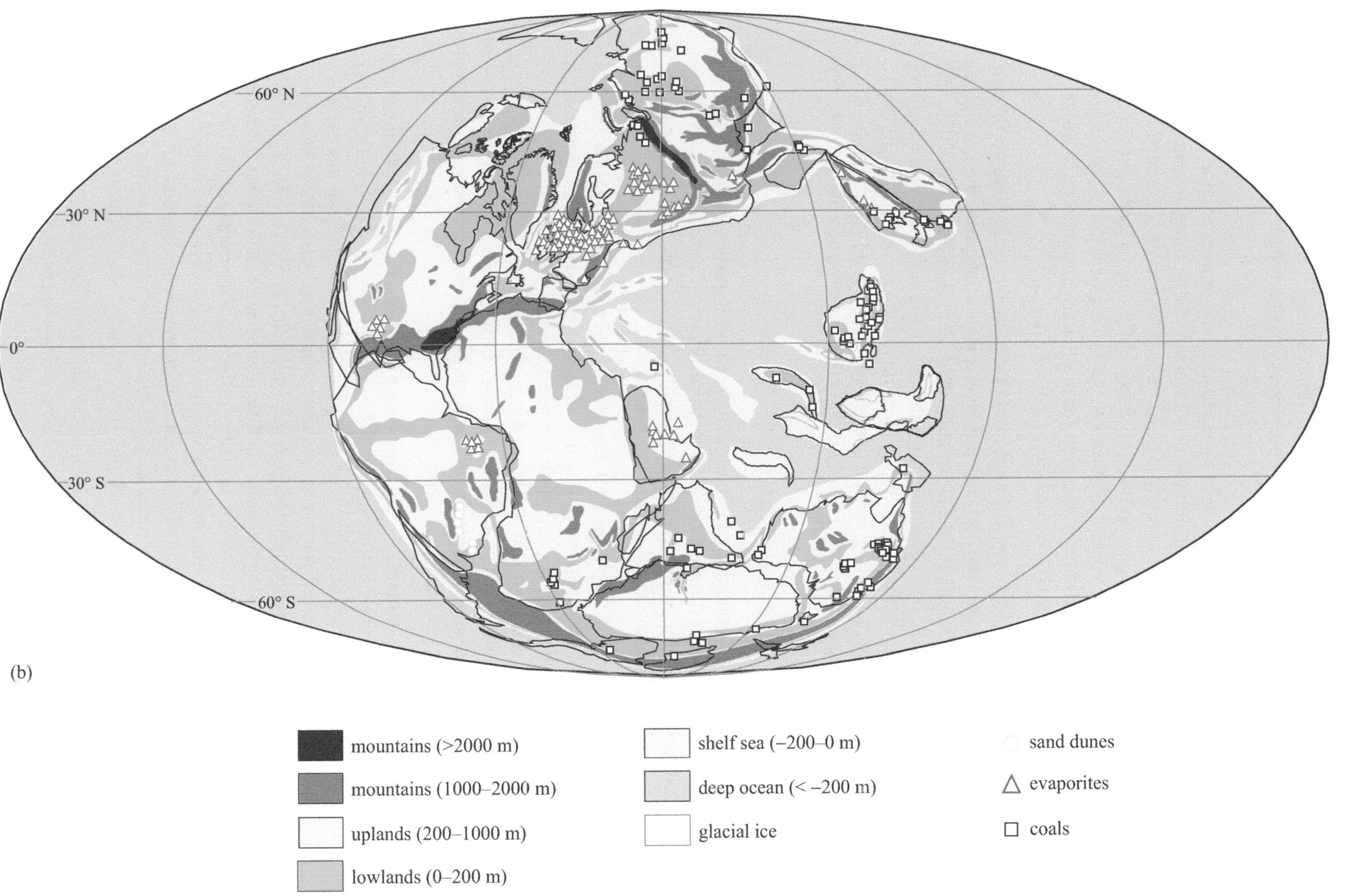

Figure 7.1(b) Pangaea in the Permian (251 Ma). (Martin, 1996)

Question 7.1

From Figure 7.1, what was the most striking aspect of the continental arrangement during the Permian compared with that today?

Figure 7.1 shows the position of Pangaea at the beginning and the end of the Permian, and illustrates the changes in location of this supercontinent and the shift from glaciated to ice-free conditions. These maps, compiled by Fred Ziegler and his colleagues at the University of Chicago, represent the most detailed palaeogeographical synthesis to date of a vast array of data. Figure 7.1a can serve as a general reference map for the Permo-Carboniferous (i.e. the late Carboniferous to early Permian) icehouse world.

Life on Pangaea was very different from that of today: there were no flowering plants and no mammals. As far as the oceans were concerned, there were no calcareous plankton, predation on shelly organisms was less intense than it is today, and sessile shelly forms lying exposed on the sea floor were commonplace. Yet there are a number of similarities in terms of the broad effects of global climate change, including the response of organisms to it, which might inform our understanding and speculations about the effects of climate change in the future.

7.1.2 Geological evidence

The distribution patterns of climatically sensitive deposits, such as evaporites, **red beds**, glacial **diamictites** (non-sorted mixtures of rock interpreted as being of glacial origin, previously known as tillites), fossilised wood and coals can provide important clues to past climates.

The formation of evaporites (salt deposits) requires that evaporation exceeds precipitation. Ideal conditions occur in arid regions, where formation can take place in enclosed basins with high temperatures and low rainfall. Modern evaporites occur mostly in subtropical regions centred around latitudes 25° to 35° in both hemispheres. A similar distribution can be seen in Figure 7.1 with respect to the inferred palaeogeography of the Permian. Much the same pattern can also be seen for desert sand dune deposits.

Red beds are sedimentary rocks containing hematite (Fe_2O_3) that formed under oxidising conditions in hot climates. The original source of iron in these rocks was often exposed igneous or metamorphic rocks that had been intensely chemically weathered. This iron was then remobilised (as Fe^{2+}) in anoxic groundwater and reprecipitated in desert sediments as iron oxide, as a result of evaporation drawing the water up towards the surface. Modern red beds form largely within 30° of the Equator (reflecting **Pleistocene** shifts between humidity and aridity), and most Palaeozoic red beds seem to have had a similar distribution, being commonly associated with evaporite deposits. Of course, such ancient red beds may be eroded to yield red soils at a later date in a different latitude, as can be seen, for example, in the countryside of Cheshire, England.

In contrast to these indicators of climatic warmth, most diamictites are considered to have been deposited by glaciers. Their widespread occurrence in southern areas of Pangaea during the Permo-Carboniferous indicates that large areas experienced glacial conditions for at least some of that time.

The formation of coal requires a net surplus of precipitation over evaporation, sufficient warmth and light for plants to grow, and isolation of buried plant material (or peat, the precursor to coal) from the oxidising atmosphere. Rainfall and plant productivity are closely linked, while periods of drought strongly affect preservation because falling groundwater levels permit oxidation of organic matter (through aerobic decomposition or burning).

Further clues come from growth rings in trees, which can also serve as climate indicators (Box 7.1). Fossil Carboniferous tree trunks from low palaeolatitudes lack or have only faint evidence of growth rings, indicating that they grew in near-constant conditions of humidity and warm temperatures, and therefore most Carboniferous coals almost certainly formed under tropical conditions at low latitudes. By contrast, thick coal deposits also formed in the Permian within 5° to 30° of the present South Pole. These coal deposits contain fossilised tree trunks with prominent growth rings, implying seasonal growth.

Box 7.1 Tree growth rings

Growth rings are formed when a tree grows at varying rates over time. Typically, variations in wood growth are caused by environmental changes such as water availability, temperature or light regime (day length). Away from low latitudes such variation is usually tied to the annual cycle of the seasons. Wood cells with large internal cavities are produced early in the growing season, when the availability of, and the demand for, water is high. These cells comprise the **earlywood**. When water is less available and demand is less (in late summer and autumn), the water-conducting space in the cell becomes constricted as the cell walls thicken. These **latewood** cells are mechanically stronger than earlywood cells, but less efficient at water conduction. No cells are produced during the period of winter 'shutdown' or dormancy of the tree. The change from large cells to small, thick-walled cells corresponds to seasonal changes in tree growth, forming a single growth ring. In non-seasonal environments, such as in tropical rainforests, there is less of a fluctuation in climate conditions through the year (although there may be some variation in rainfall) and growth is consequently more uniform, with rings being absent or only weakly developed. The ratio of wall thickness to cell cross-sectional area for water conduction is a compromise to meet both the water demand and the need for structural strength.

By combining knowledge of the present-day distributions of different climatically sensitive deposits and the climate conditions in which they form, it is possible to infer similar conditions when similar features are encountered in the geological record.

The next section focuses on some of the more important aspects of the Earth and its life within the interval spanning the late Carboniferous and early Permian.

7.2 Permo-Carboniferous glaciation and subsequent warming

7.2.1 The available record

The late Carboniferous and early Permian interval is interpreted as having been a time of pronounced cooling. This was apparently followed by a shift back towards warmer conditions during the rest of the Permian, with global climate eventually becoming warmer than at present.

So how much is really known about Permo-Carboniferous conditions? The answer is a fair amount. Even though it is often harder to work on rocks of that age than on more recent rocks – it is like walking in the dark with a light only occasionally being switched on to give an idea of what is around – a reasonable picture has been built up by studying the available evidence. Some of this is discussed in the following sections.

7.2.2 The link between atmospheric CO_2, vegetation and climate

It is now widely acknowledged that the anthropogenic build-up of atmospheric CO_2 (and other greenhouse gases) is implicated in the current rapid rise in global warming. Over a longer timescale (hundreds of thousands to millions of years), atmospheric CO_2 levels also appear to have varied throughout the Earth's history and such variations must have had an effect on global climate.

Two major routes by which CO_2 can be removed from the atmosphere are through direct uptake by plants (during photosynthesis) and by dissolution in water. As long as the rate of photosynthesis exceeds that of respiration, the longer term net effect is sequestration of atmospheric CO_2 as a consequence of the burial of organic material. On land, buried organic material is first converted to peat, then coal, so that a reservoir of organic carbon builds up, locking it away from the atmosphere. Section 3.3 describes how CO_2 is highly soluble in water, and this provides the second major route by which it is removed from the atmosphere. The overall reaction between CO_2 and water to form a weak acid, carbonic acid (H_2CO_3), can be expressed as:

$$CO_2 + H_2O \rightleftharpoons H_2CO_3 \qquad (7.1)$$

Carbonic acid dissociates to release a hydrogen ion (H^+) and a bicarbonate ion (HCO_3^-), which in turn dissociates to form the carbonate ion (CO_3^{2-}) and a further hydrogen ion. The high levels of CO_2 in vegetated soils (from respiration of plant root systems and microbial decomposition) mean that CO_2 concentrations are typically 10 to 100 times higher in the soils than in the atmosphere, and it is the acidic water in soils that is mainly responsible for the weathering of minerals in soil and rock.

Question 7.2

(a) What are the two main rock-weathering reactions involving dissolved CO_2?

(b) Which of these is considered to lead to a net drawdown of CO_2 from the atmosphere when precipitation of carbonate occurs?

A third weathering reaction that needs to be considered is that involving exposed deposits of peat and coal. In contrast to the weathering reactions just discussed, the oxidative process in this case *releases* CO_2, while using up oxygen from the atmosphere.

Plants are responsible both directly and indirectly for a considerable drawdown of atmospheric CO_2. The Permo-Carboniferous interval seems to have been outstanding in terms of the direct effects of vegetation on climate. Two significant things occurred: firstly, the spread of land plants – and in particular trees – provided a large pool of organic matter, some of which was then buried; secondly, since their first appearance in the Devonian (416–359 Ma), plants have greatly increased the rate of soil and rock weathering. The result of this weathering process was an increased transfer of CO_2 from the atmosphere to the oceans in the form of bicarbonate ions (see the answer to Question 7.2) and ultimate burial there, as limestone, then lowers atmospheric CO_2 levels. Although CO_2 is returned to the atmosphere through volcanism and weathering of exposed organic deposits, it seems that this reverse process was greatly outweighed during the late Carboniferous and early Permian by CO_2 drawdown mediated in one way or another by land plants. Models for the change in atmospheric CO_2 levels, based on the estimated balance of carbon fluxes, show a marked fall in CO_2 for this interval (Figure 7.2). This in turn would have resulted in a lowering of mean global temperatures.

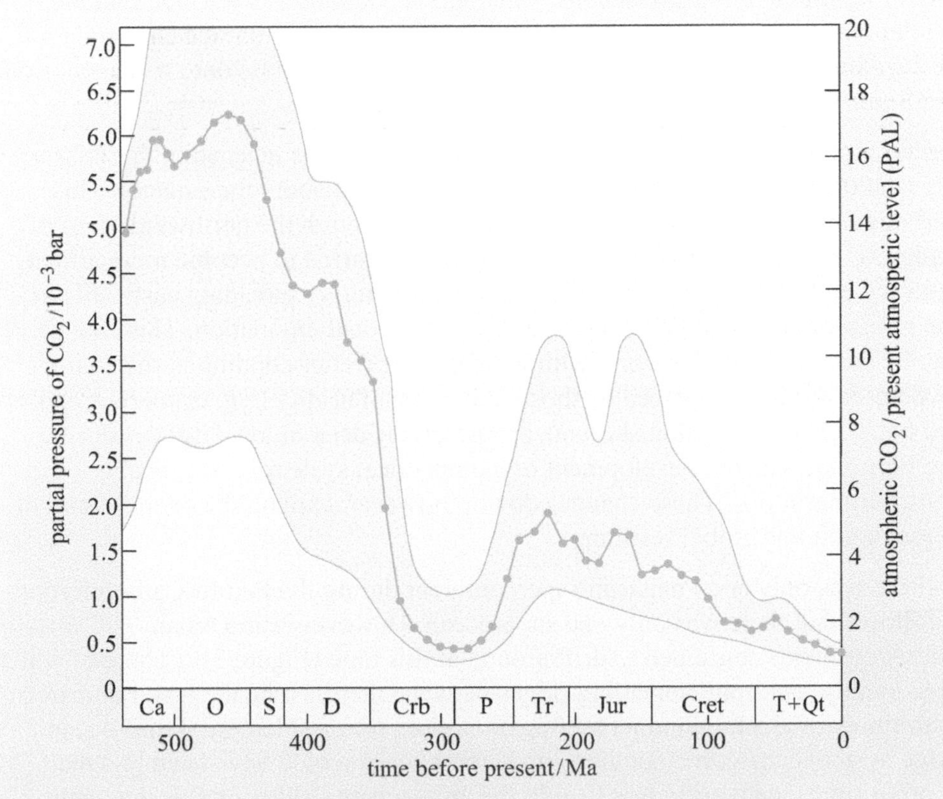

Figure 7.2 Variation in the level of atmospheric CO_2 (absolute and as a factor of present atmospheric level (PAL)) over the Phanerozoic as calculated from the GEOCARB model (Section 5.3). Note the pronounced fall in CO_2 levels around the Permo-Carboniferous boundary to levels similar to those of today. Shaded area is the uncertainty in the data. (Berner, 1994)

Other mechanisms may also have contributed to globally cool conditions during this time interval, in particular the effects of the location of the southern part of the Pangaean land mass over the South Pole, and the effect that the assembly of such a supercontinent might have had on climate. To understand this, it is important to investigate the influence of plate tectonics on global climate.

7.2.3 Continental motion and climate

The following aspects of plate tectonics are relevant to the Permo-Carboniferous icehouse world:

- the configuration and position of land with respect to the poles, and their effects on ocean circulation
- volcanic activity related to mid-ocean ridge and subduction processes
- collisional processes causing uplift and mountain belts
- eustatic changes in sea level.

As you will see shortly in the following discussion, each of these can affect climate, both locally and globally.

As noted above, the cause of the Permo-Carboniferous glaciation has, in some cases, been ascribed to the positioning of a part of the Pangaean land mass at that time over the South Pole.

Modelling of a 'cap world' (Box 3.1) configuration leads to a markedly cooler (south) polar region. If snow precipitation exceeds **ablation** (in areas of maritime influence, with moist air coming off the ocean), then its accumulation would lead to the formation of an ice cap. A counter-argument, however, is that during the Earth's history land masses were positioned over the poles at other times, including much of the Cretaceous, when global climate was warmer than today. So the onset of glaciation and subsequent melting of the polar ice cap might not be explained solely by movement of the Pangaean land mass onto and away from the South Pole.

An arrangement more reminiscent of a 'slice world' was reached in the Triassic (251–200 Ma), when the land mass was symmetrical about the Equator. This had developed gradually over millions of years through the northwards drift of Pangaea. Western equatorial regions of Pangaea started to become more arid at the very end of the Carboniferous, with arid conditions spreading eastwards as the monsoonal system developed and disrupted zonal circulation. This change can be seen in the rock record, with indicators of wetter conditions such as coals progressively replaced by those indicative of aridity (e.g. evaporites and red beds). Thus, conditions became progressively drier at low latitudes during the Permian, with the development of a monsoonal system most probably contributing to this. These changes do not, however, explain the overall trend of deglaciation and global warming.

If there was only one Pangaean supercontinent during the Permo-Carboniferous, it follows that there was only one superocean. However, some Asian microcontinents continued to drift around at this time (Figure 7.1), some of which presumably had small spreading ridges between them. None the less, compared with times of greater continental fragmentation, the total length of mid-ocean ridge systems and corresponding subduction zones would have been less than at other times, delivering less CO_2 to the atmosphere. Although this may help to explain the glaciation (through net drawdown of CO_2, as discussed earlier), it does not help to explain the subsequent retreat of glaciers during the Permian.

There is a third tectonic factor to consider in the Pangaean supercontinent and its effect on global climate, i.e. the collision of all the plates that contributed to its formation, and the consequent mountain chains this created. The assembly of

Pangaea from its constituent parts was akin to a multitude of continent–continent collisions. Numerous mountain ranges were built, although whether anything like the Tibetan Plateau formed is open to conjecture.

Question 7.3

What effect would mountain building on such a scale have had on the rate of CO_2 drawdown from the atmosphere?

Without compensation for any such increased drawdown, the consequence would have been global cooling. It is known that major mountain chains generated by such collisions did exist in the late Carboniferous and early Permian, as the eroded relics of these still form many upland areas in northwestern Europe (e.g. the Ardennes and Harz Mountains) and in the USA (the Appalachian Mountains), for example. This raises the question of whether the effect of their formation on CO_2 levels and climate can be evaluated.

High $^{87}Sr/^{86}Sr$ ratios occur in the weathering products of continental rock, whereas low ratios occur in the hydrothermal effusions associated with sea-floor spreading. Since all of this material ultimately ends up in the oceans, the values that are preserved in marine carbonate rocks containing strontium reflect the relative inputs of continental weathering and sea-floor spreading through geological time. Figure 7.3 shows a compilation of strontium isotope ratios throughout the Phanerozoic. In the long term, many factors are likely to have been involved in the overall pattern; for the time being, focus your attention on the Carboniferous, and the approaching late Carboniferous–early Permian glaciation.

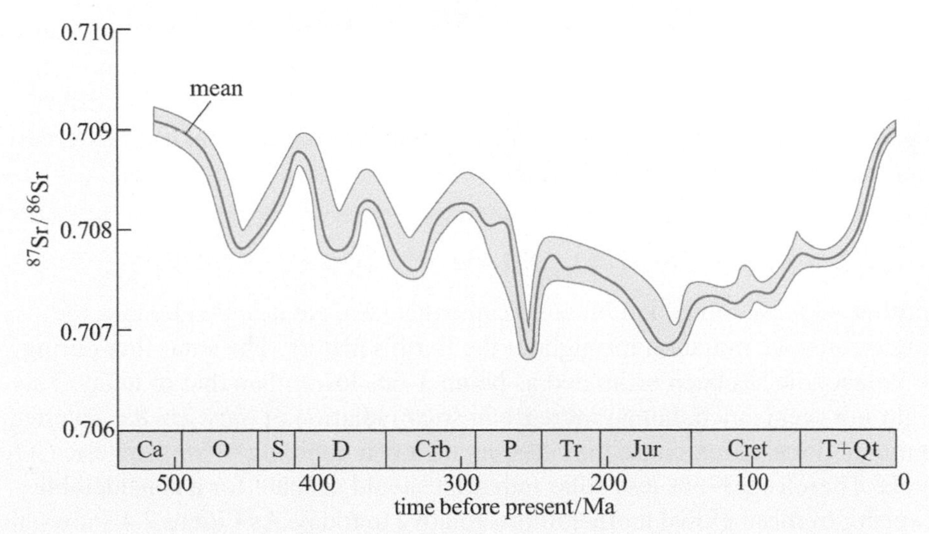

Figure 7.3 Ratios of $^{87}Sr/^{86}Sr$ in marine limestones through Phanerozoic time. The shaded area represents variation in values. (Raymo, 1991)

Note the relatively high $^{87}Sr/^{86}Sr$ ratios during the late Carboniferous–early Permian with respect to the late Devonian to early Carboniferous (before the main continental collisions involved in the formation of Pangaea), reflecting increased continental weathering relative to sea-floor spreading.

Thus, mountain building may also help to explain the globally cool conditions through increased drawdown of CO_2 during weathering of exposed rocks (see Section 5.3.1). The relationship is not quite so simple in detail as continental weathering rates also vary with temperature (as do most chemical reactions). Hence, it is to be expected that increased continental weathering rates (yielding high $^{87}Sr/^{86}Sr$ ratios) are associated with *high* atmospheric CO_2 levels and global warming. The drawdown of CO_2 would then regulate atmospheric CO_2 levels, and hence climate. The converse compensation should also occur, with lowered CO_2 levels leading to global cooling and reduced weathering rates. Figure 7.3 indicates that *increased* weathering rates have been inferred during a time of globally cool conditions. So how can this apparent paradox be explained? It seems that the sheer volume of rock made available (through major uplift) for weathering was the overriding factor. It is known from the Himalaya today that much of the *chemical weathering* of the eroded sediment eventually takes place in lowland areas flanking the mountains. If analogous to conditions during the Permo-Carboniferous, such a large area of weathering, combined with the large-scale effects of vegetation, could have resulted in abnormally high continental weathering rates in the late Carboniferous to early Permian, global cooling notwithstanding, therefore enhancing the icehouse conditions.

Finally, tectonic processes can also cause global sea levels to change, which may in turn affect global climate. Sea-level variations can occur through changes in the volume of the ocean basins. This is caused, most directly, by changes in the volume of mid-ocean ridges. Low sea levels would result in the emergence of areas of former continental shelf, and lowland swamp forests would have expanded, with a consequent rise in sediment weathering rates and drawdown of CO_2 through peat accumulation. Global sea levels would have been lowered yet further due to the accumulation of polar ice locking up large volumes of water. Although this does not explain the initial formation of the polar ice cap, it may have been a factor in maintaining icehouse conditions.

So there are several tectonic factors intertwined with the effects of vegetation and CO_2 levels on climate, each of which may help to explain the conditions seen in the Permo-Carboniferous icehouse world. Section 7.3 focuses on the effect of vegetation on the climate, but first it is important to briefly take an 'external view' of the Earth and climate change.

7.2.4 Extraterrestrial causes of climate change

Another factor involved in climate change that needs to be considered is the increase in solar radiation throughout the Earth's history. The solar flux during the Palaeozoic has been estimated as being 3–5% lower than that of today. This might not seem much, but an increase in solar radiation of only 2% has an effect on mean global temperature that is equivalent to a doubling of atmospheric CO_2 levels. Therefore, 3–5% less solar radiation should account for a considerable lowering of mean global temperatures, relative to today. As Figure 7.4 shows, the combination of low atmospheric CO_2 levels (derived from the GEOCARB model in Figure 7.2) and reduced solar radiation results in an even sharper contrast between the Permo-Carboniferous icehouse world and other geological intervals (over the Phanerozoic). So, lower levels of solar radiation can also be invoked to explain, in part, the Permo-Carboniferous climate. If, however, solar radiation has increased only gradually through time, then it cannot fully explain both the marked global cooling of the late Carboniferous and early Permian *and* the relatively rapid subsequent warming during the Permian.

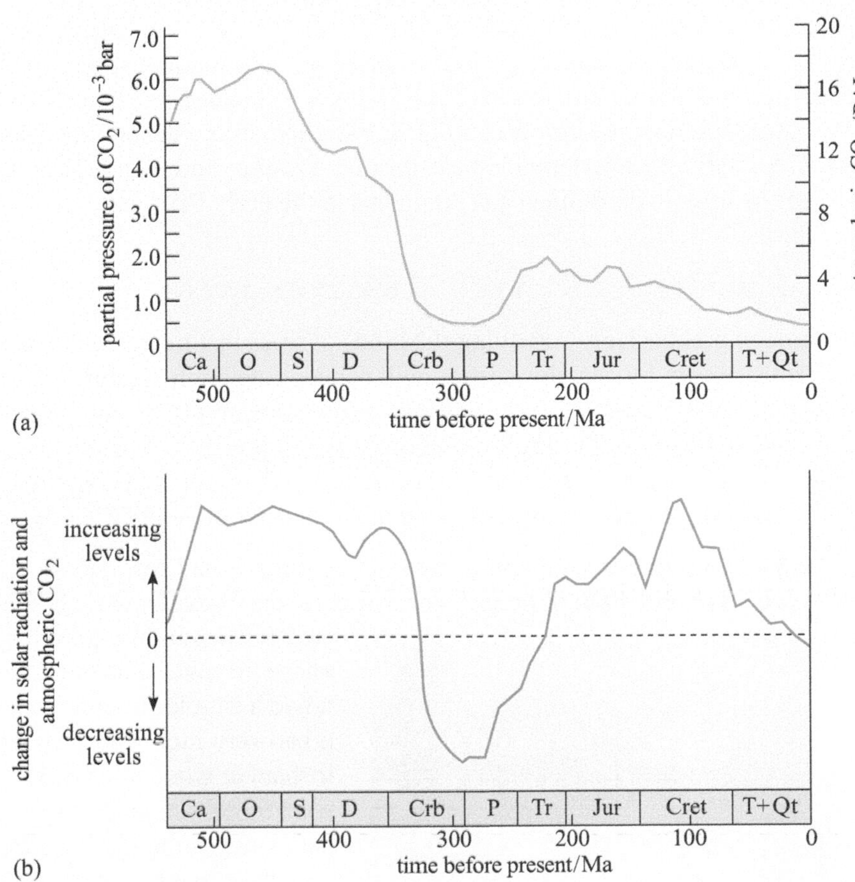

Figure 7.4 (a) Phanerozoic atmospheric CO_2 levels derived from the GEOCARB model shown in Figure 7.2. (b) Combined plot showing net forcing effect of CO_2 and increasing solar radiation on climate through the Phanerozoic relative to the present level. Note the significantly lower levels around the Permo-Carboniferous boundary compared with those of today. (Crowley, 1993)

7.3 The impact of land vegetation

7.3.1 Land plants in the Carboniferous and Permian

Given the striking effect that Permo-Carboniferous land vegetation appears to have had on CO_2 levels and global climate, it is now time to look at this in more detail. Much diversification and the appearance of extensive low-latitude forests dominated by plants that resembled very large versions of living clubmosses (lycophytes) and horsetails (sphenophytes) occurred during this time. Many of these grew as tall trees dominating the low-latitude swamps.

- Why can height be advantageous to a plant?

- Height enables a plant to compete successfully for access to light and to disperse its spores by wind over a wider area. Consequently, a tall plant would have a distinct advantage over its shorter competitors.

There was a plentiful supply of water at these low latitudes, and many plants evolved and thrived. As a result, thick peat deposits formed in the late Carboniferous low-lying wetland areas, eventually producing the major coal reserves exploited today across the Northern Hemisphere.

Towards the end of the Carboniferous, the lycophyte trees were dominant in only a few areas, with seed plants such as conifers taking over in a number of habitats. Seeds are able to survive periods of drought, germinating when external conditions become favourable: in a seed, an embryo is shielded and nourished by nutritive tissues within a protective coating. These changes in vegetation signalled the gradual loss of most low-latitude, peat-forming swamps and their widespread replacement by drier, well-drained environments in the early Permian.

7.3.2 *Lepidodendron*: a typical swamp dweller

Rather than comprehensively surveying the kinds of trees that made up the late Carboniferous swamp forests, this section will concentrate on the features of one in particular, namely *Lepidodendron*, which is a typical example of a lycophyte tree and is one of the most thoroughly studied of all Carboniferous plants. Not surprisingly, it figures prominently in Carboniferous swamp forest reconstructions (Figure 7.5), and there are examples of these in the Fossil Grove in Glasgow, UK.

Lepidodendron formed the dominant forest canopy during the Carboniferous and grew between 30 m and 40 m in height when mature, only developing a branched crown when in its reproductive phase. In spite of its large size, it had a simple structure in that it had very little woody tissue. Instead of wood, the trunks were largely supported by bark, suggesting that each tree may have lived for only one or at most a few seasons before falling over and perhaps being buried. The vigorous growth and short lifespan of these plants help to explain why so much peat accumulated in these swamps, eventually to become coal. Peat accumulates if the rate of production exceeds that of decay. Studies of modern trees show that bark contains chemicals that make it more resistant to biological decay; they can also inhibit decay of other substances, such as carbohydrates and forest litter. Consequently, there is little sign of decay in late Carboniferous plants, allowing the short-lived *Lepidodendron* trees to be buried in the swamps; some examples of their fossilised remains are shown in Figure 7.6.

(a)

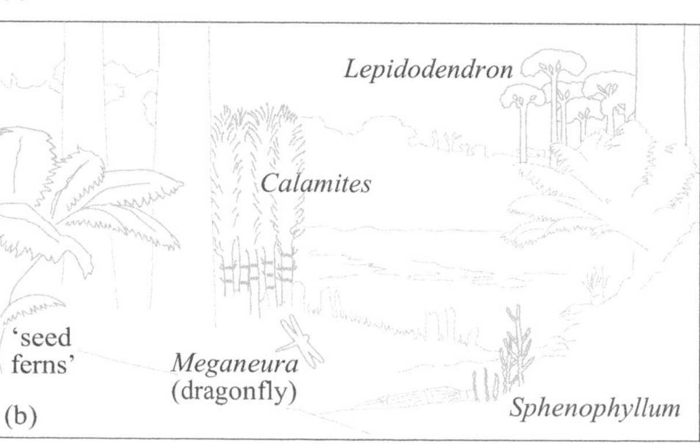

Figure 7.5 (a) Reconstruction of a late Carboniferous, low-latitude swamp forest. (b) Major plant types (and the Carboniferous dragonfly).

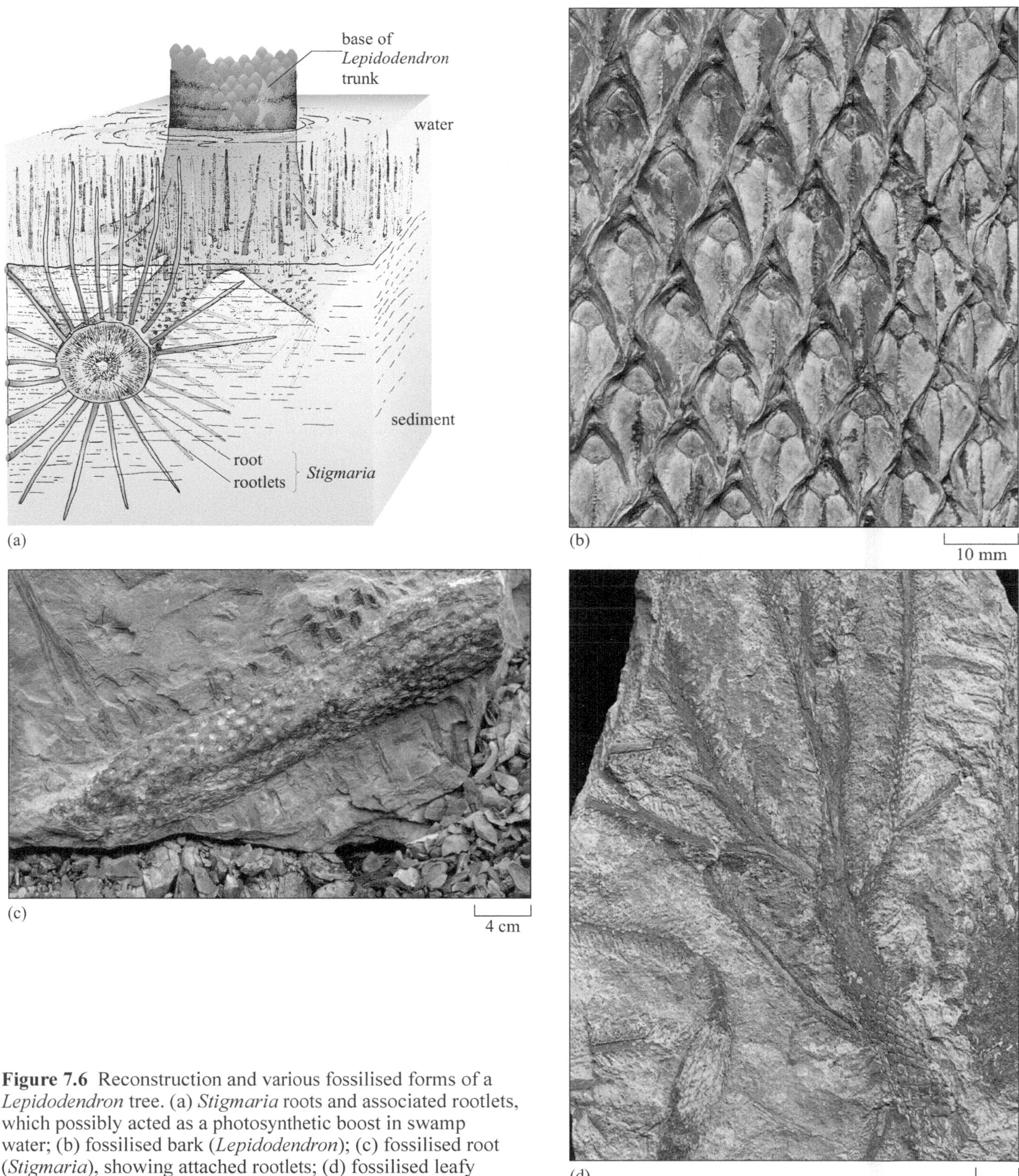

Figure 7.6 Reconstruction and various fossilised forms of a *Lepidodendron* tree. (a) *Stigmaria* roots and associated rootlets, which possibly acted as a photosynthetic boost in swamp water; (b) fossilised bark (*Lepidodendron*); (c) fossilised root (*Stigmaria*), showing attached rootlets; (d) fossilised leafy branches and reproductive organs. (Philips and DiMichele, 1992)

Near the base of the *Lepidodendron* trunk, the primary water-conducting tissue (**xylem**) was restricted to a cylinder only a few centimetres in diameter. Although there was not very much of it, the xylem must have been able to conduct water and nutrients from the roots up to the top of the tree. Compared with modern trees, *Lepidodendron* appears to have had no or extremely little **phloem** in its trunk. Phloem is the vascular tissue that enables movement of the products of leaf photosynthesis (such as sugars) to the rest of the plant, including the roots. So how did the root system (and the rest of the tree) get its food?

The rooting structure of the *Lepidodendron* tree (their fossils are known as *Stigmaria*) terminated in numerous finger-width hollow rootlets that were helically arranged on root branches (Figure 7.6a and c). In many respects, these projections appear to be more similar to leaves than roots, but leaves that have been modified for anchorage. Furthermore, as the root systems were shallow, some of them could have been exposed to sunlight filtering through water in the swamp. As the three vital ingredients for photosynthesis (H_2O, CO_2 and energy) would have been available in the shallow swamp waters, it is possible that the roots and leaves photosynthesised and nourished the plant independently.

7.3.3 The legacy of the forests

Plants like *Lepidodendron*, which were superbly adapted to and dominant in low-latitude swamp environments, were important because they enhanced CO_2 drawdown from the atmosphere. So why did they become extinct?

Although these lycophytes are only remotely related to seed plants, they had developed a reproductive structure similar to a seed. Indeed, differences between a lycophyte reproductive structure and a 'true' seed are largely technical. Referring back to the advantages mentioned earlier of seed plants and their ability to survive and even thrive in drier habitats, this suggests that the lycophytes should have been able to adapt to a changing environment as the swamps dried up and the climate changed. As discussed above, *Lepidodendron* had highly specialised rooting organs that spread out in shallow sediments and were well adapted to support a tree living in waterlogged habitats. The overly specialised adaptations of their underground organs were therefore the most likely cause of their extinction, rather than the failure of their reproductive systems. There are probably two reasons for this: growth in drier and firmer soils would seem to have been very difficult, if not impossible, for the shallow-rooted *Stigmaria* with its presumably delicate rootlet apices; and, by being buried, the rootlets would have lost the energy (sunlight) needed to nourish themselves. Thus, with the demise of lycophytes such as *Lepidodendron*, the way became open for the remaining seed plants to dominate and diversify. Some of these early seed plants developed leaves adapted to life in dry environments and became dominant elements of lowland European and North American vegetation in the early Permian. As you can see in Figure 7.7, a number of these already existed in the late Carboniferous, but were marginalised in drier habitats on the fringes of the wetlands; once the lycophytes died out, they seized the opportunity to spread across the low-latitude areas of Europe and North America.

It should be stressed that these changes in low-latitude vegetation and climate did not all occur at the same time on a global scale. Further east, in Asia, broad-scale patterns of vegetation and climate remained similar from the Carboniferous until

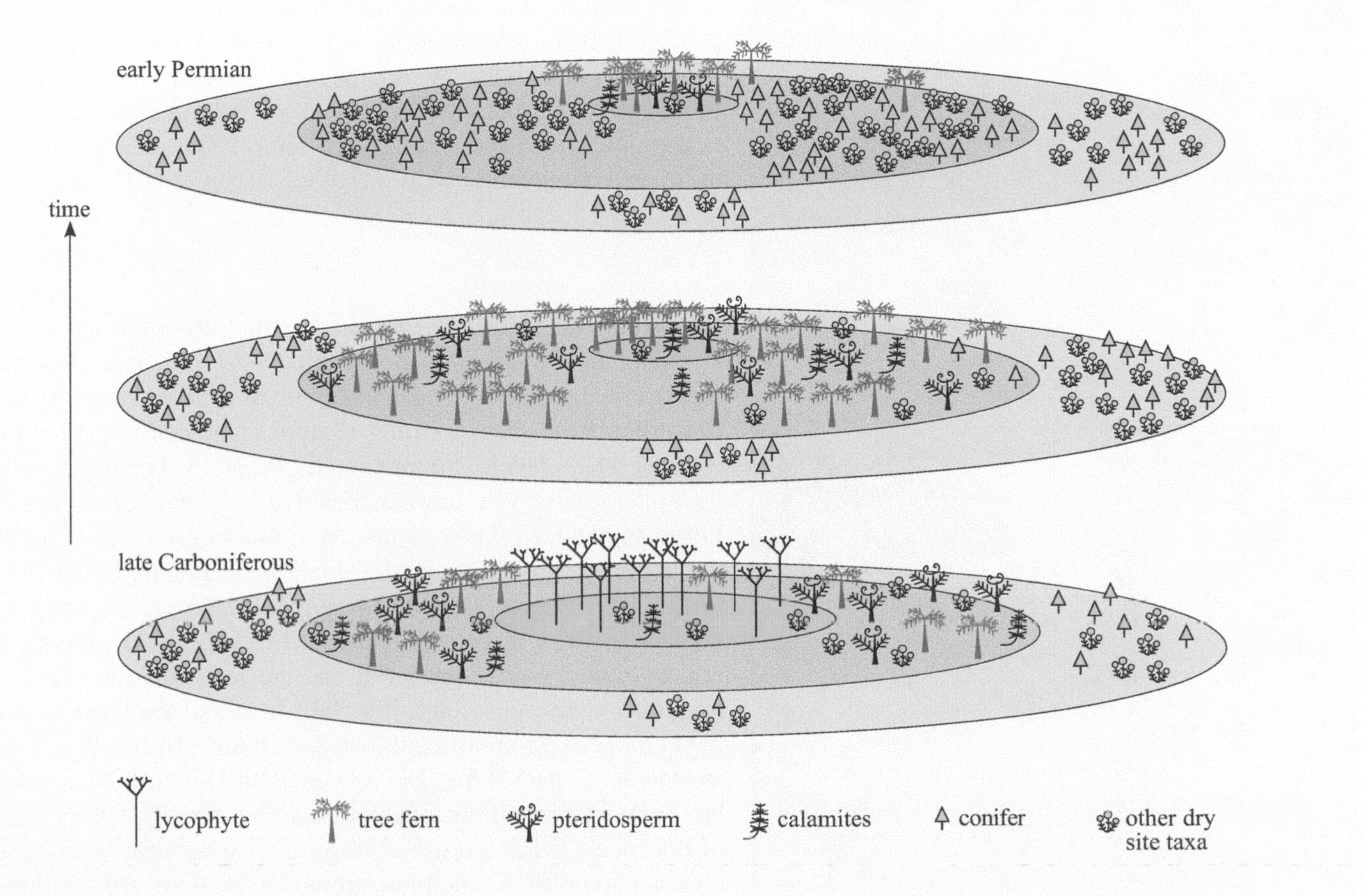

Figure 7.7 Late Carboniferous–early Permian change in equatorial European/North American plant communities. The dominant lycophytes of the late Carboniferous gave way to previously marginal plants (e.g. conifers) adapted to drier habitats. The inner circle represents the wettest environments, the middle band represents the non-swamp habitats that were progressively affected by seasonal dryness into the Permian, and the outer band represents the uplands that were rarely wet. (DiMichele and Aronson, 1992)

the end of the Permian, when conditions finally became more arid and – in contrast to Europe and North America – China, Australia and South Africa have major Permian coal reserves.

Elsewhere, extensive forests had developed in high southern latitudes during the early Permian. These were dominated mainly by the **glossopterids** (deciduous trees with distinctive tongue-shaped leaves). These leaves formed great fossilised mats where they were shed and buried annually in the swamps (Figure 7.8 overleaf). Glossopterids have a particular historical claim to fame in that they played a major role in the original reconstruction of Pangaea by Alfred Wegener, the German geologist who proposed the theory of continental drift, early in the 20th century. He noted that they were found in several areas across the present southern continents and so used them to help show the assembly of those continents in the past. The vegetation must have been abundant, since the dead remains accumulated as thick peat deposits (providing commercially mined coal reserves in Australia, for instance). From earlier discussions about the extensive southern ice sheets, it may therefore seem strange that such thick peat deposits should have accumulated shortly after the ice retreated from this area. In contrast to the warm,

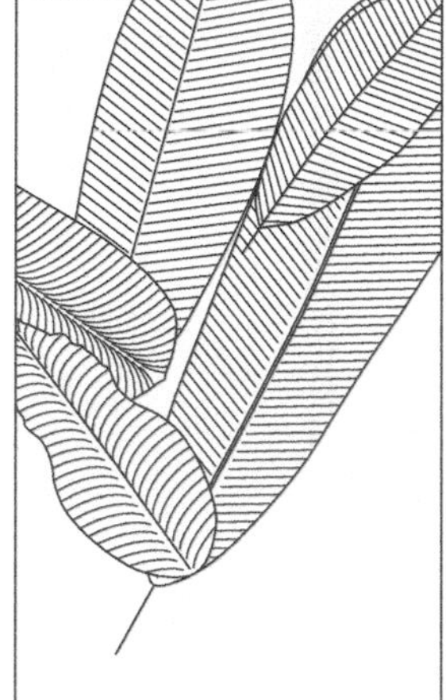

Figure 7.8 Several *Glossopteris* leaves preserved in layers of fine-grained sediment. The positions of the leaves are shown in the key. (Bob Spicer/Open University)

humid conditions at low latitudes in the icehouse world, these peats formed in cool, swampy bogs at high latitudes in the newly developing greenhouse world. (Section 7.5 returns to the question of how such productivity could have occurred at high latitudes by examining similar forests that grew some 100 Ma later in the even more extreme greenhouse world of the Cretaceous.)

7.4 A synthesis for the icehouse

Intense glacial conditions typified the late Carboniferous and early Permian, but why did they occur and persist? The role of plants seems to have been crucial. The timing of the maximum spread of Carboniferous coal-forming swamps coincided with minimum estimated CO_2 levels and maximum glaciation. The impact of land plants, however, was inextricably linked with other factors. For example, the relative reduction of mid-ocean ridge systems might have resulted in a sea-level fall, providing extensive lowlands for the coal-forming swamps. Moreover, the reduced volcanic activity might have produced less CO_2 to counteract the effects of the biotic sequestering and the accumulation of polar ice might itself have ensured further falls in global sea level, resulting in yet more exposure of continental shelf area. In addition, movement of part of the land mass over the South Pole may also have played a role in the observed cooling. Finally, the formation of Pangaea itself as a result of major collisional processes, causing continental uplift, would have subsequently increased weathering rates. The copious sediment shed from these mountains would also have contributed to the preservation of coal by continually burying the swamp peats in vast delta systems. Together, these factors established a sustained net drain on atmospheric CO_2, so reducing temperatures and maintaining icehouse conditions.

These conditions were followed by deglaciation towards the end of the early Permian, resulting in only a few small ice caps remaining on the highlands of southern Africa. Cool, humid conditions prevailed over the rest of the southern part of the land mass, enabling the growth of glossopterids and the formation of high-latitude coal deposits. There was a subsequent loss of even these remaining ice caps, with the further development of globally warm conditions in the late Permian. The movement of Pangaea away from the South Pole, so as to straddle the Equator in the Triassic, may have given rise to the development of an extremely strong monsoonal system, with pressure systems and precipitation zones swinging seasonally from hemisphere to hemisphere. Warming would have been enhanced by exposure and oxidation of some of the organic carbon sequestered during the late Carboniferous and early Permian. Indeed, the loss of most coal-swamp vegetation may have provided another positive feedback to the warming, since there would have been less capacity for CO_2 removal from the atmosphere. It seems that tectonically driven influences on climate now overwhelmed any cooling and related change due to CO_2 reduction by biotic sequestering. The result was a greenhouse world that lasted for some 250 Ma. Some of the factors that allowed for the development of a greenhouse world during the Cretaceous form the focus of attention in the next section.

7.5 The Cretaceous greenhouse world

Figure 7.9 shows the palaeogeographic arrangement of continents for the mid-Cretaceous, along with the distribution of some climatically sensitive deposits. The widespread occurrence of coals at high palaeolatitudes (>60°) demonstrates that temperatures were high enough there for plant life to flourish and that large volumes of glacial ice (prevalent during the early Permian) were now absent. At low palaeolatitudes (<30°), the extent of evaporites (especially in the Northern Hemisphere) shows that large areas were predominantly arid, with evaporation exceeding precipitation. Contrast this with the extensive tropical rainforests of today in the Amazon Basin, Africa and South East Asia. Although some coals (a sign of active plant growth and, therefore, water) formed at mid- to low palaeolatitudes (<60°) in the Cretaceous, they did so on a limited scale and only close to ocean margins.

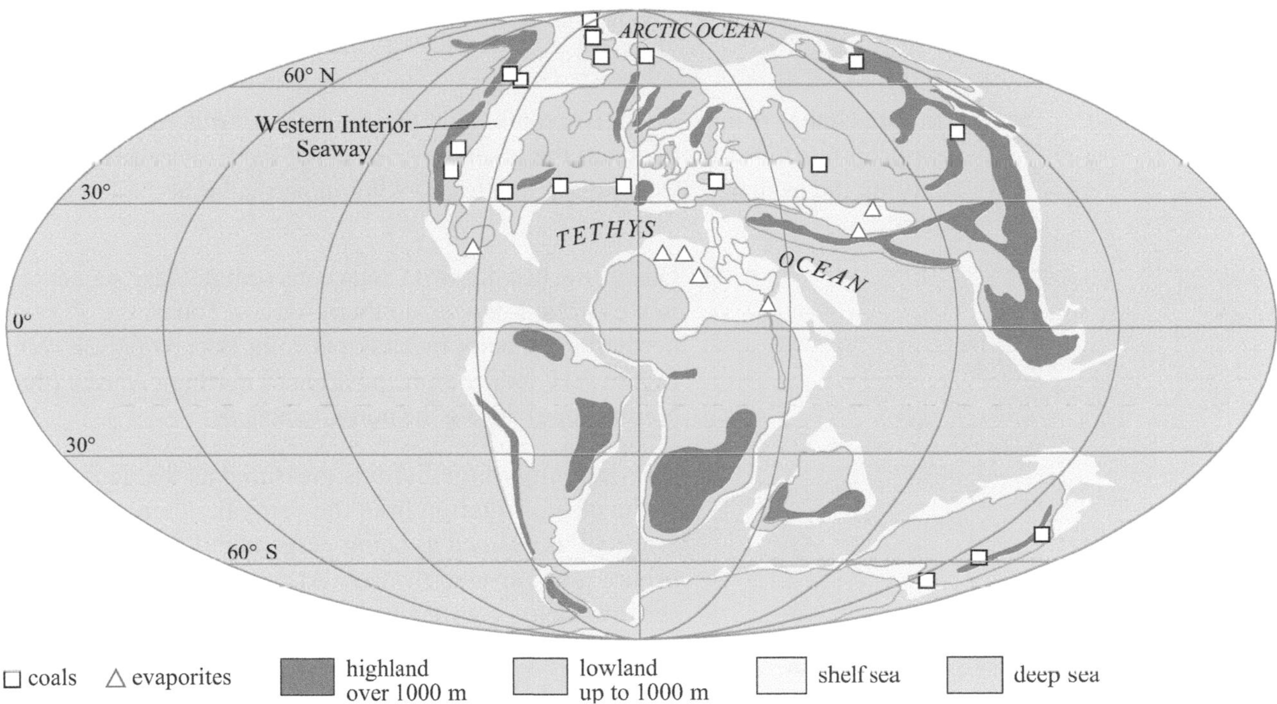

Figure 7.9 Mid-Cretaceous palaeogeography (around 95 Ma) and climatically sensitive deposits. (Parrish et al., 1982)

Another noticeable feature of the map is that the individual land masses are smaller in extent than those of today: there are no corresponding areas of land as large as the modern Eurasia or North America, for example. The reason for this is twofold: firstly, referring back to the map showing the latest Permian palaeogeography (Figure 7.1b) reveals that the Cretaceous world was a product of the rifting apart of the supercontinent Pangaea; secondly, Cretaceous sea levels were, at times, a few hundred metres higher than at present, so that large areas of the continents were covered by shallow seas.

Question 7.4

From what you now know about the differences between the Cretaceous Earth and the Permo-Carboniferous icehouse Earth, suggest two reasons why sea levels should have been relatively higher in the Cretaceous.

One of these shallow continental interior seas formed during the Cretaceous was the Western Interior Seaway in North America, which at various times during the Cretaceous connected the Tethys Ocean with the Arctic Ocean. These shallow seas were important to the world climate in that:

- they were a source of moisture in what would otherwise have been dry land areas
- they warmed and cooled more slowly than the surrounding land (because of the high heat capacity of water) and, therefore, affected heat distribution
- they acted as conduits for heat as water currents flowed from one ocean to another.

7.6 Polar climate

7.6.1 The case of Alaska

Climate change is most strongly expressed at the poles so, for this reason, it is appropriate to begin to look at the greenhouse world of the Cretaceous by studying the evidence for its polar climates, looking in particular at the High Arctic.

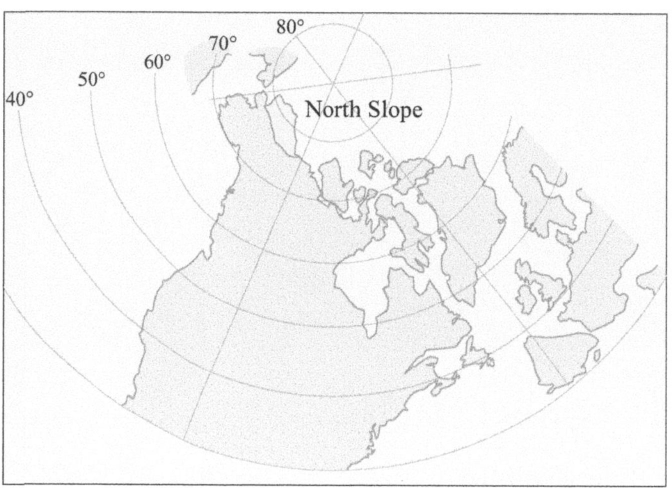

Figure 7.10 Mid-Cretaceous position of Alaska.

One of the best known Cretaceous sedimentary sequences in the Arctic is that of northern Alaska. Today, the northernmost point in Alaska is Point Barrow (71.23° N); Figure 7.10 indicates the position of northern Alaska (the North Slope) during the mid-Cretaceous.

The obliquity of the Earth is measured as the departure of the axis of rotation from the vertical, where the 'vertical' is defined as being perpendicular to the Earth's orbital plane around the Sun. At the latitude of Point Barrow today, continuous winter darkness (i.e. darkness longer than 24 h) does not occur, but there are about 2 months of twilight. For a period of almost 4 months the period of light during the day changes; during the summer there is a period of about 2.5 months with continual light.

The Cretaceous rocks found in northern Alaska are extremely rich in coal. In fact, current estimates suggest that there are 2.75×10^{12} tonnes of coal lying under northern Alaska, which represents about one-third of the total US coal reserves of all ages combined (including those of the Carboniferous). By any measure, this near-polar environment was an extremely effective long-term carbon sequestering and storage system during the Cretaceous.

Observation of tree stumps preserved in life position rooted in fossil soils indicates that the coal-forming vegetation during this time was in the form of forests rather than tundra. Their existence poses two questions: first, why was this forest ecosystem so effective at capturing carbon; and second, how was the carbon so effectively buried?

It is easier to answer the second question first. Rocks of mid- to late Cretaceous age were laid down in northern Alaska as a series of deltaic sediments that built out northwards into a trough to the north of the newly uplifted Brooks Range Mountains. Although the details are unresolved, the mountain range appears to have been generated by local plate collision. Uplift was most rapid during the last part of the early Cretaceous, but continued through into the late Cretaceous. Erosion of these mountains during the mid- to late Cretaceous supplied the deltaic sediments.

Plants colonised the delta flood plains skirting the mountains and, as the sediments containing the organic matter subsided (due to weighting and gravity), more sediments were deposited on top of them and more forests grew. Such a process, typical of many long-lived deltas, provided an ideal setting not only for the preservation of fossils but also for burying peat.

7.6.2 The Arctic forests

There are peats forming in northern Alaska today, but there are no forests. Since northern Alaska was further north of its present position in the Cretaceous, the conclusion is that climatic conditions were not the same in the Cretaceous as they are now. Conditions must have been more favourable in the Cretaceous to have allowed the growth of forests at these latitudes – but did they grow?

■ What alternative explanation might there be for Cretaceous polar forests if climate conditions then were the same as now?

□ The tolerances of plants might have changed. For example, the plants might have had a different biochemistry that allowed enzymes to function more efficiently at lower temperatures.

This is unlikely, however, because it would be expected that plants with such specialised adaptations would have survived subsequently, particularly during the Quaternary glacial events, and still be in evidence today. The fact that such plants are not present, particularly in today's cool climate, suggests they never existed.

So, if conditions (rather than the fundamental requirements for plant growth) were different, how can these conditions be deduced? To do this, it is necessary to examine the rocks and try to decode the information they contain; not only must the fossils be examined but also the sedimentary context in which they are found.

Plant fossils have been known from the Arctic for a very long time. Some finds of apparently warmth-loving plants at high latitudes can be eliminated from further consideration because they are found on tectonic plates that were at lower latitudes when the plants were alive, but which subsequently drifted to their near-polar settings of today. This still leaves the problem of northern Alaska, the palaeolatitude of which during the Cretaceous is well constrained (to within 5°). A wealth of fossils can be found there, but the fact that they represent trees is incompatible with present conditions at the latitudes where they grew.

Figure 7.11 *Nilssonia* frond. (Bob Spicer/Open University)

7.6.3 The cycad conundrum

Among the more problematical finds in North Alaska are those that are thought to be fossil cycads. These plants, which only occur today at low latitudes where frosts are absent or infrequent, are represented most commonly in the Cretaceous fossil record of Alaska by the leaf form, *Nilssonia* (Figure 7.11).

Modern cycads typically have the appearance of a squat palm tree: they have a trunk composed mostly of tissue rather like that in an apple or a potato and very little wood. At the top of the trunk is a crown of leathery evergreen fronds. The sensitivity of the modern cycad to frost and its evergreen characteristics have posed severe problems for interpreting Arctic palaeoclimates. For a start, if the biology of Cretaceous cycads was the same as that of the modern cycads, the polar regions must have been much warmer than today. It is difficult to see how such conditions could have been sustained with long periods of winter darkness.

Even more problematic, however, is the fact that modern cycads retain their leaves all the year round: they are evergreen. In addition to photosynthesising, plants must respire to release the energy stored in their food reserves. The rate at which metabolic processes (including respiration) proceeds increases with temperature. The optimum temperature range varies with the type of plant, but in general most species function best between 5 °C and about 40 °C. Below 5 °C, the metabolic rates of all living things are very slow, whereas above 40 °C some processes are inhibited or prevented as enzymes are damaged. Thus, the balance of photosynthesis and respiration is both light-dependent and temperature-dependent. In climates where temperatures never (or only very rarely) drop to near freezing, respiration rates are relatively high. In cold climates, especially in winter, respiration rates are very low, so respiration in the leaves does not use up much food. Plants in cold, dark situations, such as experienced today during winters at high latitudes, can survive even though the darkness prevents the production of food by photosynthesis. This is because they do not exhaust their food store by having a high rate of respiration.

Assuming that in the late Cretaceous the Earth's obliquity was the same as at present (Section 7.6.1), and given that the palaeolatitude of the centre of the North Slope was approximately 80° N, it follows that the *Nilssonia* plants would have experienced continuous winter darkness for some three months, and that in the spring and autumn there would have been a period of two weeks of twilight. If the *Nilssonia* plants were indeed evergreen, maintaining the leaves through a long, dark but relatively warm winter would have caused an intolerably high respiratory drain on the plant, particularly for young plants and seedlings.

It is not surprising that some scientists have suggested that if the Arctic cycads were evergreen, they must have received winter light. This would enable them to manufacture food to replace some or all of that consumed by leaf respiration. To achieve this more even distribution of light all the year round at the poles, the obliquity of the Earth must have been considerably less than the present 23.4°, with an obliquity possibly as low as 5°. This reduction is, of course, outside the range produced by the Earth's regular orbital variations and would have required a special mechanism to have produced this change.

Are there any additional factors that need to be taken into account before becoming concerned about the apparent anomaly of evergreen plants growing in northern Alaska in the Cretaceous? One factor that has been overlooked so far is the question of whether we can be sure that the palaeomagnetic pole used to determine the palaeolatitudinal positions of the continental plates was the same, or nearly the same, as the Earth's rotational pole. It is the position and inclination of the rotational pole or axis that determines the light regime. To answer this question, the symmetry of distribution of climatically sensitive deposits can be examined. These sediments should be symmetrical about the rotational pole because the rotation of the Earth controls atmospheric dynamics and, therefore, climate. If the position of the sediment-determined rotational pole agrees with that determined palaeomagnetically, it can be assumed that the rotational and magnetic poles were essentially the same.

In the latest Cretaceous, when northern Alaska was at palaeomagnetic 85° N, the rotational pole appears to have been within 4° of the palaeomagnetic pole, i.e. less than the expected error for palaeomagnetic positioning. It may be concluded, therefore, that the rotational and magnetic poles were effectively the same at that time, and that the palaeomagnetically determined palaeolatitude can be assumed to have been more or less the same as the true palaeolatitude in relation to the rotational pole. Therefore, it has to be presumed that these cycads did indeed experience prolonged winter darkness.

So, has the biology of these plants been correctly interpreted? Modern cycads exhibit what is known as a relict distribution, i.e. they are restricted to relatively few sites that are geographically isolated, with individual genera occurring at several of these separated sites. The more widespread distribution and greater taxonomic diversity of cycads in the Mesozoic implies that their present distribution must represent merely a relict of the former dominance of the group as a whole. This poses two questions:

1 Were the environmental conditions that suit cycads more widespread in the Mesozoic (i.e. was global climate more uniformly warm)?

2 Have the remaining relict forms adapted to a narrower range of conditions than that tolerated by Cretaceous forms?

To answer these questions, it is necessary to go back to the fossils and examine them without being prejudiced by the biology of the modern relicts. In 1975, two Japanese palaeobotanists, Tatsuaki Kimura and Shinji Sekido, described a Cretaceous cycad quite unlike any that are alive today; they called this plant *Nilssoniocladus* because it bore *Nilssonia* leaves branching from side shoots (Figure 7.12). It is this unique cycad that is found across North Alaska in the Cretaceous.

Figure 7.12 Reconstruction of part of the *Nilssoniocladus* plant, which grew throughout the mid- and late Cretaceous of the Arctic.

Unlike typical modern cycads, which have a squat, palm-like appearance with a thick trunk, *Nilssoniocladus* had a thin, which vine-like stem. Arranged along this stem were side shoots that bore numerous scars where leaves had been attached. This observation suggests that *Nilssoniocladus* was a deciduous cycad (i.e. one that shed its leaves seasonally throughout its life). Moreover, the plant apparently not only shed its leaves seasonally but also did so in a synchronous fashion (i.e. leaves were shed at the same time, leaving the plant devoid of all leaves for significant periods in its life). Evidence for this comes from the observation that *Nilssonia* leaves are often found as leaf mats and that individual leaves have not simply broken off the main plant, but have distinct bases that are specialised for leaf separation. Moreover, the leaves are usually found intact and therefore must have been shed intact, without damage or rotting. Leaves of different sizes are also found on single bedding planes, indicating that they all fell off the plant at the same time. As a result of this powerful evidence that *Nilssoniocladus* was deciduous, it can now be inferred that it was the biology of the Cretaceous cycads that was different and that modern cycads do indeed represent a relict subset of a group that previously had a wider range of environmental tolerances.

Notice here that a combination of sedimentological evidence, observations on leaf characteristics (size, shape and whether they display signs of decay) and shoot architecture is being used. The study of sedimentological and biological evidence relating to explanations of fossilisation phenomena is known as **taphonomy** (from the Greek *taphos*, meaning a tomb) and is of critical importance in palaeoenvironmental research.

7.6.4 A new perspective

What are the implications of this new understanding of Mesozoic cycads? Firstly, it is dangerous to extrapolate uncritically the biology and climatic tolerances of modern plants back in time. Secondly, if the cycads were deciduous, then the paradox caused by winter darkness ceases to be a problem (at least for the cycads), but was such a deciduous nature a general characteristic of Cretaceous vegetation of northern Alaska?

Similar evidence for deciduous features has been found in many other plant groups. For example, the angiosperms first appear in the middle part of the Cretaceous and have large leaves that look very much like those of modern plane trees (Figure 7.13). These are informally termed platanoid leaves and are typical of deciduous angiosperms: they tend to be simply constructed, in that they have a thin texture. This contrasts with longer-lived, thicker leaves produced by evergreens, such as rhododendron or holly. Moreover, platanoid leaves have relatively thin cuticles because they do not have to limit water loss during the winter, when the root zone might be frozen. They also have an expanded base to their stalk, which is a feature of seasonally shed leaves.

Many platanoid leaves are big, being more than 10 cm in breadth; they also have a wide size range, varying from small to large. In spite of their large size and the coarseness of the sediment, these fossil leaves are invariably complete within the rock. As with the interpretation of leaf mats, this strongly suggests that they were shed intact and were all buried together before any of them had time to decay.

Figure 7.13 Cretaceous leaf in sandstone (×0.5). (Bob Spicer/Open University)

So far, representatives of angiosperms and cycads have been considered, but these groups represent only a small fraction of the polar land vegetation. The most ubiquitous plants appear to have been deciduous conifers similar to modern redwoods, together with ginkgos, or 'maidenhair' trees.

The options of 'deciduousness' or 'evergreenness' are restricted to long-lived woody taxa; herbaceous plants do not normally shed their leaves. What did the herbaceous plants do during the winter darkness?

Much depends on whether the plants were annuals or perennials. Annuals grow from seed each year, reproduce and then die in the autumn; overwintering is therefore in the form of seeds. Perennial herbaceous plants adopt a different strategy: they die back to basal food storage organs such as rhizomes (a modern example is the iris), corms (e.g. crocus) or bulbs (e.g. tulips and onions). In the Cretaceous, there is no evidence for corms or bulbs, but many ground-cover plants had rhizomes, as evidenced by the remnants of rhizomes found in fossil soils. Many are characteristic of modern *Equisetum* (horsetails or scouring rushes), and are assigned to the fossil form *Equisetites*, whereas others represent ferns. In some instances, it is evident that ferns formed the ground cover because they are found preserved *in situ*, buried beneath volcanic ash-falls overlying fossil soils. In the winter, the above-ground parts of *Equisetites* and associated ferns, which formed two main ground-cover plants, would have died back to their rhizomes and become dormant (fossil evidence suggests that grasslands had not yet evolved).

Fossil evidence shows that the near-polar forests also supported a diverse fauna, including large herds of plant-eating dinosaurs, including the duck-billed *Edmontosaurus* and the horned ceratopsians, which were preyed upon by meat-eating dinosaurs such as *Tyrannosaurus*.

Wildfires were also common in the polar forests (as evidenced by abundant fossil charcoal) and must have released some plant carbon back into the atmosphere as CO_2, while yet more would have been oxidised through bacterial action, or converted into methane under reducing conditions in boggy areas. Other remnants of the vegetation would have been transported as partly rotted microscopic particles into sediments, where they have survived as particulate material or as organic molecules.

From the kinds of evidence discussed above, it is possible to build up a picture of the northern Alaskan Cretaceous forests as one consisting of plants that were either deciduous or died back to underground organs during the winter months. As such, there is no need to postulate year-round light, or day–night cycles.

There is further direct evidence that the polar light regime was like that of today. In addition to fossil leaves, the Cretaceous sediments of northern Alaska yield abundant fossil wood that has well-preserved internal structures. At some sites in northern Alaska, tree stumps are found rooted in fossil soils, indicating that trees did indeed grow under near-polar conditions. When sections of the fossil logs are studied under a microscope, fine preservation of the structure, down to the subcellular level, may often be seen. Interpretation of these preserved structures, however, is not straightforward, as a given pattern of ring features may be the result of genetic factors rather than environmental constraints. If a number of different taxa display the same pattern, it is probable that this pattern was environmentally produced, either through direct influence on growth or through selection for certain genetically produced characteristics that confer specific advantages in the prevailing environmental regime.

The pattern observed in many of the northern Alaskan woods of mid-Cretaceous age is distinctive. The earlywood produced in the spring is clearly visible as xylem cells with large internal cavities through which water flowed. The cell walls of the earlywood cells are thin. The latewood cells have thicker walls and smaller cavities, forming the dark bands in the wood. The uniformity of the mid-Cretaceous earlywood shows that growth conditions varied little throughout the growing season. This principally means that drought (or waterlogging) and chill rarely occurred. It also means that other events that could diminish growth, such as severe insect attack, were also of no great consequence. Towards the end of the Cretaceous, however, environmental conditions changed markedly.

Growth rings found in some fossils show interruptions during the summer growing season. These interruptions can be caused by any factor that reduces the tree's photosynthetic activity. For example, they might be caused by: sudden cold snaps that depress the rate of photosynthesis; drought, which causes the stomata to close as a water-conservation measure, thereby stopping the supply of CO_2 to the photosynthetic process; or trauma, such as massive insect damage. Such rings are called **false rings** and are a measure of environmental variations that exceed the tolerance of that particular plant. Many of the latest Cretaceous woods are characterised by such frequent false rings. Often, false rings and true (annual) rings are difficult to differentiate, and this in turn makes measuring ring thicknesses difficult. In general these later trees have narrower rings but, because the amount of latewood is about the same as that in the mid-Cretaceous specimens, the ratio of earlywood to latewood diminishes. This implies that growth conditions throughout the latest Cretaceous summer were less temperate than those in the mid-Cretaceous.

Simultaneous examination of the sediments associated with different late Cretaceous fossils reveals that drought indicators, including preserved mud cracks, are lacking, although the frequency of charcoal in the latest Cretaceous sediments is higher than in the earlier beds, implying that vegetation was drier and so more predisposed to combustion. In the absence of direct evidence for insect damage (e.g. chewed leaves or borings in the wood), the most likely explanation for the variation in summer growth of late Cretaceous vegetation was temperature fluctuation: there were periodic cold snaps. This might imply that the average summer temperatures were lower in the latest Cretaceous, such that temporary falls in temperature from the lower average value were sufficiently cold to limit photosynthetic activity.

7.6.5 Taking stock

Using qualitative sedimentological and palaeobotanical evidence only, the following points have been established.

1 The obliquity of the Earth's rotational axis was essentially the same in the late Cretaceous as it is now.

2 The rotational pole was essentially in the same location as the magnetic pole.

3 The polar light regime was similar to that of today.

4 The Arctic climate supported forests up to at least 85° N.

5 The overall polar climate must have been warmer than now.

6 There was sufficient rainfall to support abundant plant growth.

7 The vegetation must have dried sufficiently to burn from time to time.

8 The Arctic forests were deciduous.

9 Some plants had substantially different climatic adaptations when compared with their modern relatives.

By now, quite a detailed picture of the Cretaceous Arctic has been established. There is substantial evidence to show that, in many respects, the high southern latitude (Antarctic) vegetation and climate mirrored that of the Arctic.

7.7 Low-latitude vegetation and climate

7.7.1 Water supply

This section investigates the other end of the Equator-to-pole temperature gradient that powers the climate system: the low latitudes. Returning to Figure 7.9 for a moment, you may recall that one feature of the Cretaceous low latitudes was the extensive area of arid climates characterised by evaporites. To form, evaporites need a high evaporation rate, as well as a supply of water carrying dissolved minerals, so that continued salt precipitation takes place. The presence of evaporites over large continental areas, therefore, implies that those areas were not constantly dry but must have had some water supply on a frequent basis. If this were so, some evidence for vegetation might be expected. Fortunately, this is exactly what is seen, and, as before, the plants tell a great deal about the environment.

In Cretaceous sediments deposited between palaeolatitudes 40° N and 40° S, it is common to find large quantities of the fossil pollen type *Classopollis*. This is a very easily recognised pollen form that is known to have been produced by a family of extinct conifers known as the Cheirolepidiaceae (Figure 7.14 overleaf). The pollen is found in the cones of these plants, for which fossils of both foliage and wood have also been identified. Examples of cheirolepidiaceous foliage and wood can be found in many early Cretaceous sediments in southern England (which was then at ~36° N), with the sediments deposited at the margins of a seaway that connected with the Tethys Ocean. For example, on the Isle of Portland, tree stumps, logs, foliage and pollen are found associated with sediments of earliest Cretaceous age containing **halite (salt) pseudomorphs**. This and other evidence suggests the plants grew under considerable evaporative stress, where rainfall was, at best, highly seasonal. There is also a strong seasonal signal in the growth rings of the wood.

Different species of the Cheirolepidiaceae are found in the early Cretaceous sediments of lower latitudes. Again, wood, foliage, cones and pollen are all present, and the foliage form considered here, called *Frenelopsis*, shows some extremely interesting adaptations to aridity.

Figure 7.14a shows a foliage branch of a typical member of the Cheirolepidiaceae, many species of which have been reconstructed as a tree with a single main trunk. The foliage shoots are segmented with each successive segment inserted in the previous one, rather like a stack of paper cups. The leaves are not immediately obvious but are small, triangular-shaped appendages situated at the rim of each shoot segment. In the species shown here, there are three leaves per segment but some species have only one leaf per segment.

Each segment is covered in rows of tiny dots. These are, in fact, the stomata and there are large numbers of them on each segment. A single stoma of a related genus *Pseudofrenelopsis* is illustrated in Figure 7.14b. It consists of a ring of finger-like projections that point towards each other over a deep pit in the thick cuticle. At the bottom of this pit is the stomatal aperture, bounded by a pair of guard cells. The thick cuticle provides a clue as to the functional significance of this elaborate stomatal structure: it reduced water loss.

A thick cuticle is usually a good indicator that the plant producing it was exposed to considerable water stress, but this stomatal architecture represents a further strategy for resisting desiccation. The deep pits (Figure 7.14c) would have contained relatively static air that would have become water-saturated, so protecting the guard cells from drying. In effect, this adaptation is a 'captured' boundary layer. The guard cells could only have had a thin cuticle if they were to retain the ability to change shape to control the stomatal apertures and they were, therefore, particularly susceptible to desiccation. Furthermore it seems likely that, at times of extreme drought, loss of water from the cells supporting the finger-like projections would have caused them to close up, so sealing the tops of the stomatal pits and protecting the guard cells from drying.

Another adaptation to drought conditions is seen in the architecture of the junctions between foliage shoot segments. This is shown in Figure 7.14d. At the top edge of the lower segment, the cuticle was fringed with a line of hairs that lined the edge of the groove between the segments. Within the groove, the cuticle was very thin compared to that directly exposed to the atmosphere. We cannot be sure of the significance of this structure, but it is likely that the hairs would have been the sites for the nucleation of water droplets when dew formed or when it was foggy. The same phenomenon is seen when you hang a woollen sweater out overnight in damp air – the hairs of the sweater become covered in droplets of dew. If this did happen with Pseudofrenelopsis, then the droplets would have grown until they touched the cuticle of the adjacent segment, and the water would have been drawn into the thinly cutinised groove between the segments and absorbed by the plant. This plant may well have had the ability literally to extract water from the air, as mist and fog rolled in from the nearby sea.

The Cheirolepidiaceae were not the only plants adapted to this harsh environment. There were also some ferns with very thick cuticles and specialised stomata, designed to combat water loss. Additionally, parts of the fern fronds are often found as charred remains, indicating the vegetation was frequently burned and must, therefore, have been dry for this to happen.

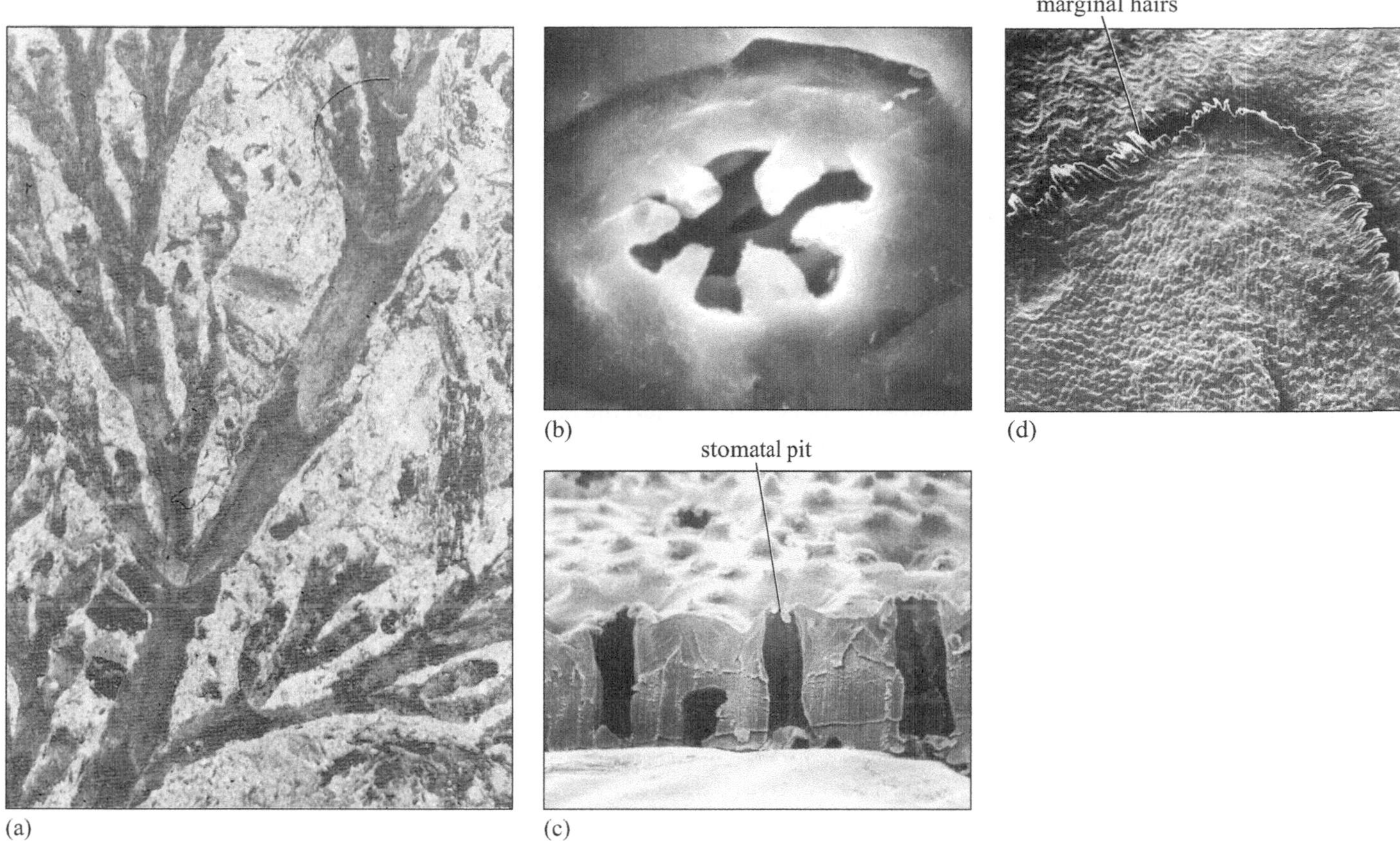

Figure 7.14 (a) *Frenelopsis ramosissima* shoot showing the typical segmented structure seen in many members of the extinct Cheirolepidiaceae (×13). (b) Scanning electron micrograph of a single stoma of *Pseudofrenelopsis* (×2600). (c) Vertical section through the cuticle of *Pseudofrenelopsis* showing the deep stomatal pits (×310). (d) Leaf of *Pseudofrenelopsis* showing marginal hairs behind which is the groove at the junction with the next shoot segment (×80). ((a, c and d) Joan Watson; (b) Bob Spicer/Open University)

7.7.2 Qualitative and quantitative inferences

The low-latitude plants around the margin of the Tethys Ocean and Tethyan margin plants were architecturally very different from those growing in the polar regions, and these differences reveal a great deal about the varying climates of those regions, and in particular the temperature, evaporative stress and frequency of rainfall. It is also possible to determine that the main sites of land-based plant productivity were at high and not low latitudes, and that the equivalent of today's tropical rainforests did not exist to any great extent during the Cretaceous. These observations are of critical importance in understanding what might happen to the future climate in general, and agriculture in particular, should the Earth return to its warm (greenhouse) conditions. Even more can be discerned when these observations are combined with those of sediments; however, as all of this information is qualitative, it is not possible to quantify what the exact temperature was, nor how much rainfall there was. Alternative methods of obtaining quantitative palaeoclimate data are therefore required.

7.8 Climate reconstructions

7.8.1 Plants as thermometers

So far in this chapter, the evidence that suggests the late Cretaceous polar light regime was more or less as it is now has been reviewed. Apart from deducing that the overall temperature regime was warmer than now, although not necessarily very warm as some frosts were likely, it has not been possible to quantify the temperature regime so far. This section considers some of the ways in which atmospheric Cretaceous polar temperatures can be determined using techniques that can be applied wherever suitable plant fossils are preserved.

To begin, it will be assumed that the range of temperature within which growth could have occurred is much the same as it is today, i.e. somewhere between 5 °C and 40 °C. This is a rather large temperature range and is little help in trying to understand the dynamics of polar (and global) climate. As such, some additional criteria must be considered. In relation to this, palaeobotanists have long recognised that land plants are excellent climate indicators and, broadly speaking, two approaches are commonly adopted:

- nearest living relative technique
- physiognomy.

7.8.2 The problem with relatives

The first technique is based on the climatic tolerances of nearest living relatives (NLRs) and is known as the **NLR technique**. This assumes that ancient plants and plant communities lived under similar conditions to those of their nearest living relatives. The success of the technique, which is widely used in Quaternary studies, depends on the correct identification of the fossil and, of course, limited evolutionary change. Where the fossils are in the form of reproductive structures, such as pollen or seeds, NLR techniques are particularly useful because the taxonomy of living plants is based mostly on their reproductive characteristics; correct identification, however, is less difficult than it is when working with vegetative organs, such as leaves. For *pre-Quaternary* fossils that may represent extinct taxa with very different environmental tolerances from their living relatives, and for vegetation for which there are no living counterparts, such as the greenhouse polar forests, a different approach has to be adopted. As such, the NLR technique is not suitable to investigate Cretaceous flora as climate indicators.

7.8.3 Following form

The second method for using plants as climate indicators is based on the architecture, or **physiognomy**, of the plant or community, and is applied when the fossils are in the form of vegetative organs, particularly leaves. As plants cannot move around once they have taken root, they must be well adapted to their local environment or they will die, either because an environmental tolerance is exceeded or because they are out-competed by better-adapted plants. Either way, there is a selective premium on being as efficient as possible in the local circumstances with respect to water conservation, gas exchange and light interception. In the course of evolution, most plants have become honed for successful exploitation of particular environmental niches, and many display specially adapted physiognomies. One extreme example is seen in desert plants, which have a low surface area to volume ratio so as to conserve water, low **leaf-area indices** and thick cuticles.

- List some examples of plants that are adapted to a desert environment.
- Cacti and their 'Old World' counterparts, euphorbias, are but two that illustrate these traits well.

Rainforests, by contrast, are characterised by plants with large leaf-area indices, forming a vertical succession of layers within the forest.

As a result of physiognomic adaptations to environment being so consistent, quantitative comparisons for determining pre-Quaternary climates can be made, within certain limits. One of the most successful applications of this approach was devised as long ago as 1915, when two American botanists, Bailey and Sinnott, noted that the leaves of modern woody, broadleaved flowering plants (such as alder, willow and figs) tended to have smooth (entire) leaf margins in warm climates, but toothed, jagged margins in cool climates (Figure 7.15). Jack Wolfe, another American palaeobotanist, developed this methodology further in the late 1970s and, by using modern species growing in drought-free environments in South East Asia, plotted the mean annual temperature against the percentage of species with entire-margin leaves (Figure 7.16).

This technique of leaf margin analysis can be applied to the Alaskan fossils. It is known that, in mid-Cretaceous times, northern Alaska rarely experienced drought because there are numerous thick coal seams; and, although charcoal is present, it is scarce. In addition, tree rings show that growth was uniform throughout the summers; therefore water was not limiting. The sediments have also yielded a large number of leaf fossils and so far 67 different species of woody broadleaf angiosperms have been distinguished, of which 22 species have entire leaf margins preserved. Referring to Figure 7.16, this indicates that the mean annual temperature of mid-Cretaceous northern Alaska was similar to that of the southern British Isles today. However, bearing in mind the polar light regime, it is likely that the mean

(a)

(b)

Figure 7.15 Leaf margin types. (a) A variety of tropical rainforest species all displaying smooth (entire) leaf margins. Some also show an extended leaf apex ('drip tip') for shedding water. Drip tips are characteristic of plants growing under very wet conditions. (b) A typical temperate tree bearing leaves with toothed margins. (Bob Spicer/Open University)

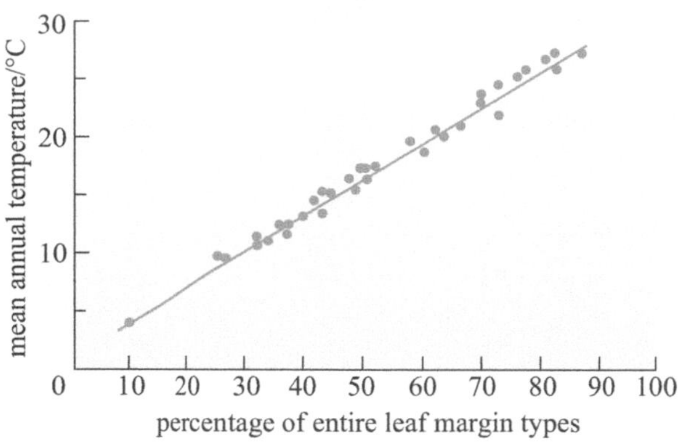

Figure 7.16 Plot of mean annual temperature against the percentage of entire leaf margin types. This graph was constructed using plants in drought-free environments in South East Asia. (USGS Printing Office)

annual *range* of temperature in the mid-Cretaceous was much greater.

A number of assumptions have been made using leaf margin analysis in this instance. First, there is an assumption that there is no change in slope of the graph (Figure 7.16) over time, i.e. that the slope determined for the plants of today is applicable to the mid-Cretaceous, which was very early in angiosperm evolution. Second, there is an assumption that the relationship between margin characteristics and temperature was not affected in any way by the polar light regime. As there is no obvious break in Figure 7.16 in the slope of the relationship for modern-day plants above or below latitude 66° N (the likely position of the Arctic Circle then), it can be presumed that the polar light regime has little effect.

Physiognomic analysis can be applied to different vegetation types. Instead of relying on the response of one group of plants (the woody broadleaves) to climatic variables, all taxa, including gymnosperms, can be involved. This is an advantageous approach to use because when a broad spectrum of taxa is used, it is unlikely that they will all relate by chance to the climate in the same way, making this technique particularly robust.

A discussion of how ancient vegetation can be reconstructed accurately from dispersed fossilised remains is beyond the scope of this book; however, suffice to say that a considerable amount of detective work is required to link leaves, wood and reproductive structures into complete plants of known form, ecological setting, community association and abundance. This detective work combines detailed observation of plant form with occurrences in a wide range of sediments, in much the same way as that described earlier when examining the leaf-shedding habits of the plants of the polar forests.

Figure 7.17 Reconstruction of northern Alaskan forests of the mid-Cretaceous.

Figure 7.17 depicts a reconstruction of the mid-Cretaceous forests of northern Alaska, predominantly composed of conifers. The most dominant plant was the deciduous conifer, which is related to the bald cypress and dawn redwood; however, other taxa such as river-margin angiosperms and ginkgos were also prevalent, with ferns and horsetails commonly forming ground-cover plants. The mid-Cretaceous vegetation was not a pure conifer forest then, but one mixed with other components of a broadleaved nature.

7.9 Geographical framework

7.9.1 The marine environment

The previous sections concentrated largely on the Cretaceous vegetation at high latitudes, as this reflected most strikingly the contrast with today's climate. In the marine realm, however, the main differences between then and now were most apparent in low to mid-latitudes associated with the Tethys Ocean, which separated the northern and southern continents (Figure 7.18), as well as in the equatorial Pacific.

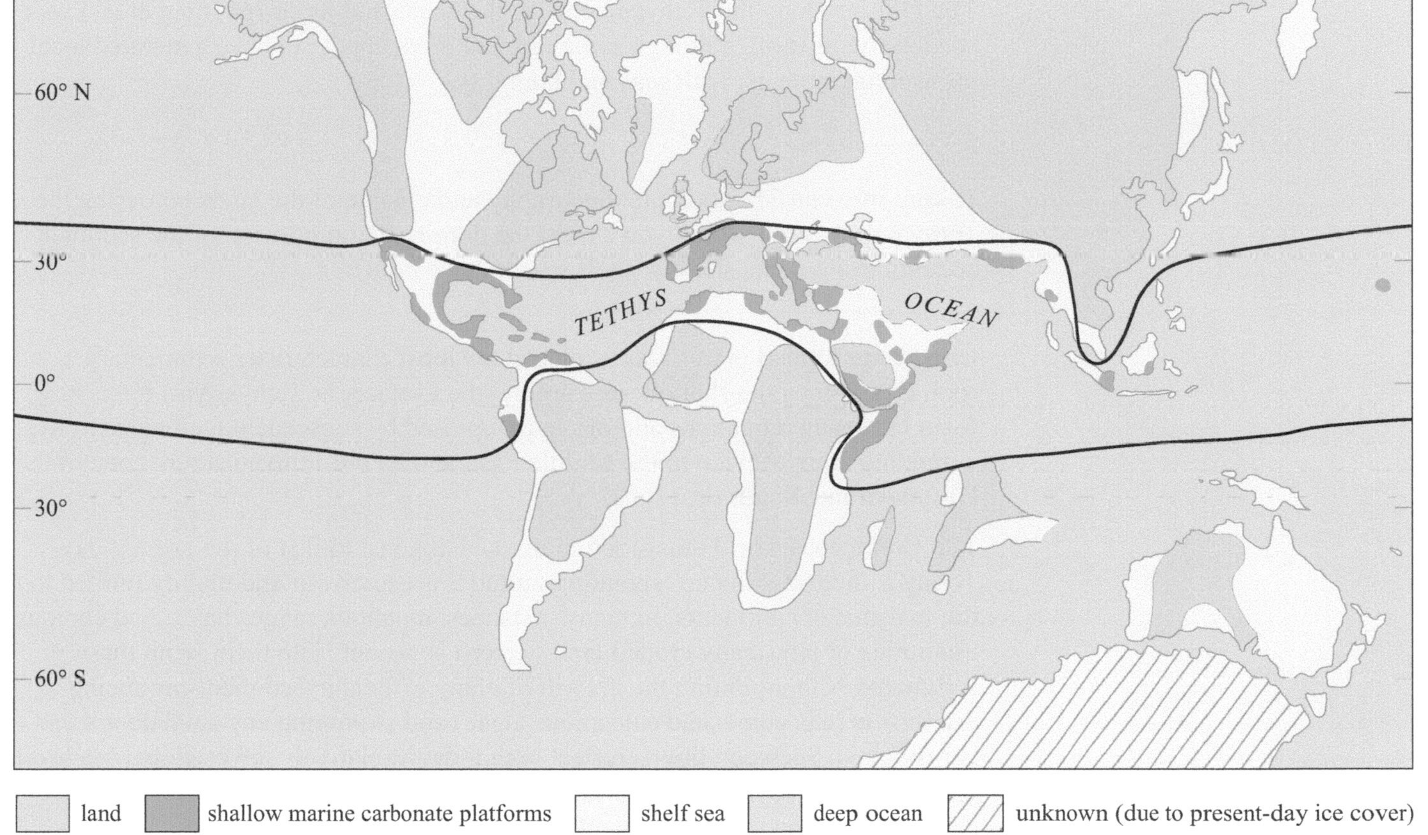

Figure 7.18 Geography of the earlier Late Cretaceous, showing the zone of major carbonate platforms formed in and around the Tethys Ocean. The distribution of land and sea in Antarctica is not well known because of the present-day ice cover. (Sohl, 1987)

- Study the palaeogeographic map in Figure 7.18, noting, in particular, those parts of the continents shown between the bold lines (i.e. between ~10° N and ~30° N). Which covered a relatively greater area – shallow sea or land?

- Within this palaeolatitudinal belt (which included the Tethys Ocean), shallow sea covered by far the larger proportion of the area than was covered by continental crust. Indeed, some continental extensions from the North African margin are shown to have formed isolated, broad, shallow marine platforms, straddling the central part of the ocean.

Elsewhere, seaways spread across the continental interiors, a consequence of the exceptional rise in global sea level during the Cretaceous Period.

Geological evidence reveals that topographical relief in the land areas bordering the shallow shelf seas was generally subdued. The main zones of uplift generated by plate tectonic activity around the Tethys Ocean were limited to volcanic island arcs, associated with the subduction of oceanic crust. These lay predominantly along the northern Tethyan margin, where they formed chains of islands and submerged ridges similar to the Sunda Archipelago of South East Asia today. One of the few mountain ranges that was present in the Cretaceous was situated away from the Tethys Ocean at higher latitudes in northern Alaska (the Brooks Range).

The climate along the Tethyan belt (i.e. the ocean and its surrounding areas) was predominantly arid, although some uplifted areas appear to have generated local monsoonal systems, with seasonal humidity.

Question 7.5

Taking into consideration the topography and climate of the lands bordering the Tethyan belt, what would you expect the dominant composition of the sediment accumulating in the shallow shelf seas to have been?

Limestone formed in this way is indeed the most characteristic sedimentary rock type left by the shallow seas around the Cretaceous Tethys. Vast tracts now form imposing mountains and plateaux, uplifted by subsequent tectonic activity, extending from Mexico to the Middle East, and on the Tethyan suture zone of the Himalaya into South East Asia.

The Cretaceous world thus stood in marked contrast to that of the present day. Today's shelf seas in corresponding latitudes are narrower and mainly limited to the margins of continents. In many instances, mountain ranges have shed copious quantities of physically eroded land-derived sediments into them, with these sediments both inhibiting the growth of many carbonate-sediment-producing organisms (e.g. corals and calcareous algae) and swamping any shell debris that is produced. Such conditions reflect both today's relatively lower global sea level and the extensive mountain building that has occurred in low and mid-latitudes since the Cretaceous. Consequently, shallow tropical carbonate provinces are now much less widespread than those of the Cretaceous.

7.9.2 Tethyan carbonate platforms

Limestones are a crucial component of the carbon cycle and, hence, of the climate system. The limestones formed around the margins of the Cretaceous Tethys are, therefore, worth a closer look, first to see how they came to be so extensive and, second, in order to estimate their quantitative contribution to carbon burial during this period.

Characteristically, the Cretaceous Tethyan limestones accumulated so as to form broad, shallow, submarine plateaux, termed **carbonate platforms**, the outer margins of which sloped away into neighbouring basins. The depth of the basins varied greatly, according to tectonic setting (i.e. the tectonic regime of a region) but, within most areas of continental crust, it usually ranged from only some tens to a hundred or more metres.

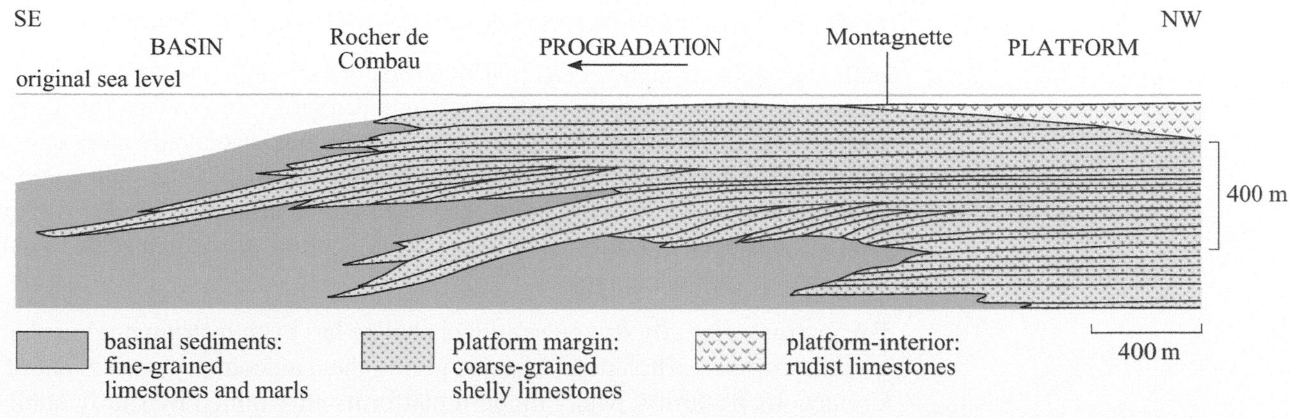

Figure 7.19 Cross-section of the platform to basin transition of the Vercors Platform in eastern France, showing the geometry of the constituent beds.

In recent years, very detailed studies have been carried out on the geometry and constitution of the beds composing such platforms, with the aim of reconstructing how the platforms grew. Figure 7.19 is a vertical section across the edge of one such platform – the Vercors Platform in eastern France, dating from the early Cretaceous. The platform shows a pattern seen very frequently in Cretaceous Tethyan carbonate platforms: sets of inclined beds of relatively coarse sediments (mainly shell debris), derived from the top of the platform, project obliquely down like overlapping tongues into the finer sediments of the basin. This indicates that the platform margin gradually built out laterally into the basin (**progradation**), as the coarser sediment spilt down its flanks. The transport of material off the top of the platform was a consequence of more shell debris being produced than could be accommodated in the shallow waters covering the platform; in other words, there was an overproduction of sediment by the platform 'carbonate factory', which was swept off by waves, tides or storm currents onto its flanks. Between periods of progradation, occasional subsidence of the platform due to a rise in sea level allowed basinal sediments to build up the flanks again until progradation resumed. These **regressive** episodes yielded the thin wedges of finer material that separate the progradational tongues.

There was, inevitably, much variation on this basic depositional theme, as a result of differences in the rates of deepening (due to either subsidence or eustasy) and in the multitude of factors affecting sediment production and transport on the platforms. For example, where contemporaneous faulting at the platform margin accentuated the depth of the neighbouring basin, the sediment spilling off the platform had a larger space to fill, so marginal slopes became steeper and outward growth less marked. On the other hand, the platforms themselves were sometimes 'drowned' when their carbonate factories failed to keep pace with rising sea levels because of adverse water conditions, e.g. storm action and/or too rapid deepening.

Nevertheless, for much of their development, the platforms tended to build up and outwards in fits and starts, as space to accommodate the sediment produced was provided by successive deepening events. At this point, it would be justified to ask why there was so much repeated deepening, and why there were not as many episodes of emergence, leading to weathering and dissolution of the limestones. This leads to a crucial distinction between the Cretaceous greenhouse world, in which these platforms developed, and today's icehouse world.

Although some evidence (e.g. marine dropstones, which are stones picked up by glacial ice sheets and then dropped at some distant location when the ice sheets melt) has been proposed to suggest that a few minor glacial advances caused sea level to fall in the early Cretaceous, this evidence is in dispute, with some critics claiming the boulders could have been rafted on floating tree trunks. Any brief glaciation that there might have been was minor and never on a scale to compare with that of the Quaternary.

By contrast, glacially driven sea-level change has been instrumental in shaping current tropical carbonate platforms, giving them a geometry unlike that of the Cretaceous platforms. Many modern platforms are rimmed by sturdy coral reefs, which reach up to sea level and form prominent barriers to incoming ocean waves. It would be tempting to suppose that this modern barrier-reef topography was entirely created by the vigorous upward growth of the coral reef itself at the platform margin. In many cases, however, the reef has been installed on a pre-existing topography that already had a raised margin. How might that have got there? The **glacio-eustasy** mentioned above provides the answer to this question. Commonly, reefs grow upon older carbonate platforms, which would have been emergent during the last glacio-eustatic fall. Dissolutional weathering of these older platforms often left behind projecting edifices around their edges, rather like castle walls, periodically cut across by gullies. When these weathered platforms became submerged once again, these prominences provided the hard foundations for the re-establishment of reefs around the margins.

The Cretaceous world lacked such glacio-eustatic fluctuations of sea level, at least on any comparable scale. Hence, significant emergence and weathering of the carbonate platforms was infrequent, except where local tectonic activity intervened. Instead, the platforms tended to experience only successive minor deepening events brought about through a combination of the overall eustatic rise of the period and regional subsidence of the crust. Between these sporadic deepening events, carbonate sediment overproduction led to the cycles of progradation discussed above.

How did this characteristic pattern of growth of the Cretaceous platforms affect the organisms that dwelt on them and, ultimately, created the platforms? The common lack of inherited topographical relief sculpted by emergence and weathering, and the frequent fluxes of carbonate sediment across the prograding margins, meant that there was little opportunity for the stable establishment of barrier reefs. The platforms thus lacked the sturdy protective rims created by barrier reefs today, which take the brunt of incoming waves from the open sea and limit the flux of water and sediment across the platform margins. Instead, the outer parts of the Cretaceous platforms tended to be dominated by migrating banks or sheets of current-swept shell sand and debris, which graded into expanses of muddier lime sediments in the platform interiors. These broad sediment surfaces supported myriads of bottom-dwelling shelly organisms, the growth and death of which returned more sediment to the system. Foremost among the larger skeletal sediment-producers, particularly in the late Cretaceous, were some oddly shaped gregarious bivalves called **rudists** (Figure 7.20). These grew within or lying upon the loose sediment as vast 'meadows' of clustered shells, rather like oyster beds

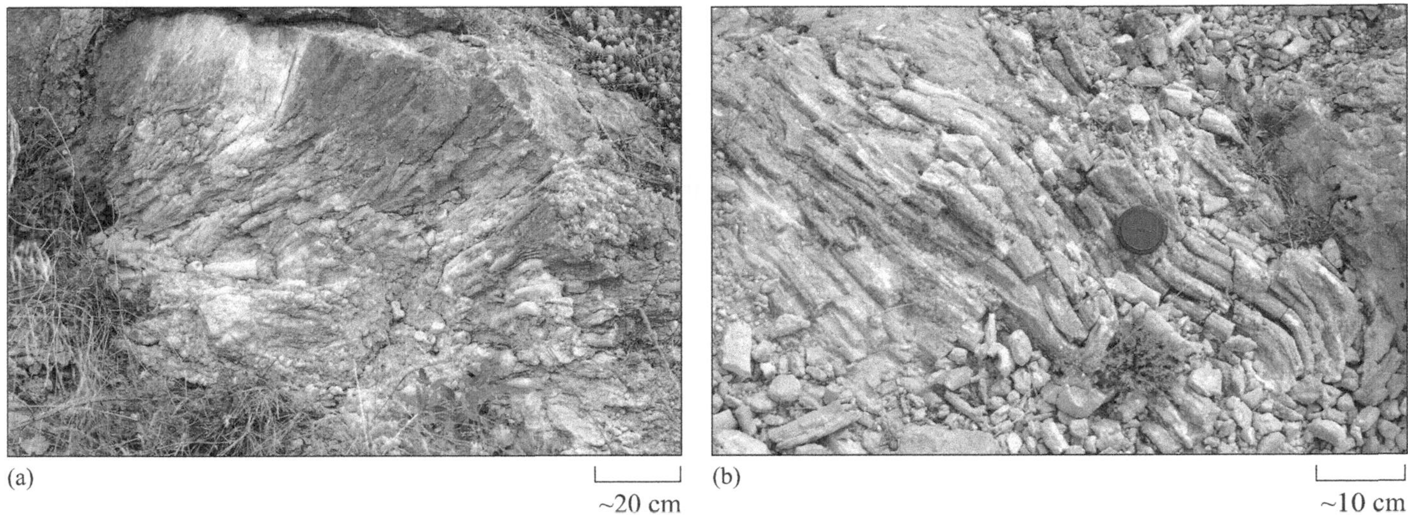

Figure 7.20 Clusters of tubular rudist bivalve shells, preserved mostly in life position, implanted in the sediment in which they grew, from the Upper Cretaceous of (a) Provence, southeastern France and (b) the Spanish Pyrenees. (Peter Skelton/ Open University)

today. Frequent disruption of these congregations, usually by storm currents, together with breakage and perforation of the shells by burrowing and boring organisms, helped to turn them into shell sand and mud. The upwards and outwards growth of the platforms was thus self-sustaining, as shell debris was harvested over huge areas and spread both inwards and out onto the marginal slopes.

Sometimes these rudist-dominated formations are referred to in textbooks as 'rudist reefs' but, as the description above shows, this is quite a misleading term as far as the anatomy and growth of the carbonate platforms were concerned. Rather, the rudists were essentially bottom sediment dwellers, forming shelly 'meadows' and associated carpets of redistributed debris.

7.9.3 Quantifying the growth of the platforms

This section considers some quantitative aspects of these Cretaceous Tethyan carbonate platforms, the areal extent of which naturally varied throughout the Cretaceous, with long periods of expansion punctuated by brief episodes of decline, when carbonate production was temporarily reduced or even halted on many platforms due to adverse conditions. During periods of adverse conditions, large tracts of the platforms subsided. This in turn necessitated regrowth by progradation from shallower areas once conditions improved. A major study by a team of French geologists has investigated various quantitative aspects of the history of a number of carbonate platforms across France. They found that one of the main phases of global platform growth was approximately half-way through the Cretaceous, between about 94 and 92 Ma (contemporaneous with some of the northern Alaskan forests). The Tethyan carbonate platforms at that time are estimated to have covered a total area of some 9.68×10^6 km^2 (an area approximately equivalent to that of China or Europe). Individual platform thicknesses, of course, varied according to the amount of accommodation space available due to subsidence, itself dependent on the local to regional tectonic setting. Growth rates calculated for several examples of this age from around the

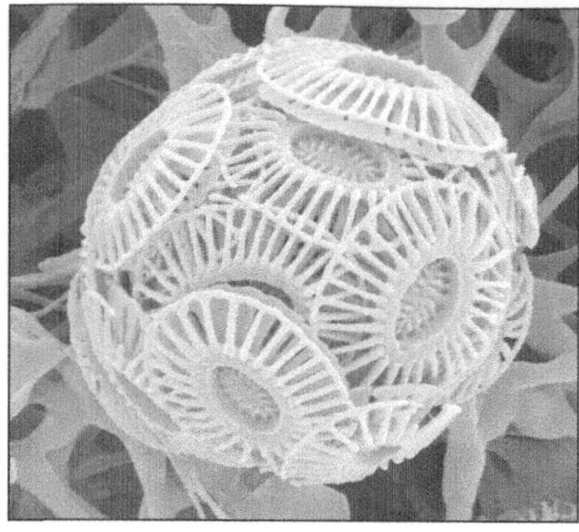

(a)

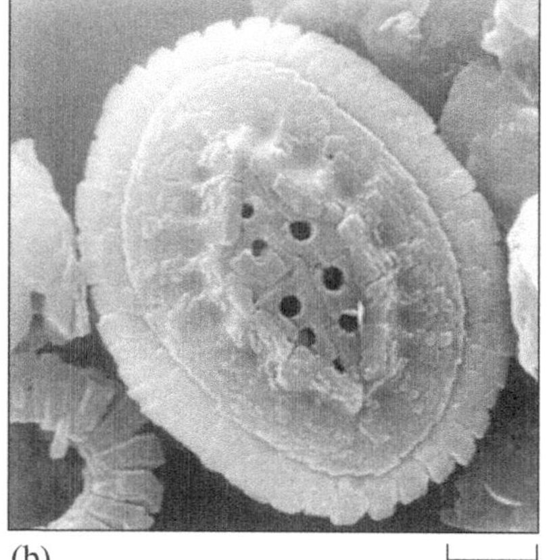

(b)

Figure 7.21 Scanning electron micrographs of coccoliths of the late Cretaceous: (a) a complete cell coating (coccosphere) of coccolith plates; (b) an isolated coccolith from a different species. ((a) J. Young; (b) J. A. Lees)

Middle East and the Mediterranean, however, commonly range between 10 and 100 m Ma^{-1}, with most closer to the lower end of the range. A value of 30 m Ma^{-1} is therefore a reasonable estimate of the average vertical rate of accumulation for the purposes of calculating burial rates, and it is important to note that, at this time, the Tethyan carbonate platforms *alone* were burying carbon (in the form of carbonate) at nearly half the total rate for *all* marine carbonate sediments today.

7.9.4 Other sinks for carbon

Tethyan platforms were by no means the only sites of carbonate deposition in the Cretaceous seas. Neighbouring basins on subsiding areas of continental crust around the Tethys Ocean frequently accumulated fine-grained carbonate sediment, or mixtures of carbonate mud and clay (marl), with the volumes involved also considerable, although less easily quantified.

At mid-latitudes, vast areas of the continents were blanketed by deposits of **chalk**, a distinctively fine-grained limestone. The white cliffs of Dover in England are probably the best-known example of such deposits, but similar thicknesses of chalk of the late Cretaceous may be traced across much of northern France and the Low Countries, into Germany, Poland and Russia. Similar deposits are also present across large tracts of the American Midwest, as well as in western Australia. Chalk is mainly composed of the minute skeletal plates of planktonic unicellular algae called coccolithophores (Figure 7.21), accompanied by the remains of other calcareous planktonic forms, such as foraminiferans (protists), and the remains of various bottom-dwelling organisms. Although calcareous-shelled plankton first evolved some time earlier (at least by the late Triassic), their abundance increased dramatically in the late Cretaceous, creating the extensive thick chalk deposits mentioned above.

As a crude approximation, the European chalks alone have been estimated to have buried carbon at a rate equivalent to the Tethyan carbonate platforms (i.e. 30 m Ma^{-1}). Still more carbonate was deposited in the oceanic realm. What is clear, therefore, is that carbon was being buried in Cretaceous limestones at a rate significantly greater than that of today. Exactly by how much is not yet known because precise estimates have still to be calculated. In addition, no compensating decline in other carbon sinks has been readily detected and instead they too seem to have increased their rates of sequestration.

In addition to the prolific burial of coal on land in high latitudes throughout the Cretaceous, huge amounts of organic carbon were also buried under the sea due to a combination of factors. In contrast to the cold saline bottom waters of today's oceans, the Cretaceous oceans tended to acquire deep reservoirs of warm saline water produced by intense evaporation at the surface (especially in the arid low latitudes). Whereas today's cold bottom waters are relatively well oxygenated, their warm counterparts in the Cretaceous were much less so for a variety of reasons, one being simply the declining solubility of gases as water temperature rises. From time to time, extensive water masses within the Cretaceous oceans became anoxic, allowing abundant organic material to be deposited without being oxidised on the way and, therefore, to accumulate in the bottom sediments. Moreover, the complex geometry of the Tethys Ocean and its surrounding seas created numerous basins with relatively restricted flow, the bottom waters of which were particularly prone to stagnation and anoxia. Indeed, many of the giant oil fields of the Arabian Peninsula contain oil from Cretaceous source rocks that accumulated in such basinal areas within the continent, and which migrated to and became trapped in shallow marine platform limestones that flanked the basins. A detailed analysis of $\delta^{13}C_{carb}$ values in the English Chalk has shown a strong positive shift in values within Upper Cretaceous strata corresponding to the end of the mid-Cretaceous episode of carbonate platform building (see Box 5.1 for a discussion of carbon isotope measurements). A similar change in Italian sequences of the same age suggests a widespread influence from briefly enhanced rates of organic carbon burial.

From the values of $\delta^{13}C_{carb}$ for the Cretaceous, it has been estimated that the rate of marine burial of organic carbon may at times have reached up to three times its present value. The timing of these episodes of organic burial in relation to the growth of the carbonate platforms is by no means simple, and is still not well resolved; there may have been a degree of counter-balancing, with one increasing as the other declined. It is however evident that, at times, both organic burial and carbonate platform growth were occurring synchronously, although in different regions.

Notwithstanding the need to resolve the details of timing, it is already clear that the overall rate of burial of carbon in the Cretaceous greatly exceeded that of today. Where all the extra carbon may have come from poses an interesting problem. Such a high rate of burial of carbon could not possibly have been sustained throughout the period from the Earth's surface reservoirs of carbon alone, without compensation from elsewhere. The atmospheric reservoir (taking today's figure of 0.76×10^{15} kg of carbon) would have been exhausted in a matter of thousands of years; clearly it was not, as the loss of this greenhouse gas would have caused a drop in global temperature, in contrast to the notably warm climatic conditions of the period. Likewise, given the rates of carbon sequestration, the total carbon reservoir in the soil and oceans would only have lasted some tens of thousands of years; that too was not detectably depleted, as there is no shortage of fossil evidence for the healthy continuation of life throughout the Cretaceous. Hence, another source of extra carbon must be sought that could sustain the excessive supply over time throughout the Cretaceous.

7.10 A surfeit of carbon: the key to the Cretaceous greenhouse

7.10.1 *De profundis*

The only remaining large-scale source of carbon available is volcanic outgassing of CO_2. During the Cretaceous, there is widespread evidence in the western Pacific, in particular, for massive volcanism associated with a **mantle superplume**. Unlike rift volcanism, the volcanoes associated with this intraplate source would have reached sea level, where their gases would have been released directly into the atmosphere. The volcanic activity in the Pacific was spread over a long time interval, largely between about 125 and 80 Ma, with a major initial phase in the first 15 Ma followed by a later phase peaking during the last 10 Ma. Carbonate platforms developed extensively, although with episodic drowning, throughout the interval in question. The anomalously high sea levels of the time, particularly of the late Cretaceous, can also be linked with the widespread thermal doming of the ocean floor associated with the volcanism.

7.10.2 The greenhouse atmosphere

If there is a plausible explanation for the hyperactivity of the Cretaceous global carbon cycle, as well as the elevated sea levels of the period, in the form of the 'Pacific superplume', then one vital piece of the jigsaw remains to be put in place: what effect did the enhanced flux of carbon have on atmospheric levels of CO_2 and hence on climate? This is a tricky question because it cannot be assumed that a simple connection exists between the rate of through-flow of carbon in the atmosphere and its level in that temporary reservoir. Yet, it is this temporary reservoir that would have exerted an influence on climate through its greenhouse effect. For example, think of the analogy of running water from a tap into a bucket that has some holes in the bottom. Given a sufficient rate of supply, the water in the bucket is going to fill up to a certain level at which the head of water in the bucket causes the rate of outflow to match that of the inflow. At this point the water level will remain balanced (i.e. in equilibrium). If the rate of inflow is increased, the level in the bucket will rise to a higher level, but increase the size of the holes as well and the rate of outflow will increase, causing the water to drop back to a lower level again if the same rate of inflow is maintained. The rate of through-flow would therefore have increased without necessarily raising the actual water level. The fluctuations in level in this analogy suggest a possible model for the Cretaceous carbon cycle.

In the Cretaceous world, volcanoes may be considered as equivalent to the tap in the above analogy, ultimately supplying the carbon; changes in volcanic activity would have caused fluctuations in the rate of supply. Continuing the analogy, the Tethyan carbonate platforms may be regarded as one of the larger 'holes' for the outflow of carbon and, as you saw in Section 3.3, the carbon would have been deposited as calcium carbonate produced by increased rates of weathering, with the calcium carbonate deposited as a by-product of weathering. The overall surface area of the platforms therefore corresponds to the size of the 'hole'. As described above, the areal extent of the platforms did fluctuate quite markedly with time, with short episodes of widespread platform drowning, punctuated by longer periods of re-establishment and progradational expansion. This pattern implies a degree of necessary lag between any increases in carbon supply (from volcanic CO_2 emissions) and compensatory expansion of the platforms.

What about the reverse, i.e. a decrease in volcanic emissions? Reversing the argument above, it could be expected that CO_2 levels in the atmosphere would drop as carbon was still being deposited. The ensuing climatic cooling (with associated effects) is therefore likely to have been detrimental to the growth of the carbonate-precipitating organisms, allowing atmospheric levels of CO_2 to start climbing. Oxygen isotope studies suggest that some of the major episodes of platform extinctions were associated with climatic cooling. When volcanic emissions rose again, there would have been undercompensation from the platforms, and a new cycle of platform growth would have ensued.

Thus, the lag between volcanic emissions and platform growth (a mismatch between the rates of change of the fluxes supplying and burying carbon) could have led to increased levels of atmospheric CO_2 throughout much of the Cretaceous. Such increased levels would certainly have contributed to climatic warming. The ecosystems that arose in response (such as the Tethyan carbonate platform communities and the polar forests discussed in this chapter) reflect the collective adaptive responses of organisms to the conditions that ensued. They have no equivalents in today's icehouse world, which has its own, equally distinctive, ecosystems.

7.11 Conclusion

Both the greenhouse world of the Cretaceous and the icehouse world of the Permo-Carboniferous reflect the complex process of continuous re-equilibration between the feedbacks of evolving life and the changing Earth. Conditions at the Earth's surface have clearly fluctuated, primarily in response to changes within the Earth itself (e.g. controlling palaeogeographical organisation and superplume activity), but also as a consequence of evolutionary innovations with major environmental impacts (largely upon atmospheric composition, but also upon weathering rates and sedimentation). Other influences have included changes in solar radiation. Overall, no consistent stable state is evident, with each age finding its own unique balance of interacting influences.

Summary of Chapter 7

1 The Carboniferous icehouse world comprised one major continent – Pangaea, part of which lay over the South Pole, creating a 'cap world' configuration with extensive ice sheets in high southern latitudes. The dominant type of plants and animals present at this time were different from today, although the distribution of Carboniferous vegetation was similar, with plant productivity highest in low latitudes. Extensive CO_2 sequestering and coal formation occurred in these equatorial regions.

2 Other factors, such as extensive mountain building, relatively fewer mid-ocean ridges, marine regression, and even lowered levels of solar radiation, contributed to the onset and persistence of globally cool conditions during the Carboniferous, either directly or by reducing atmospheric CO_2 levels.

3 The relative importance of different factors described above, along with the interplay and feedbacks between different processes, must be considered in relation to global cooling. For example, marine regressions occurred in the late Carboniferous and latest Permian, but resulted in two different effects. The first enabled vegetation to expand and colonise newly exposed land, enabling further CO_2 sequestering and cooling, whereas the second led to weathering of previously buried coal, release of CO_2 and enhanced global warming. This can be explained by changes in the relative importance and role of other factors, such as the development of a monsoonal system and changes in the kinds of plants able to colonise these environments.

4 Interpretations are further complicated by uncertainties concerning the timing of events and processes. A complex interplay of interactions and feedbacks is suspected to have been responsible for the icehouse–greenhouse transition.

5 In the Cretaceous, the Equator-to-pole temperature gradient was much smaller than at present, with the difference largely due to much warmer poles.

6 Low latitudes were seasonally arid and, compared with today, there were few areas where rainforest could develop. The low-latitude plants of this period also typically display special adaptations to conserve water.

7 The polar light regime during the Cretaceous was similar to that of the present, with prolonged periods of winter darkness.

8 The Arctic was devoid of permanent ice and supported luxuriant forests dominated by deciduous conifers, with ferns, ginkgos, cycads and some angiosperms also present. These forests were effective carbon-sequestering systems.

9 In high latitudes, delta flood plain accumulations of peat were buried as a result of subsidence and sediment shedding from nearby mountains (e.g. Alaska).

10 Palaeoclimatic determinations cannot be based reliably on relict species such as the cycads, whose modern representatives reflect a small extent of past biological diversity and climatic tolerances. More time-stable techniques based on physiognomy need to be used.

11 Flooding of the continents during the mid-Cretaceous provided broad, shallow seas around the equatorial Tethys Ocean, flanked by predominantly arid lands of low relief. These seas proved favourable sites for the development of extensive carbonate platforms.

12 Cretaceous carbonate platforms differ in structure from present-day tropical platforms because of the effects on the latter of significant glacio-eustatic fluctuations. Unlike today's platforms, the Cretaceous examples typically lacked marginal reefs, with their outer zones dominated by migrating banks of current-swept shell sand and debris. The prolific growth of shelly organisms on these surfaces fuelled the massive carbonate sediment production of the platforms.

13 The rate of burial of carbon in the form of carbonate on the Tethyan platforms alone at times equalled approximately half that in all carbonate sediments (of deep, and shallow, water origin) today. In addition to these platform carbonates, further carbon was also sequestered in the Tethyan basins and mid-latitude chalks. Given the high rates of burial of organic carbon throughout the Cretaceous as well, it has been estimated that carbon was being buried at a significantly higher rate than it is today.

14 The only likely source for this excess carbon in the Cretaceous is increased volcanism associated with the 'Pacific superplume'. Rising levels of CO_2 in the Cretaceous atmosphere can be attributed to increases in volcanic emissions and compensatory growth in the carbonate platforms. Episodic extinction and drowning of the carbonate platforms may have been associated with climatic cooling, brought about by temporary net excesses of CO_2 drawdown.

15 The icehouse and greenhouse case studies highlighted in this chapter reflect the complex process of continuous re-equilibration between the feedbacks of evolving life and the changing Earth, a process that is continuing today.

Learning outcomes for Chapter 7

You should now be able to demonstrate a knowledge and understanding of:

7.1 How a combination of physical, geological, palaeontological and isotopic evidence can be used to determine how and why the Earth's climate has varied over geological time, between the extremes of icehouse and greenhouse conditions.

7.2 The complex processes and interactions that result in the transition between icehouse and greenhouse conditions, as well as an appreciation that the same feedback mechanisms can result in vastly different outcomes in relation to the direction of global climate change (e.g. marine transgressions during the Late Carboniferous and Late Permian periods).

7.3 Why changes in physiognomy rather than the nearest living relative technique, is a more reliable method of determining palaeoclimatic conditions in pre-Quaternary sequences, and how this in turn has been used to investigate global climatic conditions during the Cretaceous greenhouse.

7.4 The processes that permitted the formation of extensive carbonate sequences during the Cretaceous, as well as how and why these differ structurally from modern-day carbonate reefs.

7.5 The roles of, and interplay between, plume-related volcanism and carbonate platform formation in altering atmospheric CO_2 levels, which in turn resulted in climatic warming during the Cretaceous.

Chapter 8

End-of-book summary

Returning to the general theme of this book: what is the fundamental nature of the relationship between evolving life and the Earth – benign partnership, or a chaotic system lurching from one temporary state of balance to another?

8.1 Possible worlds

There are several possible answers to this question. At one extreme, you might postulate that the role of life has been purely passive, with no significant effect on conditions at the Earth's surface: organisms have merely adapted to changes dictated by the Earth's physical and chemical state. At the other extreme, the Earth, together with the life it supports, might be seen as a kind of 'superorganism' that has evolved a tightly coupled system of feedbacks ensuring that 'the Earth's surface environment is, and has been, regulated at a state tolerable for the biota'; this is the 'Gaia hypothesis' of James Lovelock (see Box 2.4). Between these two extremes is the view that, while life has significantly altered conditions at the Earth's surface relative to what a lifeless Earth would have been like, the feedbacks concerned are not tightly coupled in any manner akin to the homeostatic mechanisms of an individual organism: at most, they may contribute, along with the Earth's abiotic processes, to temporary equilibria. Nevertheless, these equilibria are not stable over the long term, being subject to alteration due to changes in the balance of feedbacks both from the Earth and from evolving life itself, including the effects of sporadic major environmental perturbations.

Before reading further, you may wish to ponder these three possibilities in the light of what you have learned so far, to see which you consider is the most plausible view. Consider, first, whether or not the presence of life has had any significant effect on conditions at the Earth's surface. If so, consider whether or not such influences have consistently tended to regulate conditions in a state 'tolerable for the biota'.

8.2 Review of the options

The first option – life as a purely passive passenger on Earth's voyage through time – can probably be discounted as the scale of life's impact on the Earth has been repeatedly stressed throughout this book. That leaves the nature of such feedbacks between life and the Earth to be considered – is the interaction Gaian, or is life merely a contributor to complex, but essentially chaotic, biogeochemical cycles?

Consider, first, the expectations of the theory of evolution by natural selection. This theory emphasises selection at the level of the individual, not at the level of entire systems, let alone the global biosphere.

If natural selection theory thus provides no grounds for the Gaia hypothesis, what, then, of the empirical evidence? Does the geological record support the proposition that 'the Earth's surface environment is, and has been, regulated at a state tolerable for the biota'? Might such self-regulation have come about by means other than evolution by natural selection? The first problem in tackling this question is the woolliness of the proposition itself. Which 'biota' is being talked about? You have seen that the Earth's biota has changed quite considerably over time. If you had been able to question, for example, the anaerobic denizens of the Archean oceans, to whom today's levels of molecular oxygen would have been toxic, or the high-latitude forests of the Cretaceous Period, whose domains are now icy wastes, on the issue, their responses would not have been wholeheartedly supportive of the Gaia hypothesis. Nor, apparently, were these life forms the unfortunate victims of mass extinctions, which might be blamed on extraneous perturbations: their environments simply changed beyond their limits of tolerance and they were eventually written out of the drama (or, in the case of the anaerobic microbes, literally went underground, or into the guts of animals including humankind, where their descendants continue to thrive in anoxic regimes). Obviously, then, the reference to the biota in the proposition is not intended to be comprehensive: presumably *any* manifestation of life is intended.

The observation that life in general appears to have persisted at least from early Archean times is trivial, in this respect, as it is merely consistent with, but does not necessarily confirm, the Gaia hypothesis: it does not disprove the alternative model that life has managed to continue participating in a chaotic system that lurches between temporarily equilibrating states. It could simply be the case that the 'lurching' has not, so far, transgressed the limits of tolerance of all life forms: some form of life has always been able to survive, furnishing successive ages with appropriately adapted organisms. In other words, there has always been somewhere, so far, where DNA has proved robust.

So, does the geological record suggest any *predominant* tendency towards the establishment and stabilisation of optimal conditions? Demonstration of the mere existence of some feedbacks having this effect would not be enough to settle the issue, as some such effects could be expected in the chaotic model – by chance alone. The Gaian model explicitly proposes a coherent integration of feedbacks, together yielding homeostatic self-regulation. A suitable test, therefore, is to see whether such a pattern has been predominant over geological time. If it has not – with the feedbacks being sometimes stabilising, sometimes not so – then the Gaia hypothesis should be rejected.

Consider, for example, the various feedbacks involved in regulating Permo-Carboniferous atmospheric composition and hence climates, which were discussed in Chapter 7. Do these appear to have consistently stabilised a particular set of conditions? Or were their effects variable – sometimes maintaining a given state, and sometimes helping to install a different regime?

Again, in the Cretaceous, the extensive carbonate platform biota, as well as the high-latitude land flora, did not enjoy uninterrupted exploitation of the enhanced supply of atmospheric CO_2 and the associated warm climatic conditions. It seems that they were poorly suited to respond to hiccups in the supply, as they continued to draw down carbon at excessive rates, and so perhaps contributed to episodic crises involving climatic cooling and eventual drowning of the platforms themselves. Thus, far from consistently stabilising optimal conditions for the incumbent biota, the Cretaceous Earth–life system became implicated in important environmental fluctuations, which were attended by extinctions on a variety of scales. The same story emerges for any geological period when investigated in sufficient detail.

Of the two models that have been considered, that of a complex, essentially chaotic, system seems to be the more plausible. The Gaia hypothesis is supported neither by evolutionary theory nor by the empirical evidence of the geological record.

However, it is worth pondering how the fluctuations in conditions remained within the limits of tolerance of living organisms, and why present-day conditions seem to be so well suited to the living biota. It is, in fact, easier to answer the second question first, as it is really a trick question, but tackling it helps a little way towards answering the first, and it also raises the crucial issue of rates of change.

While the 'rapid' fluxes of biogeochemical cycles that are largely mediated by ecological interactions may balance out over the short term, slower geological and evolutionary processes alter such equilibrium states over longer timescales. Atmospheric oxygen is a case in point, the nice mutual adjustment of major sources and sinks (e.g. photosynthesis and respiration) over the timescales of human observation providing a telling contrast to the dramatic long-term changes of the past. As long as these slower changes do not outstrip the limits of tolerance conferred on individuals by available genes, evolution can deliver suitably adapted organisms. Thus, to marvel at the appropriateness of conditions for life is simply to put the cart before the horse: it is the evolutionary responsiveness of organisms that has ensured the fit. With a little bit of adaptation, today's problem can even become tomorrow's manna. This is known as the 'helpful stress effect'. In recent times, rats have provided a good example of such a change of status in their evolution of resistance to the pesticide warfarin. Warfarin is poisonous to normal rats because it suppresses blood-clotting (leading to fatal bleeding) and interferes with the uptake of vitamin K. Soon after its introduction, rats resistant to the pesticide began to appear and to replace the ordinary warfarin-sensitive strains. However, the frequency of the resistant forms rapidly declines when use of the pesticide is stopped. Thus warfarin-resistance evidently bears a cost to fitness under normal circumstances (probably related to an increased vitamin K requirement), such that the warfarin-resistant strains positively depend on the presence of the pesticide in order to survive.

Stark testimony to the importance of this principle in the history of life is provided by instances where rates of environmental change have evidently outstripped the capacity for evolutionary response of organisms: the results have been dramatic – extinction, and in cases where such perturbations were widespread and pervasive in their effects, mass extinction. Thus the *rate* of an environmental change, perhaps rather more than its nature, is crucial to its effect: be it slow enough, and it provides new opportunities for adaptation; be it too fast, and it becomes a catastrophe, leaving a trail of extinctions in its wake. Emissions of CO_2 associated with major volcanic episodes seem to illustrate the contrast between these two effects. Brief but massive eruptions of flood basalts may well have been implicated in some mass extinctions. Yet those associated with the postulated Pacific superplume of the Cretaceous, which were erupted on an even greater volumetric scale, but over a longer period, helped to create the conditions in which the spectacular carbonate platform ecosystems and high-latitude forest ecosystems of the period thrived.

The other question raised above was: how did the fluctuations in conditions remain within the limits of tolerance of living organisms? To some extent, the progressive evolutionary tracking of changing conditions discussed above can help to explain this. The limits of tolerance have themselves shifted in some respects. Had the physical and chemical conditions of today's world been miraculously switched on in, say, the Archean or the early Proterozoic, the effect on life then would certainly have been devastating, perhaps even fatal. Nevertheless, that is certainly not the whole story: the limits within which mean global surface temperatures appear to have remained, for example, are impressively modest. It is a curious but characteristic feature of complex systems, involving myriads of feedbacks between component parts, that they tend to settle within relatively narrow limits of physical and chemical states. The numerical modelling of such behaviour involves a difficult branch of mathematics – chaos theory – which is beyond the scope of this book. An important point to note, however, is that such sets of conditions ('strange attractors', as they are known) are far from inviolable. As you have already seen, they lack the protection of co-adapted homeostatic mechanisms characteristic of individual organisms. They are indeed quite sensitive to initial conditions, and unpredictable shifts to new states can result from small but critical changes in the components of the system. Possibly, then, mean global temperature may have been broadly constrained in this manner, although evidently it fluctuated between the contrasting greenhouse and icehouse modes. Perhaps, moreover, we have just been lucky, in that the early removal of carbon from the Earth, with the formation of the Moon, followed by the sustained drawdown of CO_2 from the atmosphere, removed the risk of a runaway greenhouse effect as the Sun warmed. There is no satisfactory answer to this question at present.

8.3 Conclusion

In summary, it is clear that evolving life has profoundly influenced conditions at the Earth's surface over time. Yet it is equally apparent that the feedbacks between Earth and life have interacted over the long term in a chaotic, undirected manner, lacking the kind of tightly coupled self-regulation that may be admired in individual organisms. Life on Earth is a risky game, then, but with so many players around, there have always been sufficient winners to keep it going, so far. Some puzzlingly 'benign' aspects nevertheless still stand out, such as the relative stability of mean global surface temperatures. It is still too soon to say whether this can be put down to luck, or to some, as yet poorly understood, system of global stabilisation.

Answers to questions

Question 1.1

(a) (i) At 50° N, the incoming solar radiation on 21 March (the spring equinox) is about 10^7 J m^{-2}. It increases to a maximum of 2.3–2.4 × 10^7 J m^{-2} in late June (around the time of the summer solstice/longest day), then declines to about 0.3 × 10^7 J m^{-2} towards the end of December (around the time of the winter solstice or shortest day), after which it begins to rise again.

(ii) Watts are joules per second; therefore, to convert J m^{-2} (incoming solar energy per unit area per day) to W m^{-2} (average solar power) you must divide the contour values by the number of seconds in a day, i.e. 8.64 × 10^4 s. The contour values would therefore range from:

$$\frac{0.4\times10^7\ \text{J m}^{-2}}{8.64\times10^4\ \text{s}} \text{ to } \frac{2.5\times10^7\ \text{J m}^{-2}}{8.64\times10^4\ \text{s}} \text{ or } \sim46 \text{ to } 289\ \text{J s}^{-1}\ \text{m}^{-2}$$

which is 46–290 W m^{-2} (to 2 sig. figs).

(b) Mid-latitudes (between ~30° N and S) receive the most solar radiation at any one time: > 2.5 × 10^7 J m^{-2} in summer. This is because of the long days at this time of year, when the noonday Sun is high.

Question 1.2

(a) (i) Approximately 10–20% of visible radiation is absorbed in the atmosphere, with absorption at the red end of the spectrum being greater than that at the blue/violet end. The gas responsible is ozone.

(ii) About half of incoming infrared radiation is absorbed by the atmosphere. The gases mainly responsible in this case are water vapour and carbon dioxide.

(b) Outgoing longwave radiation is absorbed mainly by water vapour and carbon dioxide, with lesser absorption by methane, nitrous oxide and ozone.

Question 1.3

(a) The main and most obvious reason is that more solar radiation reaches the Earth's surface at low latitudes than at high latitudes, for reasons demonstrated by Figure 1.4. The second reason is that, of the solar radiation that reaches the Earth's surface, much more is reflected at high latitudes than at low latitudes, particularly because of the high albedos of ice and snow (Table 1.1).

(b) The Earth–atmosphere system has a net gain of heat between about 40° S and 35° N (Figure 1.45 overleaf); polewards of those latitudes, there is a net loss of heat.

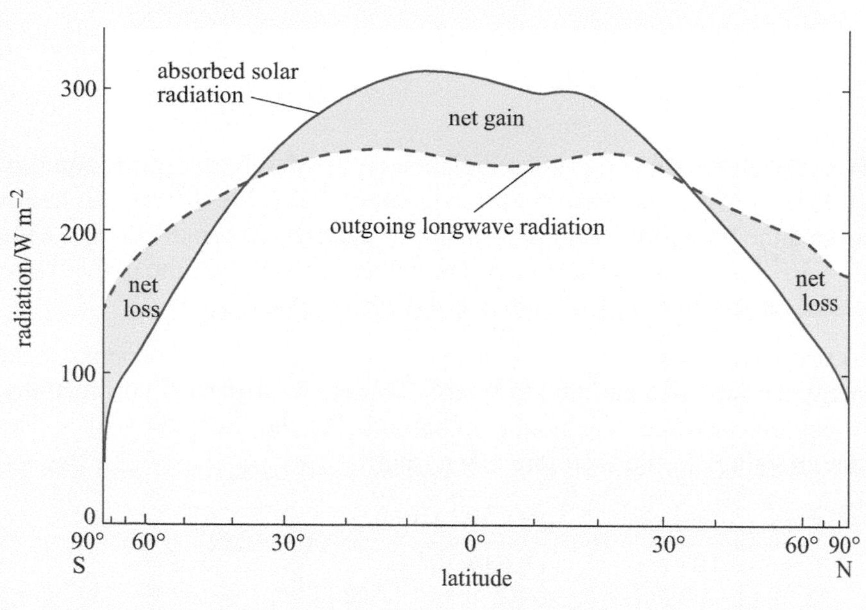

Figure 1.45 Variation with latitude of the solar radiation absorbed by the Earth–atmosphere system (solid curve) and the outgoing radiation lost to space (dashed curve). Values are averaged over the year, and are scaled according to the area of the Earth's surface in different latitude bands.

Question 1.4

If during the Little Ice Age average global temperatures were about 1 °C lower, and this were attributable entirely to a decrease in solar luminosity, there would have to have been a reduction in incoming solar radiation of about:

$$\frac{1\,°\text{C}}{0.75\,°\text{C per W m}^{-2}} \approx 1.3\ \text{W m}^{-2}$$

because the climate sensitivity is given as the change in temperature per unit change in forcing – thus the change in forcing is the change in temperature divided by the sensitivity.

Expressed as a percentage of the present-day average value:

$$\frac{1.3\ \text{W m}^{-2}}{343\ \text{W m}^{-2}} \times 100 = 0.389\% \approx 0.4\%\ \text{(to 1 sig. fig.)}.$$

Question 1.5

(a) Generally, winds blow from regions of high pressure to regions of low pressure. This is most clearly seen in the case of the Trade Winds blowing towards the zone of low surface pressure along the Equator. In addition, winds blow clockwise around high-pressure regions in the Northern Hemisphere and anticlockwise around them in the Southern Hemisphere; such flow is referred to as anticyclonic. Winds blow in the opposite direction (i.e. are cyclonic) around low-pressure regions.

(b) In regions of high surface pressure, vertical air motion will result in the air sinking whereas in regions of low pressure the air will be rising.

Question 1.6

(a) The ITCZ is (mostly) further north in July and further south in January; in other words, it moves into the hemisphere experiencing summer, with this being particularly prevalent over large land masses. This would be expected because the ITCZ, which is an area of vigorous upward convection, tends to be located over the warmest parts of the Earth's surface and, as discussed in connection with Figure 1.14, continental masses heat up much faster than the oceans in summer and cool down much faster than the oceans in winter.

(b) In July, when the ITCZ is at its most northerly position, the winds over tropical West Africa are mainly southwesterly (i.e. from the southwest). In January, when the ITCZ is at its most southerly position, the winds over the region are mainly easterly or northeasterly (i.e. from the east or northeast).

(c) The north–south shift in the position of the ITCZ is most marked over the northern Indian Ocean and over the western tropical Pacific and South East Asia.

Question 1.7

The value needed from Box 1.4 is the specific heat of water, 4.18×10^3 J kg^{-1} °C^{-1}; and 1300 km^3 is 1300×10^9 m^3, which has a mass of 1300×10^{12} kg. The total amount of heat given up each day, on average, is:

$$\begin{aligned}\text{heat given up each day} &= \text{specific heat} \times \text{mass of water} \times \text{temperature drop}\\ &= (4.18 \times 10^3\ \text{J kg}^{-1}\ °\text{C}^{-1}) \times (1300 \times 10^{12}\ \text{kg}) \times 11\ °\text{C}\\ &= 5.9774 \times 10^{19}\ \text{J} = 6.0 \times 10^{19}\ \text{J (to 2 sig. figs).}\end{aligned}$$

Therefore, ~6×10^{19} joules of heat are released to the atmosphere over the northern North Atlantic every day.

Using the same conversion as in Question 1.1 (1 J s^{-1} = 1 W), 6×10^{19} J day^{-1} is about 0.7×10^{15} W, which is equivalent to the heat output of about half a million large power stations. In the winter months, the heat lost to the atmosphere would be even greater than this average value.

Question 1.8

(a) The percentage of the Earth's water presently in the ocean:

$$= \frac{1\,322\,000 \times 10^{15}\ \text{kg}}{1\,360\,000 \times 10^{15}\ \text{kg}} \times 100 \approx 97\%.$$

The percentage of the Earth's water presently in ice caps and glaciers:

$$= \frac{29\,300 \times 10^{15}\ \text{kg}}{1\,360\,000 \times 10^{15}\ \text{kg}} \times 100 \approx 2\%.$$

(b) Even if all the ice caps and glaciers melted, the volume of water in the oceans would increase by only about 2%, so if a quarter of them melted, it would increase by 0.5%.

This may not sound much, but it would mean a rise in sea level of about 15 m. Given that a large proportion of human habitations are concentrated around the coasts, such a sea-level rise would be disastrous.

Question 1.9

(a) Figure 1.6 shows that in December only low levels of solar radiation reach high latitudes in the Northern Hemisphere, which is insufficient to allow phytoplankton populations to grow. By May, light levels have risen considerably in the Northern Hemisphere – and the image in Figure 1.43a illustrates the burst in primary productivity, known as the 'spring bloom'.

(b) This region is the centre of the Atlantic subtropical gyre, driven by anticyclonic winds. As discussed in the text in connection with Figure 1.36, such gyres are regions of downwelling, and hence can only support limited primary productivity.

By contrast, the cyclonic subpolar gyres are regions of divergence and upwelling (Figure 1.36bii), which is another reason why primary productivity can be high there once light levels rise in spring.

Question 1.10

Increased primary productivity would mean bigger plankton blooms and a greater flux of dimethyl sulfide to the atmosphere. This in turn could lead to more cloud formation, resulting in an increase in the Earth's albedo.

Question 2.1

(a) In Equation 1.2 (photosynthesis), oxygen is being liberated by breaking C—O bonds in CO_2. In addition the carbon in CO_2 is linked to two oxygen atoms, whereas in organic matter (glucose in Equation 1.2) each carbon is linked to other carbon atoms and hydrogen atoms as well as oxygen atoms. This reaction is therefore a reducing reaction.

(b) In respiration (Equation 2.1) the reverse process is going on, in which the C—C and C—H bonds in glucose are being broken and the hydrogen and carbon are combining with oxygen to produce CO_2 and water. This is therefore an oxidising reaction. (However, you should note that the oxygen molecules are being broken up and recombined with C and H so, as far as the oxygen is concerned, the reaction is reducing.)

Bonds with oxygen are very stable and require a large amount of energy to break them so photosynthesis requires an input of energy from the Sun to take place – it is an endothermic reaction. By contrast, the oxidation of glucose releases energy because C—C and C—H bonds can be broken relatively easily. The reaction is therefore exothermic.

Question 2.2

(a) According to Figure 2.3, the amount of carbon in terrestrial plant material is about 560×10^{12} kgC and the net fixation of carbon from the atmosphere (i.e. the net annual input of carbon to the reservoir of living plants) is 120×10^{12} kgC y^{-1}. The mean residence time for carbon in plant biomass is therefore:

$$\frac{560 \times 10^{12} \text{ kgC y}^{-1}}{120 \times 10^{12} \text{ kgC y}^{-1}} \approx 4.5 \text{ years (to 2 sig. figs).}$$

The amount of carbon deposited in soil is also 60×10^{12} kgC y^{-1}, while the amount of carbon in the soil reservoir at any one time is about 1500×10^{12} kgC (Figure 2.3). The mean residence time for carbon in soil organic matter is therefore:

$$\frac{1500 \times 10^{12} \text{ kgC}}{60 \times 10^{12} \text{ kgC y}^{-1}} = 25 \text{ years.}$$

(b) You could expect short residence times because small numbers divided by large numbers give even smaller numbers.

Question 2.3

(a) (i) The annual flux of carbon into tropical rainforests is the total amount of carbon fixed in them annually, i.e. their total net primary productivity, which is $(17.0 \times 10^6 \text{ km}^2) \times (0.9 \text{ kgC m}^{-2} \text{ y}^{-1}) = 15.3 \times 10^{12} \text{ kgC y}^{-1}$. (Having now calculated this value, you can complete the last column of Table 2.1.)

(ii) The percentage of total global net primary productivity contributed by the rainforest is:

$$\frac{15.3 \times 10^{12} \text{ kgC y}^{-1}}{48.3 \times 10^{12} \text{ kgC y}^{-1}} \times 100 = 31.68\% = 31.7\% \text{ (to 3 sig. figs).}$$

As you will probably have noticed in working out your answer, this high percentage is partly due to the large area of the globe occupied, even now, by tropical rainforest. (Note, however, that in doing this calculation, it is assumed that the area of the globe covered by rainforest remains constant, whereas in reality it is decreasing.)

(b) (i) To make a comparison area for area, the column headed 'Mean NPP per unit area' must be used. The value for tropical evergreen forests is 0.900 kgC m^{-2} y^{-1}, and that for boreal forests is 0.360 kgC m^{-2} y^{-1}, so tropical evergreen forests are:

$$\frac{0.900 \text{ kgC m}^{-2} \text{ y}^{-1}}{0.360 \text{ kgC m}^{-2} \text{ y}^{-1}} = 2.5 \text{ times more productive than boreal forests.}$$

(ii) One factor to be considered is rainfall. Tropical forests grow at low latitudes where there is heavy rainfall associated with the ITCZ. However, boreal forests grow in the subpolar regions where precipitation is also fairly high, so this cannot be the main reason. Another important difference between the two areas is the amount of light they receive during the year: tropical regions have high daytime light levels all year, but subpolar regions have very low light levels (or are dark) for much of the year and hence have

a very short growing season. (*Note*: because of length of the growing season, even temperate savannah and tall grassland are more productive than boreal woodland and boreal forest.) The most important reason, however, is that growth rates are greatly affected by temperature.
At the present time, this is the factor that limits terrestrial primary productivity at high latitudes.

(iii) Residence time in living plant material is given by the total plant biomass divided by the total net primary productivity. Therefore, according to Table 2.1, the residence time of carbon in swamps and marshes is:

$$\frac{13.6 \times 10^{12}\ \text{kgC}}{2.2 \times 10^{12}\ \text{kgC y}^{-1}} = 6.18\ \text{y} \approx 6\ \text{years.}$$

In boreal forests the residence time is:

$$\frac{108 \times 10^{12}\ \text{kgC}}{4.3 \times 10^{12}\ \text{kgC y}^{-1}} \approx 25\ \text{years}$$

which is effectively the average lifetime of a tree. The residence time in boreal forests is approximately four times that in swamps and marshes.

(Again, for this question, it is assumed that the areal extents of swamps and marshes, and of boreal forests, remain constant.)

Question 2.4

(a) The reverse reaction for photosynthesis is that for respiration (Equation 2.1).

(b) The reverse reaction for carbonate weathering is carbonate precipitation (Equation 2.8).

Question 2.5

(a) The values on the right-hand axis are all negative and, as explained in the caption, negative values correspond to CO_2 concentrations in the atmosphere being higher than those in surface waters. This means that the situation is out of equilibrium, with a concentration gradient across the air–sea interface and a net flux of CO_2 *into* the ocean (Figure 2.8).

(b) As it was the time of the spring bloom, the phytoplankton were multiplying and fixing carbon (i.e. there was high net primary productivity). Where there were more phytoplankton in the surface water (i.e. chlorophyll concentrations were high), more carbon was being fixed, causing more CO_2 gas to enter surface waters from the atmosphere; where there were fewer phytoplankton, the reverse was true.

(As you might have realised, the phytoplankton – and indeed zooplankton feeding on them – would also have been respiring, releasing CO_2 into the water; but at times of high primary productivity, sufficient organic debris would be falling out of the surface layers to drive a net flux of CO_2 into the ocean.)

Question 2.6

(a) In the northeastern North Atlantic (as well as at high southern latitudes), the net flux of CO_2 is into the ocean. This is a region where deep-water formation is occurring as a result of intense cooling of surface water (and brine-rejection) (Figure 1.33). Cold water can take up a relatively large amount of CO_2 before becoming saturated and, on sinking down from the surface, would allow yet more CO_2 to dissolve.

(b) At both seasons of the year there is a flux of CO_2 into the ocean in the central North Atlantic, partly because surface waters converge and sink there (Figure 1.37). By April–June, however, the water is being warmed and so the region where there is a net flux of CO_2 out of surface water is beginning to spread northwards. In addition, at this time of year there is high net primary productivity in the northern part of the ocean (Figure 1.43a), which also contributes to a net flux of CO_2 from air to sea.

(c) In low latitudes, water is upwelling to the surface and warming. Both the decrease in pressure and the increase in temperature force CO_2 out of solution, so that there is a net flux from sea to air (despite any high primary productivity along the Equator that would tend to draw CO_2 down into the ocean).

Question 2.7

(a) According to Table 2.2, the total mass of carbon in marine plant material is 1.76×10^{12} kgC, i.e. about 2×10^{12} kgC. This is a small percentage of the standing stock of biomass on land, whether the value of 560×10^{12} kgC (from Figure 3.3) or 827×10^{12} kgC (from Table 2.1) is used. In the first case, the answer is ~0.36%, in the second it is ~0.24%.

(b) If an average of the input and output fluxes shown in Figure 2.19 is used, the residence time for carbon in living phytoplankton is:

$$\frac{2 \times 10^{12} \text{ kgC}}{38 \times 10^{12} \text{ kgC y}^{-1}} \approx 0.05 \text{ years or about 19 days.}$$

By contrast, the residence time of carbon in the terrestrial biomass reservoir is about nine years (Question 2.2a), so carbon cycles much faster through the marine biomass reservoir than through the terrestrial biomass reservoir.

Question 2.8

The statement is only partly true. Both limestone and shells or skeletal remains (which eventually become limestones) are forms of inorganic carbon. They are precipitated from dissolved inorganic carbon in seawater (mainly HCO_3^-), and are not composed of large organic molecules made up of carbon, hydrogen and oxygen.

Question 2.9

(a) Your answer may include any two of the following ways in which greenhouse warming could indirectly affect the global carbon cycle.

If the Earth's surface became warmer, evaporation of water (from both the sea-surface and the land) would increase, there would be more water in the atmosphere, and so precipitation would increase; in other words, the hydrological cycle would become more active. (To some extent, this is being seen at the present time, in the form of more droughts at lower latitudes and increased rainfall in subpolar latitudes.)

As a result of warmer, wetter conditions, terrestrial biological productivity would increase (consider how you encourage plants to grow). This would fix more carbon from the atmosphere, supply more to soil in organic debris and perhaps increase the rate at which carbon is preserved on land (e.g. in swamps). Marine productivity might also increase as a result of higher sea-surface temperatures (although other factors might counteract this). (There is also some (inconclusive) evidence that an increase in the concentration of atmospheric CO_2 increases net primary productivity.)

Warmer, wetter conditions would also increase rates of chemical weathering, particularly if plant growth were increased. Increased weathering of silicates would result in an increase of the net flux of carbon to the ocean (Figure 2.21).

(b) Increased weathering of silicates and increased primary production on land and in the sea both result in increased rates of sedimentation of calcareous and organic sediments in the ocean. However, although an increased flux of organic-rich material is likely to result in more rapid burial and preservation of carbon, the same cannot necessarily be said for inorganic carbon. This is because an increase in the concentration of CO_2 in the atmosphere and increased primary productivity would both lead to an increase in dissolved inorganic carbon in the deep ocean, and hence an increase in the acidity of deep waters and an increase in the rate of dissolution of calcareous remains (i.e. a deeper carbonate compensation depth).

You were not expected to think of the last aspect; it has been included to illustrate the complexity of the global carbon cycle and how difficult it is to make predictions about the results of changing fluxes in any part of the cycle.

Question 2.10

(a) The fluctuations in atmospheric CO_2 concentration are a result of the uptake of CO_2 by plants during photosynthesis in spring and summer, i.e. removal of carbon from the atmosphere and its fixation in living plant material. (Note that it is the lows that correspond to spring and summer, and the highs that correspond to winter.)

(b) It is a very evocative description, but rather misleading. Breathing – effectively *respiration* – involves the uptake of oxygen and the release of carbon dioxide (Equation 2.2). Respiration of biomass continues all year round (although more so in the spring and summer), but high rates of primary production and hence *net* primary productivity (production minus respiration) *only* occur in spring and summer. Thus (as discussed in (a)), the pattern is primarily a manifestation of photosynthesis rather than of respiration.

(c) The pattern is dampened in the Southern Hemisphere because primary productivity per unit area in the ocean is much less than on land (e.g. compare the columns for 'Mean NPP per unit area' in Tables 2.1 and 2.2), and the Southern Hemisphere is largely ocean.

Question 3.1

(a) As shown in Figure 1.20, the tropospheric wind systems of the two hemispheres are to a large extent separate, meeting at the Intertropical Convergence Zone (ITCZ). For example, air being carried northwards across the Equator by the Trade Winds is likely to rise at the ITCZ and then be carried southwards again, rather than be carried northwards to the Northern Hemisphere.

(b) As shown schematically in Figure 1.20, the tropopause (i.e. the boundary between the troposphere and the stratosphere) is higher at low latitudes than at high latitudes. As discussed in the caption for Figure 1.19, it is generally at a height of about 17 km in low latitudes but only 8–10 km high at high latitudes. As a result, a volcano is more likely to be able to send gases (and aerosols) into the stratosphere at high latitudes than at low latitudes.

Question 3.2

If the gases produced by an eruption do not get into the stratosphere, any aerosols are formed in the troposphere and are soon washed out. Even the biggest explosive volcanic eruptions – such as that of Mount Pinatubo in 1991 – are short-lived, and their aerosols remain in the stratosphere for only a year or two. Thus, although their short-term effects may be profound, the climate forcing they cause is not sustained for sufficiently long for the climate system to respond. For this to happen, feedback loops (e.g. involving growth of vegetation or spread of ice sheets) have to have time to become established, over periods of tens or hundreds of years.

Flood basalt eruptions, however, may be exceptional in that they involve the steady effusion of sulfur-rich magmas over periods of decades or more.

Question 3.3

Table 3.3 Answer to Question 3.3.

Flood basalt or large igneous province	Mean age (Ma)	± (Ma)	Stage or epoch boundary at which extinction occurred	Age (Ma)	± (Ma)
Emeishan Traps	259	3	Guadalupian–Lopingian (M)	260.4	0.7
Siberian Traps	250	1	Permian–Triassic (M)	251	0.4
Central Atlantic Magmatic Province	201	1	Triassic–Jurassic (M)	199.6	0.6
Karoo Ferrar	183	2	Pliensbachian–Toarcian (E)	183	1.5
			Bajocian–Bathonian	168	3.5
			Tithonian–Berriasian (E)	145	4
Parana Etendeka	133	1	Valanginian–Hauterivian	136.4	2
Ontong–Java 1	122	1	Early Aptian (E)	125	1.0
Rajmahal/Kerguelen	118	1			
Ontong–Java 2	90	1	Cenomanian–Turonian (E)	93.5	0.8
Caribbean Plateau	89	1			
Madagascar Traps	88	1			
Deccan Traps	65.5	0.5	Cretaceous–Palaeogene (M)	65.5	0.3
North Atlantic 1	61	2			
North Atlantic 2	56	1	Paleocene–Eocene (M)	55.8	0.2
Ethiopia and Yemen	30	1	Oil event	30	2.5
Columbia River	16	1	Early Miocene–Mid Miocene (E)	16.0	0.1
			Serravallian–Tortonian	11.6	0.3
			Pliocene–Pleistocene (E)	1.81	0.02

The eight extinction events and continental flood basalt–large igneous province (CFB–LIP) ages highlighted in red show a high degree of correlation in that their ages are within the uncertainties of each other. Other CFB–LIPs are not associated with any extinction events and some extinctions are not closely associated with known CFB–LIPs. These coincidences suggest a link between basalt eruptions and some extinctions but, clearly, extinctions can have other causes.

Question 3.4

(a) Given that they occur at intervals of hundreds of millions of years (and the whole globe is affected), ice ages would plot in the upper part of the right-hand bubble (Figure 3.21).

(b) If major ocean basins open and close on timescales of 100–200 millions of years, then the timescale on which the disposition of continents over the surface of the globe changes significantly is of the same order as the timescale at which ice ages come and go.

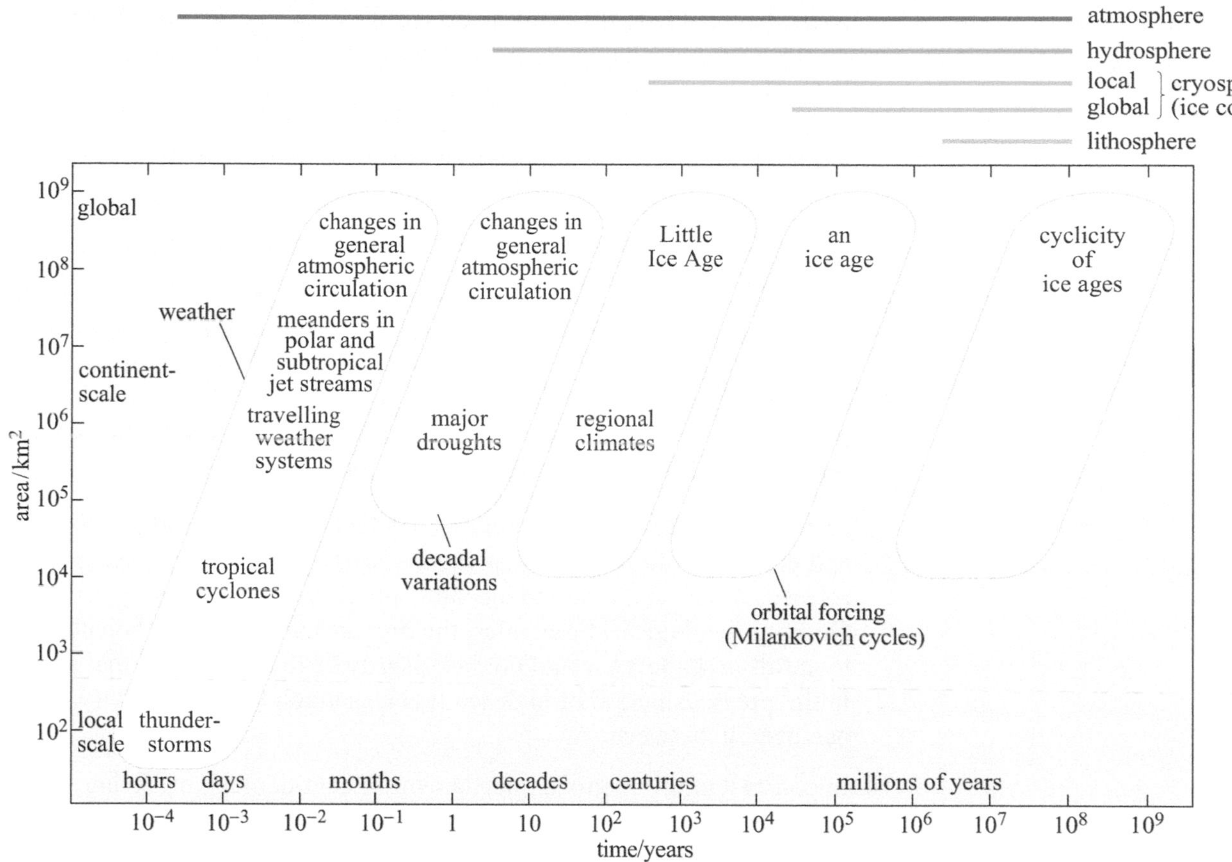

Figure 3.21 Answer to Question 3.4a.

Question 3.5

(a) Changing sea levels may open or close oceanic gateways to surface (or even deep) currents. This could influence how much heat is carried polewards and thus affect the climate as a whole.

(b) Generally speaking, during periods of high sea level, the Earth's average albedo is reduced, as sea-surfaces generally have a lower albedo than land. More solar radiation is absorbed, which would reinforce any global warming trends. Conversely, falling sea level exposes more land that has a higher albedo and would encourage global cooling.

Note, however, that this argument is somewhat oversimplified, as both land and (especially) sea have a greater albedo for low sun elevations (i.e. at high latitudes).

Question 3.6

(a) (i) As continental shelves have a flat topography, relatively small changes in sea level can result in relatively large areas being flooded or exposed. Assuming sea level is starting from a low level, a rise in sea level would initially result in a significant increase in the area of low-lying, swampy coastal land, where organic remains (e.g. peat and coal) could be preserved. As discussed in Chapter 2, preserved organic carbon is a long-term sink for carbon, which would remove carbon dioxide from the atmosphere on geological timescales. Thus, an increase in sea level could lead to a decrease in the concentration of atmospheric carbon dioxide and global cooling.

Conversely, falling sea levels could expose organic remains to oxidation and erosion, returning carbon dioxide to the atmosphere and encouraging global warming.

(ii) A rise in sea level would increase the area of shallow seas, where shallow-water carbonate-secreting organisms (such as certain algae, bivalves and corals) could flourish. Accumulation of these carbonate remains represents a long-term sink for carbon, removing CO_2 from the atmosphere on geological timescales and leading to global cooling.

Carbonate rocks exposed by falling sea levels would be subject to oxidation and erosion in the same way as organic carbon and with the same result: release of carbon dioxide to the atmosphere and subsequent global warming. On long (geological) timescales, the organic carbon and carbonates exposed by a fall in sea level would not necessarily be the same ones that accumulated in the previous period of high sea level, as much tectonic activity might have occurred in between.

(b) In contrast to the chemical mechanism of continental weathering, the effects described for (i) and (ii) (both of which involve parts of the biosphere) are mechanisms for negative feedback, and therefore act to bring about stability in the climate system.

Question 4.1

Condensation of moisture to form rain releases latent heat originally taken up from the ocean as latent heat of evaporation. Thus, air that is rising because it has been warmed by the underlying continent (Figure 1.29b) has an extra heat source, and rises more vigorously, intensifying the low pressure over the region and drawing in moist air from the south even more strongly.

Question 4.2

(a) There have been a number of large fluctuations in the relative abundance of *G. bulloides*, but the first main change occurred between about 9 Ma and 7.5 Ma, when the relative abundance increased rapidly from <1% to >50%. In more recent times, the proportion has fluctuated, but has never been reduced to less than what it was at 8.5 Ma (with the exception of the extreme low at about 5.5 Ma).

(b) The increase in the relative abundance of *G. bulloides* at 7.5–9 Ma could well indicate an increase in upwelling of nutrient-rich water. As shown in Figure 4.10b, at present, upwelling and high levels of primary productivity occur during the southwest monsoon. Therefore it is reasonable to interpret Figure 4.11b as an indication of a strengthening of the winds of the southwest monsoon at 8–9 Ma.

Question 4.3

(a) (i) The total flux of dissolved material from the rivers listed in Table 4.1 is:

$$\frac{679\times10^{6}\ \mathrm{t\ y^{-1}}}{2130\times10^{6}\ \mathrm{t\ y^{-1}}}\times100=31.9\%\text{ of the global total (to 3 sig. figs).}$$

(ii) The total flux of suspended material from the rivers listed in Table 4.1 is $3518\times10^{6}\ \mathrm{t\ y^{-1}}$, which is:

$$\frac{3518\times10^{6}\times\mathrm{t\ y^{-1}}}{20\,000\times10^{6}\times\mathrm{t\ y^{-1}}}\times100=17.6\%\text{ of the global total (to 3 sig. figs).}$$

(b) Approximately 30% of the global dissolved products of weathering and ~18% of the global suspended material is derived from just 5% of the available land area by the rivers in Table 4.1. This suggests unusually high weathering rates in the Tibet/Himalaya region.

Question 4.4

According to Figure 4.13, there was a period of relatively fast cooling starting at about 80 Ma, followed by a second period with a high rate of cooling starting at ~50 Ma, then a sharper increase in the rate of cooling commencing at about 15 Ma, and a final even sharper decline starting at about 5 Ma.

(a) In relation to the GEOCARB model, Figure 3.10 indicates that the rate of production of oceanic crust gradually decreased between 110 Ma and 80 Ma, and then decreased more sharply over the next 1 Ma. After that, it remained roughly constant. This general decline in ocean crust production over the course of the past 110 Ma or so is consistent with the GEOCARB model. In addition, the model is also supported by the sharp temperature drop at ~85–70 Ma, with the decrease in ocean-floor production occurring over roughly the same period. According to Figure 3.10, however, the rate of production of ocean floor since 50 Ma has been fairly stable, and so the decrease in temperature over the past 50 Ma cannot be accounted for on the basis of the GEOCARB model.

(b) The Himalaya and the Tibetan Plateau began to rise some time after 50 Ma, but presumably the high rates of weathering would have been established some time after this, when the topography had become high and steep. The phase of global cooling over the past 50 Ma, therefore, is consistent with the mountain-forcing model.

Question 4.5

(a) If the flux of organic carbon from the continents increased significantly, then the average $\delta^{13}C$ value of marine sediments would decrease.

(b) The largest increase in the flux of organic carbon to the ocean will be indicated by the largest decrease in $\delta^{13}C$ in marine carbonates, which according to Figure 4.16 occurred at about 58–56 Ma. (You can assume that the $\delta^{13}C$ plot in Figure 4.16 is determined mainly by the flux of organic carbon from *land* because that from marine primary and secondary production is very small by comparison.)

Question 5.1

(a) The lunar cratering record suggests that 4.0 Ga ago, the Earth was experiencing a storm of impacts that might have repeatedly boiled off any incipient oceans. It would be ill-advised to leave behind the thermally insulated capsule on this trip, and indeed your time-travel agent might well ask you if you had some alternative destinations in mind. If you risked a visit and succeeded in spotting any standing water through the window of your capsule, you might find it fringed with greenish slime.

(b) By contrast, you could probably alight from your capsule on the trip back to 2.5 Ga, but not without the breathing apparatus because of the scarce molecular oxygen in the atmosphere, and you would need lashings of suncream because of the fierce ultraviolet radiation let in by the lack of an effective ozone screen. Your sandwiches would be preferable to the slimy greenish-grey mats you might encounter along the coast, and the book could be handy unless you really like looking at barren landscapes.

(c) Breathing might be manageable, if somewhat strenuous, on the trip back to 500 Ma, so you could possibly leave the suncream behind now as an effective ozone layer has been developed at this stage. Unless you took some sandwiches, you would need the crabbing net to catch your lunch at the seaside (but with no guarantees as to what it would taste like). Meanwhile, the book would still be useful, as you would have to wait another 100 Ma or so for the appearance of complex and visually interesting life forms on land.

(d) Top on the list for the trip back to 100 Ma would be your weapons to fend off dinosaurs, if not to help you catch your lunch.

(e) The weapons would again prove useful for your trip back to 100 000 years, the threat now being from large mammals or aggressive earlier members of your own species, resenting your presence.

Question 5.2

The long-term burial of organic material allowed molecular oxygen to build up in the atmosphere. If all the material produced through photosynthesis remained available for respiration (or combustion), it would have consumed the same amount of oxygen as had originally been released during its production. With the removal of such organic material, however, photosynthesis could yield a net excess of oxygen.

Question 5.3

The genetic material (DNA) in a eukaryote cell is contained in a nucleus, whereas the DNA in a prokaryote cell is confined to a single loop within the cell.

Question 5.4

The diagnostic features most likely to be detected are:

- cell size (eukaryotes tend to be larger than prokaryotes)
- presence of organelles
- cellular organisation (differentiated cells in multicellular forms).

Other molecular (e.g. genetic) attributes are unlikely to be preserved.

Question 5.5

The ability to engulf other cells, which is the first step towards endosymbiosis, would require loss of the constraining, rigid prokaryotic cell wall.

Question 5.6

The most likely time for mitochondria-bearing eukaryotes to proliferate is in the early Proterozoic, when the deposition of banded iron formations began to decline and red beds started to appear in the rock record, e.g. around 2.0 Ga. These changes in the sedimentary record are thought to reflect the appearance of molecular oxygen in the surface waters of the oceans.

Question 5.7

Single mutations whose effects were hidden by an unaffected partner would still accumulate through the generations, increasing the eventual probability of double mutations at matching sites on the paired chromosomes. (To return to the aeroplane analogy, this is rather like the risk of continuing to make further flights in a twin-engined aeroplane with only one functioning engine.)

Question 5.8

The observation implies that sexual reproduction evolved only after there was diversification between the asexual forms and the new sexual ones from a common ancestral group. It must have arisen in the common ancestor of the remaining eukaryote groups. The main burst of eukaryote evolutionary history may, therefore, have been triggered by the appearance of sexual reproduction.

Question 5.9

The molecular clock theory supposes a constant average rate of divergence between corresponding sequences. Therefore extrapolating a straight line in Figure 2.13 to the maximum divergence value known among animals (~190%) allows the latest time of the origin of the animals to be estimated at around 1.0 Ga.

Question 5.10

No because, as discussed above, the focus of natural selection is on the individual genetic entity, which is the only *necessary* beneficiary of any adaptations. What happens above that level (e.g. at the population level) as a consequence of adaptations of individuals is dependent on whether changes are either beneficial or disadvantageous to the maintenance of the system under consideration.

Question 6.1

A mini-glaciation.

Question 6.2

The most obvious example is that of the human population explosion, itself a consequence of a decrease in the individual death rate (i.e. increased survivorship of individuals), due to social and medical advances. Alternatively, if you are a gardener, you might have thought of the unwelcome rise in aphids each summer: this is unleashed by the onset of rapid asexual reproduction, easily outstripping the death rate of individuals, as they feed on new plant growth.

Question 6.3

Following on from that in the late Ordovician, comparable sharp falls in diversity occurred in the:

- late Devonian
- late Permian
- late Triassic
- late Cretaceous.

At first glance, it might seem that the late Triassic extinction was not on a par with the others. However, the relative drop in diversity was still considerable, given the depleted numbers of families that remained after the particularly severe late Permian extinction.

Question 6.4

After each mass extinction, the diversity of organisms increased relatively rapidly. Such a consistent response accords well with the hypothesis discussed in relation to the Cambrian Explosion, i.e. that of unrestrained evolutionary radiations filling vacated habitats.

Question 6.5

Following the trends proposed, it is possible that rates of origination could decline and extinction rates increase until they eventually matched one another, as diversity simultaneously rose. As a result of this balanced budget of origination and extinction, diversity would stay at the same level (or at least hover around it, allowing for small random fluctuations).

Question 6.6

No. Each successive fauna appears to have been proportionately less drastically affected by mass extinction than its predecessor. This is most clearly seen in the case of the mass extinction at the end of the Permian. At this time, family diversity in Palaeozoic fauna plummeted (the Cambrian fauna being already of negligible diversity at this time), whereas the modern fauna were much less profoundly affected. (This is also apparent in Figure 6.9.) In terms of reduction in numbers, the late Ordovician extinction almost halved the diversity of the Cambrian fauna, compared with the loss of under a third of the families in the Palaeozoic fauna. Thus, each mass extinction (particularly that at the end of the Permian) left a biased set of animal groups among the survivors.

The reasons for such differential survival are not well understood and may not just reflect differences in the fitness of individuals in their perturbed environments, according to the normal principles of natural selection. It is also possible that certain properties of whole groups, such as their geographical distribution, may have enhanced survival and/or extinction potentials. For example, there is some evidence that genera containing species distributed across different regions stood a better chance of surviving a mass extinction than those limited to a single area. This form of hierarchical filtering can be readily explained by the environmental perturbations affecting different regions unequally (see Box 6.2).

Question 6.7

As the calcareous plankton died and sank to the sea floor, they would have added to the carbonate sedimentation found further offshore and in deeper water, continuing the trend noted earlier for the Cambrian radiations.

This offshore deposition enhanced the carbonate contribution to the carbon cycle, representative of a time when these planktonic groups underwent major radiation.

Question 6.8

It can be supposed that sufficient CO_2 to meet photosynthetic requirements (limited by other factors) could have entered the plants through relatively few stomata. If water were the limiting factor for photosynthesis, then having only a few stomata would have been advantageous because it would have reduced water loss to the atmosphere.

As plants colonise the land, rising percentage levels of carbon were transferred from the atmosphere and fixed in the soil, with increasing amounts ending up buried in sediments. Land plants, therefore, added a new component to the biological removal of CO_2, which prior to their evolution had only taken place in the marine environment.

With atmospheric CO_2 decreasing, it, rather than water, eventually became the limiting factor of photosynthesis, and plants therefore had to adapt by increasing their stomatal numbers. The presence of stomata, and the fact that cuticles and waxes are not totally impervious to water inevitably led to greater water loss from plants back to the atmosphere; to compensate, more water had to be supplied to the leaf. This, in turn, led to natural selection of a more efficient root system and better vascularisation within plants. Thus, with increasing branching systems and leafy photosynthetic areas, there is a corresponding increase in vascular complexity in late Devonian plants. Alongside this, the diameters of the water transport cells also increased in many taxa (Figure 6.21). Early ferns provide some good examples of the developments described above (Figure 6.22).

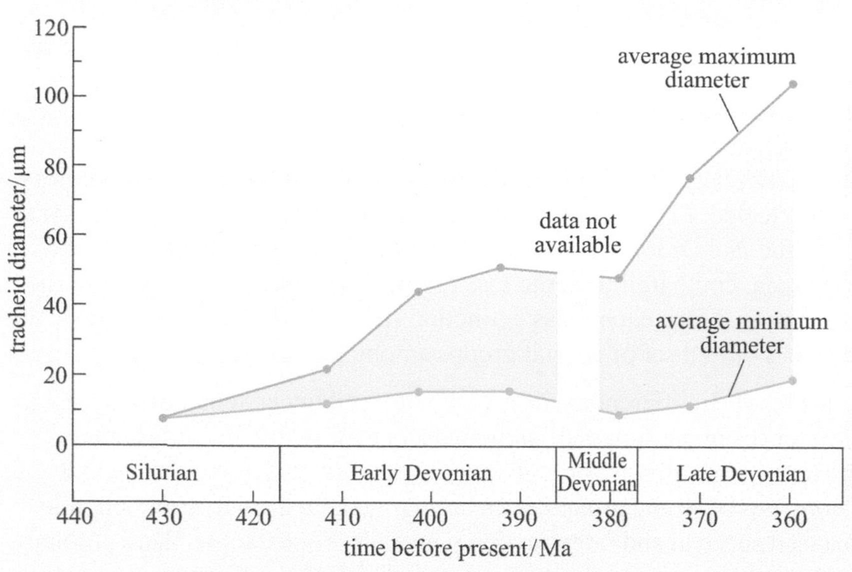

Figure 6.21 Changes in water transport cell diameter in land plants over time.

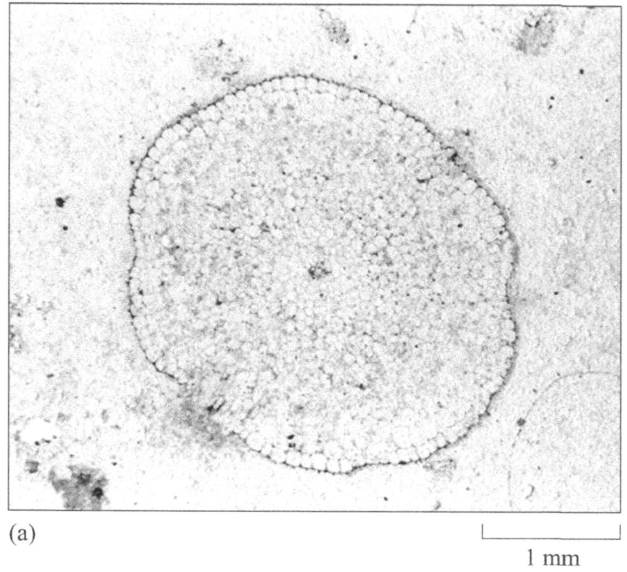

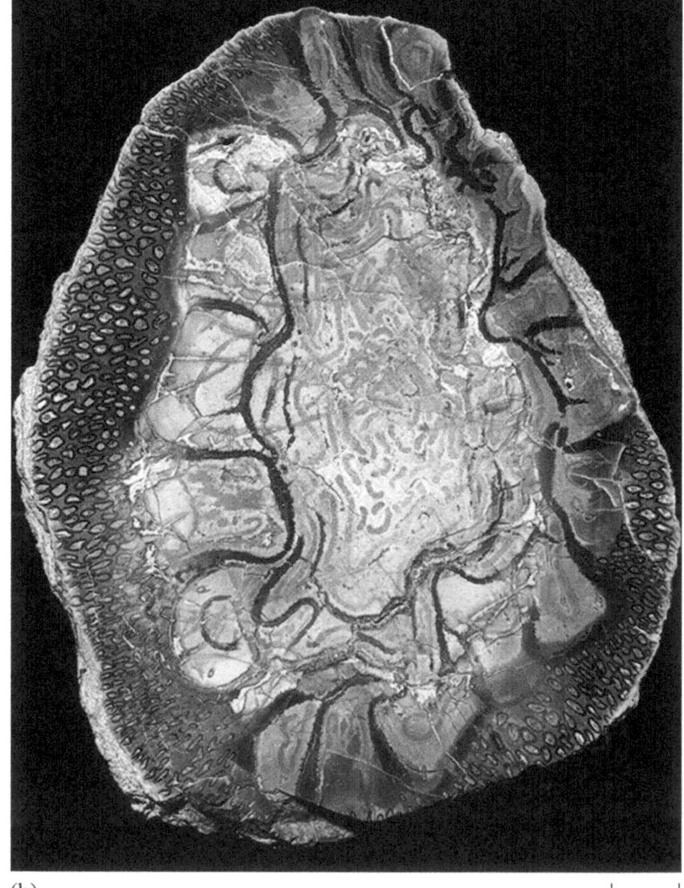

Figure 6.22 Early plants and their vascular systems. (a) Cross-section through a stem of the early Devonian *Rhynia* plant, the dark patch in the centre being the single vascular strand. (b) Cross-section through the stem of a Carboniferous fern, *Psaronius*, showing a far more complex vascular system with many arching worm-like strands in the central region (the circular structures around the edge are cross-sections through roots). (Bob Spicer/Open University)

Question 7.1

In contrast to today, Figure 7.1 shows there was one enormous supercontinent (Pangaea), which incorporated all bar some of the Asian microcontinents and which extended from the South Pole to high northern latitudes.

Question 7.2

Referring to Section 2.3.2, the two main reactions considered involve the weathering of calcium carbonate ($CaCO_3$) and silicate ($NaAlSi_3O_8$) rocks (Equations 2.7a and b). The first reaction takes in one molecule of CO_2 for each molecule of $CaCO_3$ weathered but, because the precipitation of carbonate releases it again, there is no net drawdown of CO_2:

$$CaCO_3 + CO_2 + H_2O \underset{\text{carbonate precipitation}}{\overset{\text{carbonate dissolution}}{\rightleftharpoons}} Ca^{2+} + 2HCO_3^-$$

The second reaction – silicate weathering – removes two molecules of CO_2 from the atmosphere for every silicate molecule weathered:

$$\underbrace{CaSiO_3}_{\text{silicate rock}} + \underbrace{2CO_2 + 3H_2O}_{\text{from rainwater}} \rightarrow \underbrace{Ca^{2+} + 2HCO_3^- + H_4SiO_4}_{\text{stream/river water}}$$

As the precipitation of carbonate releases one molecule of CO_2 into the atmosphere (see the carbonate precipitation reaction above), a net drawdown of CO_2 from the atmosphere occurs when silicate rocks are weathered.

Question 7.3

Such mountain building would have resulted in an increase in the amount of rock available to be eroded. This in turn would have led to more CO_2 being sequestered from the atmosphere as rock silicates were weathered, ultimately ending in the oceans as $CaCO_3$.

Question 7.4

First, the rifting and break-up of Pangaea would have created several new mid-oceanic ridge systems by Cretaceous times, the combined volume of which would have displaced corresponding amounts of water from the ocean basins. Second, the water that in late Carboniferous to early Permian times was tied up in the south polar continental ice sheet would have returned to the oceans in the Cretaceous.

Question 7.5

The subdued relief, together with the aridity of the hinterland, would have meant that supplies of land-derived sands and muds carried by streams and rivers would have been temporarily and spatially limited, associated only with localised zones of uplift. By contrast, the breakdown of calcareous shells produced by the myriads of organisms thriving in the broad expanses of warm shallow sea (the 'shallow water carbonate factory') would have generated copious amounts of carbonate sediments, which would eventually form limestone. Higher temperatures leading to the thermal expansion of water also contribute to higher sea level.

Appendix A

The elements

Atomic number, *Z*	Name	Chemical symbol	Atomic number, *Z*	Name	Chemical symbol	Atomic number, *Z*	Name	Chemical symbol
1	hydrogen	H	32	germanium	Ge	63	europium	Eu
2	helium	He	33	arsenic	As	65	terbium	Tb
3	lithium	Li	34	selenium	Se	66	dysprosium	Dy
4	beryllium	Be	35	bromine	Br	67	holmium	Ho
5	boron	B	36	krypton	Kr	68	erbium	Er
6	carbon	C	37	rubidium	Rb	69	thulium	Tm
7	nitrogen	N	38	strontium	Sr	70	ytterbium	Yb
8	oxygen	O	39	yttrium	Y	71	lutetium	Lu
9	fluorine	F	40	zirconium	Zr	72	hafnium	Hf
10	neon	Ne	41	niobium	Nb	73	tantalum	Ta
11	sodium	Na	42	molybdenum	Mo	74	tungsten	W
12	magnesium	Mg	43	technetium	Tc[a]	75	rhenium	Re
13	aluminium	Al	44	ruthenium	Ru	76	osmium	Os
14	silicon	Si	45	rhodium	Rh	77	iridium	Ir
15	phosphorus	P	46	palladium	Pd	78	platinum	Pt
16	sulfur	S	47	silver	Ag	79	gold	Au
17	chlorine	Cl	48	cadmium	Cd	80	mercury	Hg
18	argon	Ar	49	indium	In	81	thallium	Tl
19	potassium	K	50	tin	Sn	82	lead	Pb
20	calcium	Ca	51	antimony	Sb	83	bismuth	Bi
21	scandium	Sc	52	tellurium	Te	84	polonium	Po[a]
22	titanium	Ti	53	iodine	I	85	astatine	At[a]
23	vanadium	V	54	xenon	Xe	86	radon	Rn[a]
24	chromium	Cr	55	caesium	Cs	87	francium	Fr[a]
25	manganese	Mn	56	barium	Ba	88	radium	Ra[a]
26	iron	Fe	57	lanthanum	La	89	actinium	Ac[a]
27	cobalt	Co	58	cerium	Ce	90	thorium	Th[a]
28	nickel	Ni	59	praseodymium	Pr	91	protoactinium	Pa[a]
29	copper	Cu	60	neodymium	Nd	92	uranium	U[a]
30	zinc	Zn	61	promethium	Pm[a]			
31	gallium	Ga	62	samarium	Sm			

[a] No stable isotopes.

Appendix B

SI fundamental and derived units

Quantity	Unit	Abbreviation	Equivalent units
mass	kilogram	kg	
length	metre	m	
time	second	s	
temperature	kelvin	K	
angle	radian	rad	
area	square metre	m^2	
volume	cubic metre	m^3	
speed, velocity	metre per second	$m\ s^{-1}$	
acceleration	metre per second squared	$m\ s^{-2}$	
density	kilogram per cubic metre	$kg\ m^{-3}$	
frequency	hertz	Hz	(cycles) s^{-1}
force	newton	N	$kg\ m\ s^{-2}$
pressure	pascal	Pa	$kg\ m^{-1}\ s^{-2}$, $N\ m^{-2}$
energy	joule	J	$kg\ m^2\ s^{-2}$
power	watt	W	$kg\ m^2\ s^{-3}$, $J\ s^{-1}$
specific heat capacity	joule per kilogram kelvin	$J\ kg^{-1}\ K^{-1}$	$m^2\ s^{-2}\ K^{-1}$
thermal conductivity	watt per metre kelvin	$W\ m^{-1}\ K^{-1}$	$m\ kg\ s^{-3}\ K^{-1}$

Appendix C

The Greek alphabet

Name	Lower case	Upper case	Name	Lower case	Upper case
alpha	α	Α	nu (new)	ν	Ν
beta (bee-ta)	β	Β	xi (cs-eye)	ξ	Ξ
gamma	γ	Γ	omicron	ο	Ο
delta	δ	Δ	pi (pie)	π	Π
epsilon	ε	Ε	rho (roe)	ρ	Ρ
zeta (zee-ta)	ζ	Ζ	sigma	σ	Σ
eta (ee-ta)	η	Η	tau (torr)	τ	Τ
theta (thee-ta; 'th' as in theatre)	θ	Θ	upsilon	υ	Υ
iota (eye-owe-ta)	ι	Ι	phi (fie)	ϕ	Φ
kappa	κ	Κ	chi (kie)	χ	Χ
lambda (lam-da)	λ	Λ	psi (ps-eye)	ψ	Ψ
mu (mew)	μ	Μ	omega (owe-me-ga)	ω	Ω

Additional figures

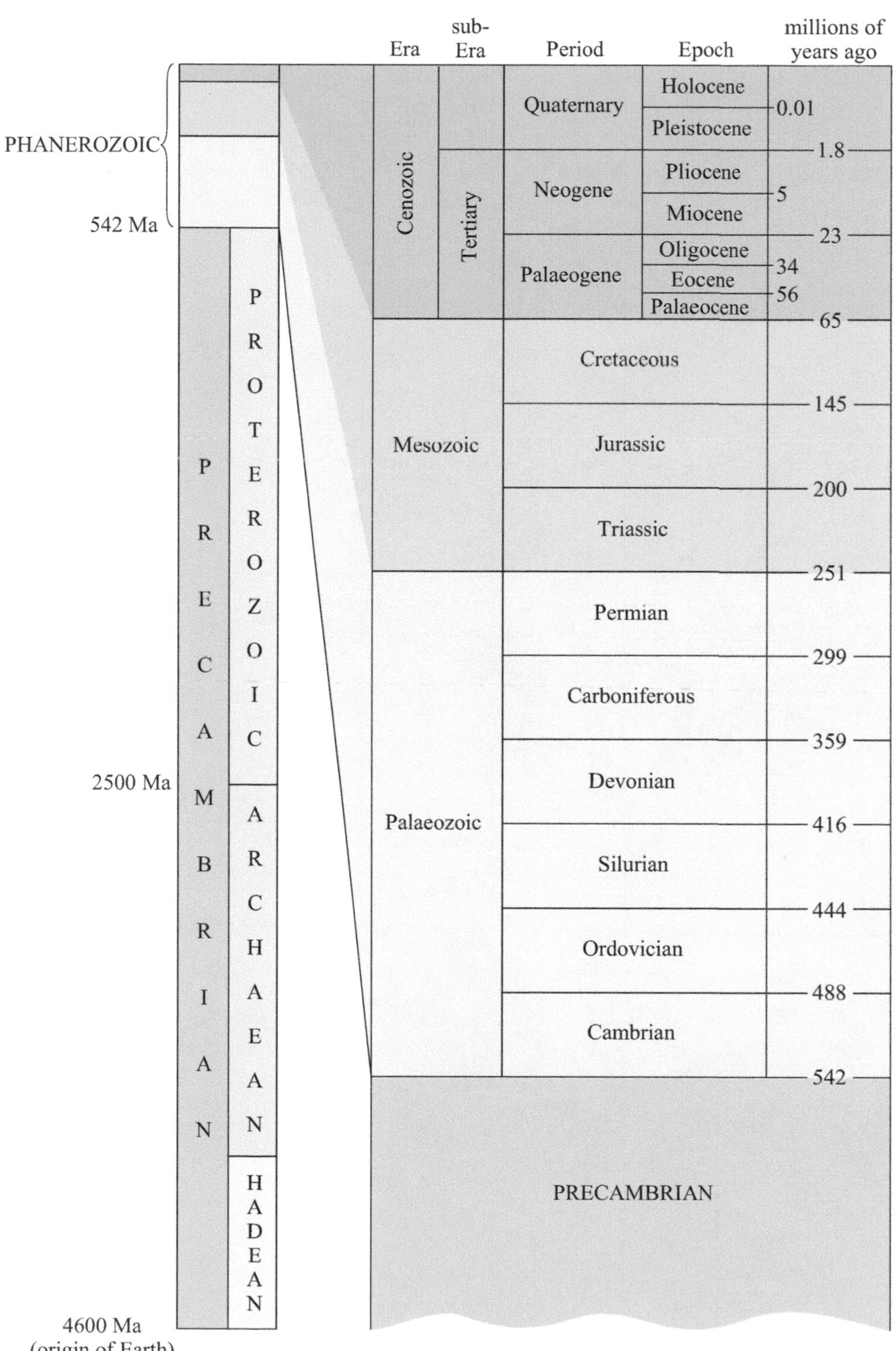

Note that time intervals are not drawn to scale. You may see other versions of this timescale with minor differences, as the subdivisions and radiometric age dates are subject to revision.

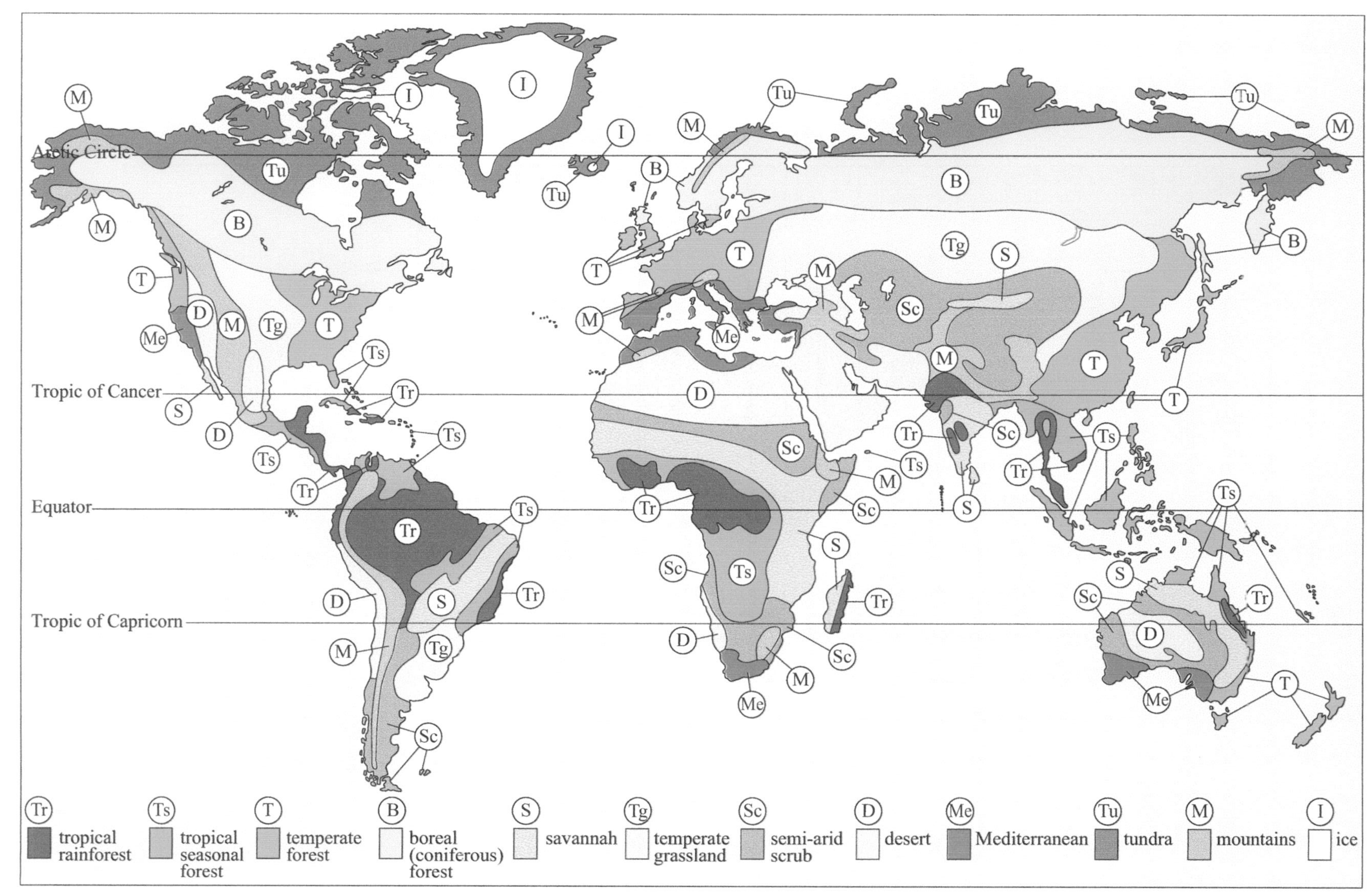

Figure 2.4 Geographical distribution of the major natural regional ecological communities or biomes.

ablation Loss of ice from a system, usually due to melting.

acritarch Any small non-calcareous, non-siliceous organic structure that cannot otherwise be classified.

adiabatic From the Greek for 'impassable', a reference to the fact that heat passes neither from nor to a substance undergoing an adiabatic process.

advection Transfer of heat by physically moving material.

aerosols Airborne solid or liquid particles.

albedo The fraction of incoming solar radiation that is reflected from a surface.

alluvial fans Fan-shaped deposit formed where faster flowing water flattens, slows, and spreads, e.g. at the exit of a canyon onto a flatter plain.

angiosperms Flowering plants.

anoxic Without oxygen, or oxygen poor.

anthropogenic Processes, objects or materials derived from human activities.

aphelion The point on an orbit that lies furthest from the Sun.

Archaea Single-celled, prokaryotic microorganisms.

atmospheric window The range of longwave radiation leaving the Earth system between 8 μm and 13 μm, where there is relatively little absorption.

autotroph An organism that manufactures its own food.

back-scattered incoming radiation Radiation from the Sun that is scattered back out into space.

bacteria Very small, single-celled life forms: the most primitive living beings.

base pairs Two nucleotides on opposite complementary DNA or RNA strands that are connected via hydrogen bonds.

bicarbonate ion (HCO_3^-).

bilaterian All animals having a bilateral symmetry, i.e. they have a front and a back end, as well as an upside and a downside.

binary fission The form of asexual reproduction in single-celled organisms by which one cell divides into two cells of the same size, used by most prokaryotes.

biogenic Produced by life processes.

biogeochemical cycle Global cycling of an element used by organisms.

biolimiting Nutrients that can become completely used up by **phytoplankton**, thus preventing the growth of further organisms.

biological pump The transfer of carbon from surface waters to the deep ocean.

biomarker A substance used as an indicator of a biological state, e.g. any kind of molecule indicating the existence of living organisms.

biomes Distinct ecosystem types.

biota All the plant and animal life of a region.

bioturbation The displacement and mixing of sediment particles by benthic fauna (animals) or flora (plants).

boundary layer The relatively still and humid layer of air next to the ground.

C3 plants Organisms using the common metabolic pathway for carbon fixation in photosynthesis. This process converts carbon dioxide and ribulose bisphosphate into 3-phosphoglycerate.

C4 plants An advancement over the simpler and more ancient C3 carbon fixation mechanism. C4 fixation requires more energy input than C3 in the form of adenosine triphosphate. These species are concentrated in the tropics and make up about 5% of plant biomass.

Cambrian Explosion The diversification of modern **bilaterian** life at the beginning of the Phanerozoic.

carbohydrates Sugars, starches and cellulose, which contain carbon, hydrogen and oxygen and form chains of simple sugars, and which function primarily in energy storage, energy transport, and plant structure.

carbon fixation A process found in **autotrophs**, usually driven by **photosynthesis**, whereby carbon dioxide is converted into organic compounds.

carbonate compensation depth (CCD) The depth in the ocean at which the proportion of calcium carbonate remaining falls to less than 20% of the total sediment.

carbonate ion (CO_3^{2-}).

carbonate platforms Broad, shallow, submarine plateaux.

carbonate system The reactions that occur between gaseous carbon dioxide and its various aqueous forms.

chalk A distinctively fine-grained kind of limestone.

chemical equilibrium A dynamic balance between forward and reverse reactions.

chemical weathering The change in the composition of rocks, often leading to a breakdown in its form.

chemosynthetic bacteria Bacteria that make food by the biological conversion of simple carbon molecules (usually carbon dioxide or methane) and nutrients into organic matter using the oxidation of inorganic molecules (e.g. hydrogen gas, hydrogen sulfide) or methane as a source of energy, rather than sunlight.

chloroplasts The **organelles** that effect photosynthesis in green plant cells.

chromosomes Separate strands of genes, usually in pairs, contained in the nucleus of a cell.

conduction Spread of heat energy through a solid, e.g. the heating of a saucepan handle when the pan is heated.

convection The transfer of heat by circulation through a gas or liquid.

Coriolis effect The deflection of winds relative to the surface of the Earth resulting from the rotation of the Earth below air masses.

crown groups A group of closely related organisms that includes the common ancestor plus all its descendants.

cycads Plants that may represent the evolutionary link between ferns and flowering plants.

decarbonation reactions A reaction during the heating of silicate and oxide minerals that releases CO_2.

diamictites Unsorted mixtures of rock that are of glacial origin.

diploid Doubling up of chromosomes in the cell nucleus.

dissolved inorganic carbon (DIC) Inorganic molecules of varied origin and composition within aquatic systems.

dissolved organic carbon (DOC) Organic molecules of varied origin and composition within aquatic systems.

earlywood Wood cells with large internal cavities produced early in the season.

Ediacaran faunas Life forms of the latest Precambrian.

Ediacaran Period The last 70 Ma or so of the Proterozoic (about 610–540 Ma ago).

electromagnetic spectrum The ordered series of all known types of electromagnetic radiation, arranged by wavelength, ranging from short cosmic rays through gamma rays, X-rays, ultraviolet radiation, visible radiation, infrared radiation, and microwaves, to the long wavelengths of radio energy.

enations Scaly leaf-like structures, differing from leaves in their lack of vascular tissue.

endosymbiotic hypothesis The idea that organelles were once independent prokaryotes that took up symbiotic residence inside ancestral eukaryotic host cells.

endothermic Reactions that *require* energy.

epeirogenic changes *See* **isostatic changes**.

eukaryote A cell or organism with a membrane-bound, structurally discrete nucleus and other subcellular compartments.

eustatic changes Worldwide sea-level changes that affect all oceans, e.g. due to the formation or melting of land ice.

evapotranspiration The process in plants whereby water drawn up from the soil by roots is lost to the atmosphere through pores in the leaves.

evolutionary faunas Sets of major groups of animals, each set showing a characteristic pattern of family turnover.

evolutionary radiations Phases of significant increase in numbers of species within groups of organisms.

exothermic Reactions that *release* energy.

extremophile An organism that can tolerate, or requires, environmental conditions considered extreme to people.

false rings Tree growth rings found in fossils, showing interruptions during the summer growing season.

fitness An individual's production of offspring, themselves surviving to be capable of reproduction, relative to other individuals.

'fixing' carbon The process in which free carbon in the atmosphere is incorporated into living material.

flood basalts Vast accumulations of horizontal basaltic rocks rapidly erupted from fissures over large areas.

forcing function An external input to the Earth's climate system.

forward-scattered incoming radiation Radiation from the Sun that is scattered forwards (for example towards the ground).

frontal current Ocean currents flowing along boundaries between bodies of water with different temperatures.

gametes Parental cells, e.g. eggs and sperm.

gene duplication The assignment of **Hox genes** to control the development of other body parts.

general circulation model A large-scale model of global climate.

GEOCARB model A model for the evolution of the carbon cycle and of atmospheric CO_2 over geological time.

glacio-eustasy Glacially driven sea-level change.

Global Stratotype Section and Point (GSSP) An internationally agreed upon stratigraphic section that serves as the reference section for a particular boundary on the geologic timescale.

glossopterids Deciduous trees with distinctive tongue-shaped leaves.

greenhouse effect The absorption of outgoing longwave radiation and the subsequent reradiation that keeps the Earth's average surface temperature at ~15 °C.

gross primary production (GPP) The total amount of carbon fixed via photosynthesis in plants.

gymnosperms Seed-bearing plants.

gyre A permanent, closed, large-scale circulation of water in an ocean basin.

Hadley circulation The latitudinal–vertical component of a helical circulatory system.

halite pseudomorphs Casts of sediment filling often cube-shaped voids from which the original halite crystals have been dissolved.

haploid Cells containing only a single set of **chromosomes**.

herbaceous Non-woody plants.

heterochrony A developmental change in the timing of events, leading to changes in size and shape of organisms.

heterotrophs Organisms that consume organics or other organisms, whether plants or animals, to obtain energy.

homeostasis Systems of self-regulation that maintain stable conditions within the organism in the face of a range of environmental perturbations.

Hox genes Genes responsible for 'pattern formation' – that is, the overall arrangement of appendages – on the body of an organism.

hydrological cycle The movement of water that underpins the cycling of many other constituents through the Earth–atmosphere–ocean system.

hyperthermophiles Organisms needing high temperatures (generally greater than 60 °C) to grow.

intermediate geological timescale (carbon cycle) Up to hundreds of thousands of years involving chemical, biological and physical components.

Intertropical Convergence Zone (or **ITCZ**) The place where the wind systems of the two hemispheres meet.

isostasy The state of gravitational equilibrium in which the Earth's crust is buoyantly supported by plastic materials of the mantle.

isostatic (or epeirogenic) changes Sea-level changes caused by vertical movements of the crust. Such movements may be caused by changes in the thickness and/or density of the lithosphere, and by loading or unloading with ice or sediments.

isotherm A type of contour line that connects points of equal temperature.

K/T boundary The end of the Cretaceous Period and, therefore, also of the Mesozoic Era, and the beginning of the Tertiary Period and, therefore, also of the Cenozoic Era; it is linked to a large **mass extinction**.

kerogen Dense residue enriched in carbon formed in an oxygen-poor environment.

kin selection Natural selection within a population in which larger than usual proportions of genetic material are shared between siblings.

latewood Wood cells produced late in the season when water is in short supply.

Le Chatelier's principle The re-establishment of equilibrium in a chemical system in which the system lessens the effect of an external constraint.

leaf-area index Ratio of the total area of all leaves on a plant to the area of ground covered by the plant.

leaf margin analysis A widely used method that applies present-day correlations between the proportion of species with untoothed leaves and mean annual temperature to estimate palaeotemperatures from fossil megafloras.

leaf physiognomy The shape, size and margin analysis of a leaf.

lignin The carbon in plant structural material.

long geological timescale (carbon cycle) Geological timescale of up to hundreds of millions of years, involving rocks and sediments.

macrophagy The ability to ingest and digest food extracellularly by allowing the development of the gut.

mantle plumes Upcurrents of magma, rising from the boundary between the mantle and the core, deep within the Earth.

mantle superplume A large, discrete, slowly rising plume of heated material in the Earth's mantle.

marine snow Clumps or aggregates of fluffy debris in the oceans made up of dead and dying algal cells and bacteria.

mass extinction Worldwide extinction of many species.

meiosis Cell division that halves the number of chromosomes while also exchanging segments of DNA between the matching **chromosome** pairs of the **diploid** parents.

mesoscale eddies Intermediate-sized eddies about 50–250 km across (i.e. about a quarter of the size of mid-latitude atmospheric cyclones and anticyclones).

methane hydrates A solid form of water that contains a large amount of methane within its crystal structure. Extremely large deposits of methane clathrate have been found under sediments on the ocean floors of the Earth.

microfossils Microscopic fossils.

midvein The main vein running through a leaf (also known as the primary vein).

Milankovich cycles Changes in climate resulting from seasonal fluctuations caused by changes in elements of the Earth's orbit, i.e. eccentricity, tilt of the rotational axis and longitude of perihelion.

mitochondria A eukaryote **organelle** responsible for the energy supply derived from aerobic respiration.

mitosis Cell division that conserves **chromosome** numbers in the nucleus.

molecular clock theory The idea that as evolutionary lineages diverge from a common ancestor so too do their sequences as a consequence of cumulative mutational change in each lineage.

molecular phylogenies Comparisons between organisms that use genetic sequences to determine how similar they are.

mountain-forcing model A model that suggests changes in atmospheric CO_2 concentrations are driven primarily by changes in chemical weathering rates.

nearest living relative (NLR) technique Assumes that ancient plants and plant communities lived under similar conditions to those of their nearest living relatives.

negative feedback Feedback in which the system responds in the opposite direction to the perturbation.

Neoproterozoic Era The time from 1000 Ma to 542 Ma ago, ±0.3 Ma.

net primary production (NPP) The carbon that is not released back into the atmosphere via plant respiration.

nutrients Elements necessary for growth in a form that can be used by organisms.

organelles Minute, intracellular structures serving a specific function in the life processes of the cell.

palaeo-altimeter A means of determining past altitudes.

particulate organic carbon (POC) Fragments of organic carbon, e.g. in soil.

perihelion The point on an orbit that lies closest to the Sun.

pH Hydrogen ion concentration in water, where a low pH corresponds to a high hydrogen ion concentration and vice versa.

phloem The vascular tissue that enables movement of the products of leaf **photosynthesis** (such as sugars) to the rest of the plant.

photic zone The sunlit surface waters of the oceans.

photosynthesis The process undertaken by organisms to build their own organic material from CO_2 and H_2O, also producing O_2.

phototrophs Organisms that carry out photosynthesis.

physical erosion The break-up of rocks, e.g. by the action of ice or changes of temperature, that is not due to chemical change.

physiognomy The architecture of the plant or community applied when the fossils are in the form of vegetative organs, particularly leaves. *See also* leaf physiognomy.

phytoplankton Minute floating algae: the primary producers of the oceans.

Pleistocene The period from 1 808 000 to 11 550 years before present, including repeated glacial periods.

polar front The boundary between warm tropical air and the underlying cold polar air.

polar jet stream A high-level, fast air current that in each hemisphere flows around the Earth above the **polar front**.

positive feedback Feedback in which the system responds in the same direction as the perturbation.

primary producers Organisms that fix carbon.

primary production The process of building living material by fixing carbon.

primary vein The midvein of a leaf.

progradation The outward building of a sedimentary deposit, such as the seaward advance of a delta or shoreline, or the outbuilding of an alluvial fan.

progymnosperms The woody, dominant canopy formers in the earliest forests.

prokaryotic cells Cells whose DNA is not bound into a nucleus and which lack organelles.

radiation Diversification (of organisms).

radiogenic strontium (^{87}Sr) The product of radioactive decay of one of the isotopes of rubidium, ^{87}Rb.

red beds Strata of reddish-coloured sedimentary rocks, such as sandstone, siltstone or shale, that were deposited in hot climates under oxidising conditions.

refugia Locations of isolated or relict populations of once widespread animal or plant species.

regressive Of an episode of regression, i.e. a fall in sea level relative to the land.

reservoir A reserve of a substance, e.g. groundwater in the hydrological cycle.

residence time A measure of the average length of time an individual molecule, e.g. water in the **hydrological cycle**, spends in any particular stage.

respiration The release by organisms of energy stored in organic matter.

rhizoids Small root hair-like appendages along **rhizomes** that anchored the plant to its substrate.

rhizome The prostrate stem of early land plants.

rudists A group of bivalves that peaked in abundance and diversity during the late Mesozoic and whose morphology consisted of a lower, roughly conical valve that was attached to the seafloor or to neighboring rudists, and a smaller upper valve that served as a kind of lid for the organism.

salinity The measure of concentration of dissolved salts.

secondary veins Second series veins in a leaf that branch off the **midvein**.

sedimentary succession Sequence of sedimentary strata.

sensible heat Heat which can be felt, or sensed.

sensitivity The size of the response of a system to a forcing of given magnitude.

sessile An organism that does not move itself about and is often attached to a substrate.

sexual reproduction The mixing together of genes from different individuals.

shocked quartz Quartz with microscopic deformation along planes, resulting from asteroid and comet impacts, i.e. intense pressure.

short biological timescale (carbon cycle) Biological timescale of months or years to decades.

snowball Earth The hypothesis that proposes that the Earth was entirely covered by ice in part of the Cryogenian and Ediacaran periods of the Proterozoic Era, and perhaps also at other times in Earth history.

solar flux The amount of solar energy that would fall on a surface at right angles to the Sun's rays.

sporangia Plant or fungal structures producing and containing spores.

spore A small usually single-celled reproductive body that is resistant to dessication and heat and is capable of growing into a new organism. Spores are produced by certain bacteria, fungi, algae and non-flowering plants.

steady-state model A model that assumes that the system has attained equilibrium.

steppe A plain without trees, similar to a prairie, dominated by shrubs and grasses, experiencing low rainfall.

steranes Substances formed from sterols that are sometimes used as **biomarkers** for the presence of **eukaryotic cells**.

stomata Openings or pores in leaves that allow for gas exchange.

stromatolites Attached, lithified sedimentary growth structures commonly thought to have been formed by the trapping, binding, and cementation of sedimentary grains by microorganisms, especially cyanobacteria.

symbiont An organism that is associated with another in a mutually beneficial relationship.

symbiosis A close relationship between organisms of different species in which both benefit.

taphonomy The study of sedimentological and biological evidence relating to explanations of fossilisation phenomena.

taxon A taxonomic category or group, such as a phylum, order, family, genus, or species.

taxonomic hierarchy The classification of organisms into a nested series of increasingly inclusive groups (**taxa**) from species to kingdoms.

tertiary veins Third series of veins in a leaf that branch fromthe **secondary veins**.

thermocline The layer within a body of water where the temperature changes rapidly with depth.

thermohaline The deep temperature–salinity circulation system of the oceans.

transpiration The loss of water from a plant's leaves.

trophic pyramid Layers in a food chain, starting with primary producers, such as algae, through to herbivores and to carnivores. The pyramid structure refers to the greater biomass generally associated with lower levels of the 'pyramid'.

troposphere The lower atmosphere, which makes up ~80% of the total mass of the atmosphere.

Universal Tree of Life A diagrammatic representation of all branches of living organisms.

vascularisation Amount of veining in plants.

water masses Homogeneous bodies of water formed by deep mixing.

xylem The primary water-conducting tissue in plants.

zooplankton The animal component of plankton; animals suspended or drifting in the water column including larvae of many fish and benthic invertebrates.

Further reading

Chapter 1

Darwin, C.R. (1859) *On the Origin of Species by Means of Natural Selection, or the Preservation of Favoured Races in the Struggle for Life*, London, John Murray.

Chapter 2

Berner, R.A. and Caldeira, K. (1997) 'The need for mass balance and feedback in the geochemical carbon cycle', *Geology*, vol. 25, pp. 955–956.

Chapter 3

Berner, R.A. and Kothavala, Z. (2001) 'GEOCARB III: a revised model of atmospheric CO_2 over Phanerozoic time', *American Journal of Science*, vol. 301, pp. 182–204.

Gibbs, M.T., Bluth, G.J.S., Fawcett, P.J. and Kump, L.R. (1999) 'Global chemical erosion over the last 250 My: variations due to changes in paleogeography, paleoclimate, and paleogeology', *American Journal of Science*, vol. 299, pp. 611–651.

Hay, W.W. (1996) 'Tectonics and climate', *Geologische Rundschau*, vol. 85, pp. 409–437.

Lyell, C. (1998) *Principles of Geology* (abridged edn, J. Secord (ed.)), London, Penguin Books Ltd.

Self, S., Thordarson, T. and Widdowson, M. (2005) 'Gas fluxes from flood basalt eruptions', *Elements*, vol. 1, pp. 283–287.

Winguth, A.M.E., Heinze, C., Kutzbach, J.E., Maier-Reimer, E., Mikolajewicz, U., Rowley, D., Rees, A. and Ziegler, A.M. (2002) 'Simulated warm polar currents during the middle Permian', *Paleoceanography*, vol. 17, no. 4, p. 1057.

Chapter 4

Harris, N.B.W. (2000) 'The role of the Himalaya and the Tibetan Plateau in climate control', *Science Spectra*, issue 23, pp. 24–32.

Ruddiman, W.F. (1997) *Tectonic Uplift and Climate Change*, New York, Plenum Press.

Chapter 5

Conway-Morris, S. (1999) *The Crucible of Creation: The Burgess Shale and the Rise of Animals*, Oxford, Oxford Paperbacks.

Darwin, C. (1998 [1859]) *The Origin of Species* (new edn), Gramercy Books.

Dawkins, R. (2006) *The Selfish Gene* (3rd edn), Oxford, Oxford University Press.

Knoll, A. (2004) *Life on a Young Planet: the First Three Billion Years of Evolution on Earth*, Princeton, Princeton University Press.

Lovelock, J. (2000) *Gaia – a New Look at Life on Earth*, Oxford, Oxford Paperbacks.

Chapter 6

Conway-Morris, S. (1999) *The Crucible of Creation*, Oxford, Oxford University Press.

McCall, G.J.H. (2006) 'The Vendian (Ediacaran) in the geological record: enigmas in geology's prelude to the Cambrian explosion', *Earth Science Reviews*, vol. 77, issues 1–3, pp. 1–229.

Runnegar, B. and Gehling, J.G. (2001) 'Understanding the Ediacarans in the context of early animal evolution', *American Zoologist*, vol. 41, no. 6, p. 1573.

Ward, P. and Brownlee, D. (2003) *Rare Earth: Why Complex Life is Uncommon in the Universe*, Springer.

Chapter 7

Brentnall, S.J., Beerling, D.J., Osborne, C.P., Harland, M., Francis, J.E., Valdes, P.J. and Wittig, V.E. (2005) 'Climatic and ecological determinants of leaf lifespan in polar forests of the high CO_2 Cretaceous "greenhouse" world', *Global Change Biology*, vol. 11, no. 12, pp. 2177–2195.

Rees, P.M., Gibbs, M.T., Ziegler, A.N., Kutzbach, J.E. and Behling, P.J. (1999) 'Permian climates: evaluating model predictions using global paleobotanical data', *Geology*, vol. 27, no. 10, pp. 891–894.

Skelton, P.W., Spicer, R.A., Kelley, S.P. and Gilmour, I. (2003) *The Cretaceous World*, Cambridge, Cambridge University Press.

Acknowledgements

The production of this book involved a number of Open University staff to whom we owe considerable thanks for their professional contributions and willingness to accommodate the requests and vagaries of the academic authors. Jennie Neve Bellamy managed the project, including the Open University course associated with this book, with undying optimism and consummate professionalism and ensured that the project kept to deadlines (both original and revised). Ashea Tambe efficiently styled the text for handover to Pamela Wardell, who copy-edited all the chapters with an uncanny attention to detail. Artwork was coordinated and perceptively executed by Sara Hack and design and layout undertaken by Neil Paterson. The index was prepared by Jessica Bartlett and the production process was managed by James Davies. We are grateful to Christianne Bailey (Open University) and Susan Francis (Cambridge University Press) for steering us successfully through the mysteries of co-publication.

We are grateful to Arlene Hunter for her continued vigilance in reading early drafts of the chapters and keeping us in line with learning outcomes. We also thank Nick Petford (University of Bournemouth) who acted as external assessor and constructive critic on both this book and the associated Open University course, and the anonymous reviewers appointed by CUP for their comments during the publisher's review process.

We also acknowledge and thank the authors of earlier Open University courses, notably S269 *Earth and Life*, which laid the foundations of this book. In particular we thank the following for work on the chapters of the S269 course that were modified for use in S279: Peter Skelton, Robert Spicer, Angela Colling, Nigel Harris, Peter Francis, Chris Wilson, Nancy Dise, Allister Rees, and Annemarie Hedges as course manager.

Grateful acknowledgement is made to the following sources for permission to use and reproduce material in this book.

Cover image copyright © Georgette Douwma/Science Photo Library.

Figure 1.1 NASA; *Figure 1.2* NSF/NASA; *Figure 1.8* Imbrie, I., et al. (1984) in Berger, A., et al. (eds) *Milankovitch and Climate*, Kluwer Academic Publishers; *Figure 1.11* Mitchell, J.F.B. (1989) 'The greenhouse effect and climate change', *Reviews of Geophysics*, vol. 27, issue 1, American Geophysical Union; *Figure 1.13* Vander Haar, T. and Suomi, V. (1971) *Journal of Atmospheric Science*, vol. 28, American Meteorological Society; *Figure 1.14* InterNework Inc., NASA/JPL and GSFC; *Figure 1.30b* Tony Waltham Geophotos; *Figure 1.31* Strahler, A. (1973) *Earth Sciences*, Harper and Row; *Figure 1.34* Mark Brandon; *Figure 1.35* Broecker, W. (1991) 'Great Ocean Conveyor', *Oceanography*, vol. 4, no. 2, Oceanography Society; *Figure 1.37* Xie, L. and Hsieh, W.W. (1995) 'The global distribution of wind-induced upwelling', *Fisheries Oceanography*, vol. 4, Blackwell Publishing Limited; *Figure 1.42* N.T. Nicoll/Natural Visions.

Figure 2.4 Cox, B. (1989) *The Atlas of the Living World*, Quarto Publishing Plc; *Figure 2.5* Nancy Dise; *Figure 2.10* Mike Dodd; *Figure 2.11* Williamson, P. 'Pictorial representation of processes contributing to the marine carbon cycle', NERC/BOFS projects; *Figure 2.12* Williamson, P. 'Variation in the concentration of chlorophyll in surface water along a ship's track in the N. Atlantic', NERC/BOFS projects; *Figure 2.13* Bob Spicer; *Figure 2.14a* A. Alldredge; *Figure 2.14b, c and d* NERC, National Oceanography Centre, Southampton; *Figure 2.16a* Courtesy of Jeremy Young, The Natural History Museum; *Figure 2.16b* N. Lefevre, Plymouth Marine Laboratory; *Figure 2.16c* D. Breger, Lamont–Doherty Earth Observatory; *Figure 2.17* N. Lefevre, Plymouth Marine Laboratory; *Figure 2.22* Thoning, K.W., et al. (1994) in Boden, T.A., et al. (eds) 'A Compendium of Data on Global Change', Atmospheric CO_2 Records, ORN/CDIAC-65, Oak Ridge National Laboratory; *Figure 2.23* Conway, T.J. et al. (1988) 'Atmospheric carbon dioxide', Tellus, Blackwell Publishing Limited.

Figure 3.3a http://dewey.cac.washington.edu/ken/pictures/hawaii/PB260022.html; *Figure 3.3b* G. Brad Lewis/Science Photo Library; *Figure 3.4a* Wesley Bocxe/Science Photo Library; *Figure 3.8* Steve Self/The Open University; *Figure 3.9* Coffin, M.F. and Eldholm, O. (1993) 'Exploring large subsea igneous provinces', *Oceanus*, vol. 36, no. 4, Woods Hole Oceanographic Institution; *Figure 3.10* Cogné, J-P. and Humler, E. (2006) 'Trends and rhythms in global sea-floor generation rate', *Geochemistry, Geophysics and Geosystems*, vol. 7, Q03011; *Figure 3.15* van Andel, T.H. (1985) *New Views on an Old Planet*, Cambridge University Press; *Figure 3.16* Haq, B.U. (1984) 'Palaeoceanography …' in Haq, B.U. and Milliman, J.D. (eds) *Marine Geology and Oceanography of Arabian Sea and Coastal Pakistan*, Van Nostrand Reinhold; *Figure 3.17* Kennett, J.P., et al. (eds) (1974) *Initial Reports of the DSDP*, vol. 29, United States Government Printing Office; *Figure 3.18* Valentine, J.W. and Moore, E.M. (1970) 'Plate-tectonic regulators …', *Nature*, vol. 228, Copyright © Nature Publishing Group.

Figure 4.1 M. Keller; *Figures 4.3, 4.5a and b, 4.6a and b, 4.8, 4.9, 4.14a and b* Nigel Harris; *Figure 4.7* Raymo, M. et al. (1988) 'Influence of late Cenozoic mountain building on ocean geochemical cycles', *Geology*, vol. 16, no. 7, Geological Society of America; *Figure 4.11a* R.W. Jordan and W. Smithers; *Figure 4.11b* Kroon, D. et al. (1992) 'Onset of monsoonal-related …', *Proc. Ocean Drilling Program Scientific Results*, vol. 117, Integrated Ocean Drilling Program; *Figure 4.12* Reprinted by permission from Macmillan Publishers Ltd: *Nature*, Quade, J. et al. 'Development of Asian monsoon revealed by marked ecological shift during the latest Miocene in northern Pakistan', copyright © 1989; *Figure 4.18* Richter, F. (1992) 'Sr isotope evolution of seawater', *Earth and Planetary Science Letters*, vol. 109, Elsevier Science.

Figures 5.1a and 5.2 Professor Andrew Knoll, Harvard University; *Figures 5.1b, 5.12c and d* Dr Peter Crimes; *Figure 5.3* Butterfield, N.J. and Chandler, F.W. (1992) 'Palaeoenvironmental distribution of Proterozoic microfossils, with an example from the Agu Bay Formation, Baffin Island', *Palaeontology*, vol. 35, pt 4, © The Palaeontogical Association; *Figures 5.12a and b* Simon Conway Morris, University of Cambridge; *Figure 5.13* Runnegar, B. © Norwegian University Press.

Figure 6.2 Bengston, S. and Yue Zhao (1992) 'Predatorial borings in late Precambrian mineralized exoskeletons', *Science*, vol. 257. Copyright © 1992 The American Association for the Advancement of Science; *Figure 6.5c* Simon Conway Morris, University of Cambridge; *Figures 6.6 and 6.9* Sepkoski, Jr, J.J. (1990) 'Evolutionary faunas' in Briggs, D.E.G. and Crowther, P.R. (eds) *Palaeobiology: A Synthesis*, Blackwell Publishing Ltd; *Figure 6.8* Benton, M.J. (1995) 'Diversification and extinction in the history of life', *Science*, vol. 268. Copyright © 1995 The American Association for the Advancement of Science; *Figures 6.11, 6.18 and 6.22* Bob Spicer; *Figures 6.13, 6.14a and 6.16* Stewart, W.N. and Rothwell, G.W. (1993) *Paleobotany and the Evolution of Plants*,

2nd edn, Cambridge University Press; *Figure 6.14b* Professor Dianne Edwards, University of Wales; *Figure 6.20b and c* Beck, C.B. (1962) 'Reconstructions of Archaeopteris and further consideration of its phylogenetic position', *American Journal of Botany*, vol. 49, Botanical Society of America Inc.

Figure 7.1 Martini, I.P. (1996) *Late Glacial and Postglacial Environmental Changes: Quaternary, Carboniferous–Permian and Proterozoic*, Oxford University Press Inc.; *Figure 7.2* Berner, R.A. (1994) '3 Geocarb II: A revised model of atmosphere CO_2 over Phanerozoic time', *American Journal of Science*, vol. 249, January, American Journal of Science; *Figure 7.3* Raymo, M.E. (1991) 'Geochemical evidence supporting T.C. Chamberlain's theory of glaciation', *Geology*, vol.19, no. 4; *Figure 7.4* Crowley, T.J. (1993) 'Climate change on tectonic time scales', *Tectonophysics*, vol. 222, Elsevier Science; *Figure 7.6* Philips, T.L. and DiMichele, W.A. (1992) 'Comparative ecology and life-history biology of arborescent lycopsids in late Carboniferous swamps of Euramerica', *Annals of the Missouri Botanical Garden*; *Figure 7.7* DiMichele, W.A. and Aronson R.B. (1992) 'The Pennsylvanian–Permian vegetational transition: a terrestrial analogue to the onshore–offshore hypothesis', *Evolution*, vol. 46, no. 3, Allen Press Inc.; *Figures 7.8, 7.11, 7.13, 7.14b and 7.15* Bob Spicer; *Figure 7.9* Parrish, J.T., Ziegler, A.M. and Scotese, C.R. (1982) 'Rainfall patterns and the distribution of coals and evaporites in the Mesozoic and Cenozoic', *Palaeogeography, Palaeoclimatology and Palaeoecology*, vol. 40, Elsevier Science; *Figure 7.14a, c and d* Dr Joan Watson, Manchester University; *Figure 7.16* USGS Printing Office, Washington DC; *Figure 7.18* Sohl, N.F. (1987) Presidential Address, 'Cretaceous gastropods: contrasts between Tethys and the temperate provinces', *Journal of Paleontology*, vol. 61, no. 6, The Paleontological Society; *Figure 7.20* Peter Skelton; *Figure 7.21a* Courtesy of Jeremy Young, The Natural History Museum; *Figure 7.21b* Dr J.A. Lees, University College London.

Sources of figures, data and tables

Beck, C.B. (1962) 'Reconstructions of Archaeopteris and further consideration of its phylogenetic position', *American Journal of Botany*, vol. 49, p. 373.

Bengston, S. and Yue Zhao (1992) 'Predatorial borings in late Precambrian mineralized exoskeletons', *Science*, vol. 257, p. 367.

Benton, M.J. (1995) 'Diversification and extinction in the history of life', *Science*, vol. 268, p. 52.

Berner, R.A. (1994) '3 Geocarb II: a revised model of atmosphere CO_2 over Phanerozoic time', *American Journal of Science*, vol. 249, January.

Broecker, W. (1991) 'Great Ocean Conveyor', *Oceanography*, vol. 14, p. 79.

Butterfield, N.J. and Chandler, F.W. (1992) 'Palaeoenvironmental distribution of Proterozoic microfossils with an example from the Agu Bay Formation, Baffin Island', *Palaeontology*, vol. 35, pt 4, November.

Carbon Dioxide Information Analysis Centre (CDIAC), http://cdiac.ornl.gov

Coffin, M.F. and Eldholm, O. (1993) 'Exploring large subsea igneous provinces', *Oceanus*, vol. 36, no. 4, Woods Hole Oceanographic Institution.

Cogné J-P. and Humler, E. (2006) 'Trends and rhythms in global sea-floor generation rate', *Geochemistry, Geophysics and Geosystems*, vol. 7, Q03011.

Conway, T.J. et al. (1988) 'Atmospheric carbon dioxide', Tellus 40B, pp. 81–115.

Courtillot, V.E. and Renne, P.R. (2003) 'On the ages of flood basalt events', *Comptes Rendus Geoscience*, vol. 335, pp. 113–140.

Cox, B. et al. (1989) *The Atlas of the Living World*, Marshall Editions.

Crowley, T.J. (1993) 'Climate change on tectonic time scales', *Tectonophysics*, vol. 222, pp. 277–294.

DiMichele, W.A. and Aronson R.B. (1992) 'The Pennsylvanian–Permian vegetational transition: a terrestrial analogue to the onshore–offshore hypothesis', *Evolution*, vol. 46, no. 3, p. 807.

Gradstein, F.M., Ogg, J.G. and Smith, A.G. (2005) *The Geological Timescale 2004*, Cambridge University Press.

Haq, B.U. (1984) 'Palaeoceanography', in Haq, B.U. and Milliman, J.D. (eds) *Marine Geology and Oceanography of Arabian Sea and Coastal Pakistan*, Van Nostrand Reinhold.

Imbrie, J. et al. (1984) in Berger, A. et al. (eds) *Milankovitch and Climate*, Kluwer Academic Publishers.

Kennet, J.P. et al. (eds) (1974) 'Initial Reports of the Deep Sea Drilling Programme', vol. 29, US Government Printing Office.

Kroon, D. et al. (1992) 'Onset of monsoonal-related …', *Proc. Ocean Drilling Program Scientific Results*, vol. 117, Integrated Ocean Drilling Program.

Martini, I.P. (1996) *Late Glacial and Postglacial Environmental Changes: Quaternary, Carboniferous–Permian and Proterozoic*, Oxford University Press.

Mitchell, J.F.B. (1989) 'The greenhouse effect and climate change', *Reviews of Geophysics*, vol. 27, issue 1, pp. 115–139.

Parrish, J.T., Ziegler, A.M. and Scotese, C.R. (1982) 'Rainfall patterns and the distribution of coals and evaporites in the Mesozoic and Cenozoic', *Palaeogeography, Palaeoclimatology and Palaeoecology*, vol. 40.

Philips, T.L. and DiMichele, W.A. (1992) 'Comparative ecology and life-history biology of arborescent lycopsids in late Carboniferous swamps of Euramerica', *Annals of the Missouri Botanical Garden.*

Quade, J. et al. (1989) 'Development of Asian monsoon revealed by marked ecological shift during the latest Miocene in northern Pakistan', *Nature*, vol. 342, p. 163.

Raymo, M. et al. (1988) 'Influence of late Cenozoic mountain building on ocean geochemical cycles', *Geology*, vol. 16, no. 7.

Raymo, M.E. (1991) 'Geochemical evidence supporting T.C. Chamberlain's theory of glaciation', *Geology*, vol. 19, no. 4.

Richter, F. (1992) 'Sr isotope evolution of seawater', *Earth and Planetary Science Letters*, vol. 109, p. 11.

Sepkoski, Jr, J.J. (1990) 'Evolutionary faunas' in Briggs, D.E.G. and Crowther, P.R. (eds) *Palaeobiology: A Synthesis*, Blackwell Publishing Ltd.

Sohl, N.F. (1987) Presidential Address, 'Cretaceous gastropods: contrasts between Tethys and the temperate provinces', *Journal of Paleontology*, vol. 61, no. 6.

Stewart, W.N. and Rothwell, G.W. (1993) *Paleobotany and the Evolution of Plants*, 2nd edn, Cambridge University Press.

Valentine, J.W. and Moores, E.M. (1970) 'Plate tectonic regulators', *Nature*, vol. 228, p. 106.

van Andel, T.H. (1985) *New Views on an Old Planet*, Cambridge University Press.

Vonder Haar, T. and Suomi, V. (1971) 'Measurements of the Earth's radiation budget from satellites', *Journal of Atmospheric Science*, vol. 28, pp. 305–14.

Xie, L. and Hseih, W.W. (1995) 'The global distribution of wind-induced upwelling', *Fisheries Oceanography*, vol. 4, p. 52.

Index

Entries and page numbers in **bold type** refer to where terms defined in the Glossary are printed in **bold** in the text. Page numbers referring only to figures and tables are printed in *italics*.

A

ablation 236
acritarchs 171
adiabatic changes **26**
advection 26
aerosols 29,108–112
Alaska, Cretaceous case study 250–252
albedo 5
  effects of mountains 139
alluvial fans 141
angiosperms 225
anoxic conditions **168**
Antarctic Bottom Water *42*, 43–44, *87*
anthropogenic effects on climate **19**
anticyclonic gyres 39
aphelion 9
Archaea 2, *202*
Archaeopteris 226
Asia
  evidence for climate change 148–153
  present-day climate 143–146
atmosphere
  as a protective filter 13–20
  constituent gases 15, *16*
  controls on carbon dioxide levels 91–93
  energy budget 17–18
  temperature structure *29*
  transport of heat and water 25–38
atmospheric CO_2
  and global climate 234–235
  and vegetation 234–235
atmospheric fronts 33
atmospheric window 17
autotrophic plants **173**

B

back-scattered radiation **109**
bacteria 53, *202*
base pairs in DNA **178**
bicarbonate ion 73
bilaterian life **195**
  radiation of 203
binary fission 176
biogenic silica **140**
biogeochemical cycles 47, 52–52
biolimiting constituents **53**
biological pump 81
biomarkers 168
biomes 67
  geographical distribution of *68*
biota 64
bioturbation 215
boundary layer 218
brine-rejection 43

C

C3 plants 152
C4 plants 152
calcareous sediments 90
Cambrian Explosion 195
  causes of 205–206
  origins of 203–205
  timing of 203
carbohydrates 62
carbon
  and climate 63–64
  in the oceans 78–89
  role in living world 61
carbon compounds (Box 2.1) 62
carbon cycle 61–102
  anthropogenic effects 97–100
  effects of mountain building 139–142
  effects of sediment burial *92*
  fluxes *98*
  geological 89–93
  in balance? 93–94
  marine 71–89
  terrestrial 65–71
  terrestrial and marine residence times 89–93
  timescale 64–93
carbon dioxide
  fluxes 84–89
  in water (Box 2.2) 72–74
  monitoring (Box 2.3) 95–96
  seasonal variations 96
  volcanic emissions 107
carbon fixing 4
  in the ocean 81
carbon isotopes (Box 4.1) 151
  variation in marine carbonates *158*
carbon sinks in Cretaceous 264–265
carbonaceous sediments 90
carbonate compensation depth 84
carbonate ion 73
carbonate platforms 260
  growth of 263–264
carbonate system (Box 2.2) **72**–74
carbonate weathering 77
Carboniferous swamp forests 240–242
cell division (Box 5.3) 179–181
chalk 264

chemical equilibrium 72
chemical weathering 47, 142
  of carbonates 77
chemofossils 175–176
chemosynthetic bacteria 56
chlorophyll, concentration variations *79*
chloroplasts 176
chromosomes 172
  diploid 178
classification hierarchy (Box 6.1) 207
climate, importance of water's properties (Box 1.4) 23
climate change
  Alaskan case study 246–247
  effect of aerosols 108–112
  effects on ocean currents 121–123
  extraterrestrial causes 238
  feedback mechanisms 135–136
  global change during Tertiary 153–164
climatic forcings, effect of volcanic eruptions *111*
climate models
  GEOCARB 156–160, *235*
  mountain-forcing model 156–160
  steady-state 163
  testing the models 160–164
climate reconstructions, using plants 256–258
clock analogy for history of life 196
clouds (Box 1.5) 28–29
coccolithophores *83*
  in chalk 264
colonisation of the land 217–227
conduction 14
continental drift 117–130
continental shelves 132
convection 25
  in the oceans 45–46
Coriolis effect (Box 1.6) **30**, 31–32
crown groups 203
cycads 225
  Cretaceous examples in Alaska 248–249
cyclonic gyres 39

D

δ^{13}C (Box 4.1) **150**, 151
δ^{13}C values in chalk 265
Daisyworld 190
decarbonation recations 140
Deccan Traps 112–114
density-driven currents in the deep ocean 25
diamictites 232
diatoms 83
Dickinsonia 185, 201
diploid chromosomes **178**
dissolved inorganic carbon (DIC) 67
dissolved organic carbon (DOC) 67
downwelling 45–46, *47*

E

earlywood 233
Earth
  air conditioning system 24–46
  comparisons with other planets 1
  early life 167–194
  eccentricity of orbit 9
  extremes 229–270
  heating system 24–46
  microbial life 2
  past environments 167, *171*
Earth's surface temperature pattern 20–23
eccentricity of Earth's orbit 9
Ediacaran faunas 185–187, 195, 200–202
  interpretation of 185–186
Ediacaran Period 185
electromagnetic spectrum (Box 1.3) **13**–14
  of the Sun and Earth *15*
emergence of life 167–194
enations 224
endosymbiotic hypothesis 176
endothermic reactions **61**
epiorogenic sea-level changes **131**
eukaryotes 167, *202*
  architecture and evolution 177–183
  beginnings of 174–177
  cell division (Box 5.3) 179–181
  cell structure 172, *173*
  diversification of 183–184
  divisions of *174*
eustatic sea-level changes **131**, 133–135
evapotranspiration 27
evolution
  altruism in 188–189
  rules for 187–191
evolution of life
  divergence of major animal groups *199*
  plants 217–227
evolution of the leaf (Box 6.4) 223–225
evolutionary faunas 213
evolutionary radiations 195, 212–215
evolutionary record, turnover of life 206–209
exothermic reactions **61**
extremophiles 2

F

false rings 252
feedback mechanisms, *see* positive *and* negative feedbacks
  between life and Earth 271–274
  in ecosystems 191
fitness 188

fixing of carbon **4**
flood basalts 112–117
  and mass extinctions 115–117
foraminiferans *83*
  oxygen isotope record *127*, 128, *153*
forcing function 19
forests
  Arctic 247
  Carboniferous swamp forests 240–242
  early evolution 225
  legacy of 242–244
forward-scattered radiation **109**
fossil record 195
  plants *221*
frontal currents 41

G

Gaia hypothesis (Box 5.4) 190, 271–274
gametes 179
gene duplication 205
general circulation model (GCM) for climate **144**
GEOCARB model 156–164, *235*
glacio-eustasy 262
Global Stratotype Section and Point (GSSP) 195
global surface current system *40*
global warming in Cretaceous 134
Globigerina bulloides 150
glossopterids 243
greenhouse atmosphere 266–267
greenhouse effect 17
greenhouse world
  Cretaceous 245–246
  Cretaceous volcanism 265–266
gross primary production (GPP) 62
gymnosperms 225
gyres 39

H

Hadley circulation 33
halite (salt) pseudomorphs 253
haploid cells **175**
haploid gametes 179
Hawaiian volcanism 105–107
heat transport by the upper ocean 39–41
herbaceous plants **226**
heterochrony 201
heterotrophs 171
hierarchy of classification (Box 6.1) 207
Himalaya, effect on atmospheric CO_2 154–160
homeostasis 189
Hox genes 205
hydrological cycle 47
hyperthermophiles 2

I

icehouse world 229–233
  geological evidence for 232–233
  role of plants 244
intermediate geological timescales 64
Intertropical Convergence Zone (ITCZ) 24, 33, *36, 37*
IRONEX II experiment 55
isostasy (Box 3.3) **131**–132
isostatic sea-level changes **131**
isotherms *22–23*

K

K/T boundary 169
K/T boundary extinction 212
kerogen 90
kin selection 189
kingdoms of organisms (Box 5.2) 173

L

large igneous provinces, global distribution of *114*
latewood 233
Le Chatelier's principle 74
leaf-area indices 257
leaf evolution (Box 6.4) 223–225
leaf margin analysis 148, 257
leaf physiognomy 147
Lepidodendron 240–242
life on Earth, early fossil evidence 169
lignin 66, 226
long geological timescales 64

M

macrophagy 205
mantle plumes 105
mantle superplume 265
marine snow 81
mass extinction 115–116, 195, 209–212
  consequences for ecological relationships 215–216
  the 'big five' (Box 6.2) 211–212
Mat world 170–172
meiosis 175, 179
  in eukaryotes (Box 5.3) 179–181
mesoscale eddies 40
methane, as a greenhouse gas 99
methane hydrates 99
microbial life, diversity of (Box 1.1) 2
microfossils 170
midvein 223
Milankovich cycles *9*, **10**–12
  as forcing function 20
mitochondria 176
mitosis 179
  in eukaryotes (Box 5.3) 179–181
model worlds and their climates (Box 3.1) 119–121
molecular clock theory 183, 197

molecular phylogenies 197
monsoon 35, 142–153
mountain-forcing model 156–164

N

natural selection 182, 271–272
nearest living relatives (NLR) technique 256
negative feedback 12
  in climate cycles *159*
  in carbon cycle 101
  in climate change 135–136
Neoproterozoic 167
net primary production (NPP) 63
  mean values *69, 80*
Nilssonia, Nilssoniocladus 248–250
North Atlantic Deep Water *42*, 43–44, *87*
nutrients 51

O

ocean
  role in heat transport 39–41
  role in recycling elements *51*
ocean basin volume changes 134
ocean currents *40*
  and climate change 121–123
ocean water volume changes 133
Ontong–Java Plateau 114
organelles 172
orographic precipitation 37, *38*
oxygen isotopes and the climate record (Box 3.2) 127–128
oxygen through time (Box 5.1) 168

P

palaeo-altimeter 146
palaeogeography
  Cretaceous *245*
  late Cretaceous 259–265
  Permian 229–232
Pangaea
  break-up of 124–129, 134
  Permian maps of *230, 231*
Panthalassa 124, 134
particulate organic carbon (POC) 67
pattern formation 205
perihelion 9
Permo-Carboniferous glaciation 234–238
pH 75
Phanerozoic life 195–228
Phanerozoic mass extinctions 211–212
phloem 242
photic zone 53
photosynthesis 3–4, 62
  marine environments 81
phototrophs 3
physical erosion 141
physiognomy of plants **256**–258
phytoplankton 53
  seasonal variations in
    concentration *149*
plant biology, extrapolation into fossil record 250–252
plants
  adapting to atmospheric CO_2 changes 225
  as thermometers 256
  effect on climate 239–244
  evolution of 217–227
  herbaceous 226
  structure of early land plants 218–222
plate tectonics
  and climate change, in icehouse world 236
  and life 129–130
  effects on climate and life 103–138
Pleistocene 232
polar climate 246–253
polar front 34
polar jet stream 34
positive feedback 12
  in climate change 135–136
  plant evolution 219–220, 225
precession 9–11
primary producers 4
primary production 4–5
  global distribution map *5*
  global estimates of marine *80*
  global estimates of terrestrial *69*
  gross 70
  net *69*
  rates of *69*
primary productivity, in North Atlantic *54*
primary vein 223
progradation 261
progymnosperms 225
prokaryotes 172
  cell structure *173*
Proterozoic–Phanerozoic transition 196, 198–206

R

radiation
  back-scattered 109
  forward-scattered 109
radiation of bilaterian life **203**
radiations and extinctions 206–216
radiogenic strontium 161
rain shadow 37
red beds 232
refugia 206
regressive episodes on carbonate platforms **261**
reservoir 48
  carbon 64
residence time 48, 66
respiration 62
reverse weathering 84
rhizoids 220
rhizome 220
Rossby waves 34
rudists 262, *263*

S

salinity 43
sea-level changes 130–136
  epiorogenic 131
  eustatic 131, 133–135
  isostatic 131
  Phanerozoic *132*, 133
  secondary effects on climate 135–136
seasons (Box 1.2) 6–8
secondary vein 223
sediment fluxes in Asian rivers 155–156
sedimentary succession 183
selfish gene 188
sensible heat 27
sensitivity of a system **19**
sessile plants **173**
sessile shelly organisms 232
sexual reproduction 179–183
  effect on natural selection *182*
shelf seas 132
shocked quartz 210
silicate weathering 77
siliceous sediments 90
slushball Earth 198
snowball Earth 198
solar flux 5
solar radiation 10–12
  changes in the Phanerozoic 238
sporangia 220
spores (Box 6.3) **217**
Spriggina 201
Standard Mean Ocean Water (SMOW) 128
steady-state climate model **163**
Stefan–Bolzmann law 14
steppe 144
steranes 175
stomata 62, 225
stromatolites 170
strontium isotopes (Box 4.2) 161
  in rivers 162–163
  in seawater 161
  in the Phanerozoic *237*, 238
subtropical gyres 39
Sun, energy from 5–12
symbionts 177, 186, 201
symbiosis 172

T

taphonomy 250
taxon, taxa 203
taxonomic hierarchy (Box 6.1) **207**
tertiary vein 223
Tethyan carbonate platforms 260–264
thermal energy (Box 1.3) 13–14
thermocline 39
thermohaline circulation 42, **43**–46
Tibetan Plateau 142–153
  dating the uplift 146–148
  effect on atmospheric CO_2 154–160
  effect on global average temperature 154
tilt 9–11
timescales
  intermediate geological 64
  long geological 64
  short biological 64
Trade Winds 33
transpiration 27, 223
tree growth rings (Box 7.1) 233
  false rings 252
trophic pyramids 178
troposphere 27

U

Universal Tree of Life 201
upwelling 45–46, *47*

V

vascularisation 222
vegetation at low latitudes 253–255
vein system of a leaf 223–224
Vendian Period 202
  fossils 186
Vendobionta 202
Vienna Standard Mean Ocean Water (VSMOW) 128
volcanic aerosols 108–112
volcanism 103–107
  distribution map *104*

W

water
  average chemical composition of *50*
  properties of (Box 1.4) 23
water masses 42
weathering
  of carbonates 77
  of silicates 77
  of the Himalaya 156
  effects on climate change 160
wind-driven surface currents 24
winds in the atmosphere 24–25

X

xylem 242

Z

zooplankton 53